Slotted Waveguide Array Antennas

Mario Boella Series on Electromagnetism in Information and Communication

Piergiorgio L.E. Uslenghi, PhD – Series Editor

The Mario Boella series offers textbooks and monographs in all areas of radio science, with a special emphasis on the applications of electromagnetism to information and communication technologies. The series is scientifically and financially sponsored by the Istituto Superiore Mario Boella affiliated with the Politecnico di Torino, Italy, and is scientifically co-sponsored by the International Union of Radio Science (URSI). It is named to honour the memory of Professor Mario Boella of the Politecnico di Torino, who was a pioneer in the development of electronics and telecommunications in Italy for half a century, and a vice president of URSI from 1966 to 1969.

Published Titles in the Series

Fundamentals of Wave Phenomena, 2nd Edition
by Akira Hirose and Karl Lonngren (2010)

Scattering of Waves by Wedges and Cones with Impedance Boundary Conditions
by Mikhail Lyalinov and Ning Yan Zhu (2012)

Complex Space Source Theory of Spatially Localized Electromagnetic Waves
by S.R. Seshadri (2013)

The Wiener–Hopf Method in Electromagnetics
by Vito Daniele and Rodolfo Zich (2014)

Higher-order Techniques in Computational Electromagnetics
by Roberto Graglia and Andrew Peterson (2015)

Slotted Waveguide Array Antennas

Theory, analysis and design

Lars Josefsson and Sembiam R. Rengarajan

The Institution of Engineering and Technology

Published by SciTech Publishing, an imprint of The Institution of Engineering and Technology, London, United Kingdom

The Institution of Engineering and Technology is registered as a Charity in England & Wales (no. 211014) and Scotland (no. SC038698).

First published 2018

The Institution of Engineering and Technology
Michael Faraday House
Six Hills Way, Stevenage
Herts, SG1 2AY, United Kingdom

www.theiet.org

British Library Cataloguing in Publication Data
A catalogue record for this product is available from the British Library

ISBN 978-1-61353-189-1 (hardback)
ISBN 978-1-61353-190-7 (PDF)

Typeset in India by MPS Limited

Contents

About the authors

Lars Josefsson and Sembiam Rengarajan met at the University of California, Los Angeles (UCLA), USA, where both worked on slotted waveguide array antennas together with Professor R.S. Elliott, a well-known scientific leader and teacher in this field. Since then they have kept contact and collaborated on research projects and courses several times.

Lars Josefsson was born in Norrköping, Sweden. He graduated from The Royal Institute of Technology in Stockholm (KTH) and got his PhD from Chalmers University of Technology in Göteborg. A grant from the Marcus Wallenberg Foundation and a Fulbright Scholarship enabled his stay in 1982–83 as a Visiting Scientist at UCLA, working on waveguide slotted array antennas. He became Adjunct Professor in Antenna Technology (1983–86) at Chalmers and obtained the equivalent position at KTH (1996–2003). He returned to Chalmers again as Adjunct Professor (2004–07).

For more than 40 years Lars was with Ericsson Microwave Systems AB in Mölndal, Sweden, where he became responsible for new antenna technology and systems, internal R&D projects and internal education relating to antennas. In 2004 he received the Thulin Silver Medal from the Swedish Aeronautical Society for his contributions to several generations of airborne radar antenna developments. He is fellow of the IEEE since 1999.

In 2003 he formed his own consulting company: Lars Microwave (technical consulting, research and education). Lars Josefsson is the author and co-author of more than 100 scientific publications. He is the author (together with Patrik Persson) of the book *Conformal Array Antenna Theory and Design*, published by IEEE/Wiley in 2006.

Sembiam R. Rengarajan received the PhD degree in Electrical Engineering from the University of New Brunswick, Canada, in 1980. Since then he has been a faculty member at California State University, Northridge, USA. He has held visiting appointments at Chalmers University of Technology, Sweden; University of Santiago de Compostela, Spain; University of Pretoria, South Africa; and Technical University of Denmark. In addition to teaching electromagnetics and antenna courses, he has performed significant research in topics on applications of electromagnetics to antennas, scattering, and microwave components. He has also consulted for many companies and government agencies. He has published more than 240 journal articles and conference presentations.

Much of Sembiam's research has dealt with slotted waveguide array antennas. He collaborated with Prof. Elliott in the 1980s and 1990s for more than a decade.

His research in this area was supported by the University of California, Hughes Aircraft Company, Lockheed Martin, Saab Ericsson Space, European Space Agency, Rantec Corporation, DSO Singapore, NASA/Jet Propulsion Laboratory (JPL), among others. He served as a Distinguished Lecturer for IEEE Antennas and Propagation Society, lecturing on slotted waveguide array antennas all over the world. Presently he serves as the Chair of the United States National Committee for the International Union of Radio Science (USNC-URSI). He has received more than 20 awards from NASA for his research contributions to JPL. He is a life fellow of IEEE.

Abbreviations and acronyms

AF	array factor
CEM	computational electromagnetics code
CFRP	carbon fibre-reinforced plastic
dB	decibel
DBF	digital beam forming
DOA	direction of arrival
EM	electromagnetic
FDTD	finite difference time domain
FEM	finite element method
FSS	frequency selective structure
GA	genetic algorithm
GHz	giga hertz
GO	geometrical optics
GSM	generalised scattering matrix
GTD	geometric theory of diffraction
HFSS	high-frequency structural simulator
HTCC	high temperature co-fired ceramics
IE	integral equation
LEO	low earth orbit
LMS	least mean square
LTCC	low temperature co-fired ceramics
MC	mutual coupling
MIMO	multiple input multiple output
MIT	Massachusetts Institute of Technology
MMIC	monolithic microwave integrated circuit
MM	mode matching
MoM	method of moments
MTI	moving target indication
NDT	non-destructive testing

PEC	perfect electric conductor
PMC	perfect magnetic conductor
PO	physical optics
RCS	radar cross section
RF	radio frequency
SAR	synthetic aperture radar
SIW	substrate integrated waveguide
SLL	sidelobe level
SMI	sample matrix inversion
STAP	space time adaptive processing
TACAN	tactical air navigation
TE	transverse electric
TEM	transverse electromagnetic
THz	terahertz
TM	transverse magnetic
TRM	transmit/receive module
UCA	uniform circular array
ULA	uniform linear array
UTD	uniform theory of diffraction
Wi-Fi	~WLAN
WLAN	wireless local area network

List of symbols used

Symbol		Unit
a	width of rectangular waveguide	m
A	magnetic vector potential	Wb/m
b	height of rectangular waveguide	m
B	susceptance	Siemens
B	magnetic flux density	T
d	slot spacing in the H-plane	m
D	electric displacement density	Coul/m^2
	also antenna directivity	–
E	electric field intensity	V/m
f	frequency	Hz
F	electric vector potential	Coul/m
g_{mn}	mutual coupling integral	–
G	conductance	Siemens
	also Green's function	–
	also antenna gain	–
H	magnetic field intensity	A/m
J	electric current density	A/m^2
k	wavenumber	rad/m
k_0	wavenumber in free space	rad/m
L	half length of rectangular slot	m
M	magnetic current density	V/m^2
M_s	magnetic surface current density	V/m
P	complex power	W
R	resistance	Ω
S	Poynting vector	W/m^2
	also scattering matrix	–
t	waveguide wall thickness	m
V	voltage	V
w	width of rectangular slot	m
X	reactance	Ω
Y	admittance	Siemens

Symbol		Unit
α	attenuation constant	Neper/m
β	phase constant	rad/m
γ	propagation constant	m^{-1}
Γ	reflection coefficient	
ε_0	permittivity of free space	Farad/m
ε_m	Neumann factor, $=1$ for $m=0$, $=2$ for $m>0$	–
ε_r	relative permittivity	–
ε	permittivity, $\varepsilon = \varepsilon_r\, \varepsilon_0$	
ϕ	azimuth angle in spherical coordinates	radians
η_0	impedance of free space	Ω
λ	wavelength	m
λ_0	free space wavelength	m
λ_g	guide wavelength	m
μ_0	permeability of free space	Henry/m
θ	tilt angle of a coupling slot	radians
	also the elevation angle in spherical coordinates	radians
σ	conductivity	Siemens/m
σ_{sc}	scattering cross section	m^2
ω	angular frequency	rad/s
χ^2	the coupling coefficient of a coupling slot	–
δ	Dirac delta function	–
	also slot offset from the centreline	–
δ_{ij}	Kronecker delta function	–

Foreword

The Mario Boella series is devoted to textbooks and monographs in all areas of Radio Science, with a particular emphasis on the applications of electromagnetism to information and communication technologies. The series is financially sponsored by the Istituto Superiore Mario Boella (ISMB), a research entity affiliated with the Politecnico di Torino, Italy, and is scientifically co-sponsored by the International Union of Radio Science (URSI). It is named to honour the memory of Professor Mario Boella of the Politecnico di Torino, who was a pioneer in introducing electronics and telecommunications into higher education in Italy for half a century and was a vice-president of URSI from 1966 to 1969.

This research monograph on slotted waveguide antennas is the eighth volume in the ISMB series. It contains a very comprehensive coverage on a subject that is of practical importance in antenna and radar systems for a variety of civilian and military applications. Both theory and design, as well as manufacturing aspects and future applications, are discussed in depth. The authors are well-known authorities in the field. Dr Josefsson has decades of experience as a design and project engineer, and has been honoured for his technical achievements. Professor Rengarajan has taught and done research on antennas and radars for many years; he is highly regarded in the electromagnetics community for his original contributions. The two authors have combined their complementary knowledge to produce a monograph that will be very useful to students, instructors and practitioners.

Piergiorgio L. E. Uslenghi
ISMB Series Editor
Chicago, April 2018

Preface

The purpose of this book is to present in one volume a comprehensive treatment of slotted waveguide array antennas from an engineering perspective. This includes history, concepts, design principles, current state of the art and a discussion about future applications. The book is addressed to engineers, physicists and students interested in slotted waveguide array technology. It is written on an intermediate level. It may serve as a supplementary text for graduate courses in antennas and applied electromagnetic theory.

Published papers on slotted waveguide array antennas are abundant, starting in the 1940s. Writing this book has meant a thorough study of the available literature and it has required a careful analysis of the major contributions to the field. A simple search on IEEE Xplore® gave 2059 publication hits matching the key word 'slotted waveguide array antenna' for the period 1946 to February 2016. Most striking is a continued increase in the number of publications per year; see the figure below. This number has an exponential growth. While a detailed analysis and interpretation of this statistic might not be useful, it is clear that the subject remains important and is of continued interest for modern applications.

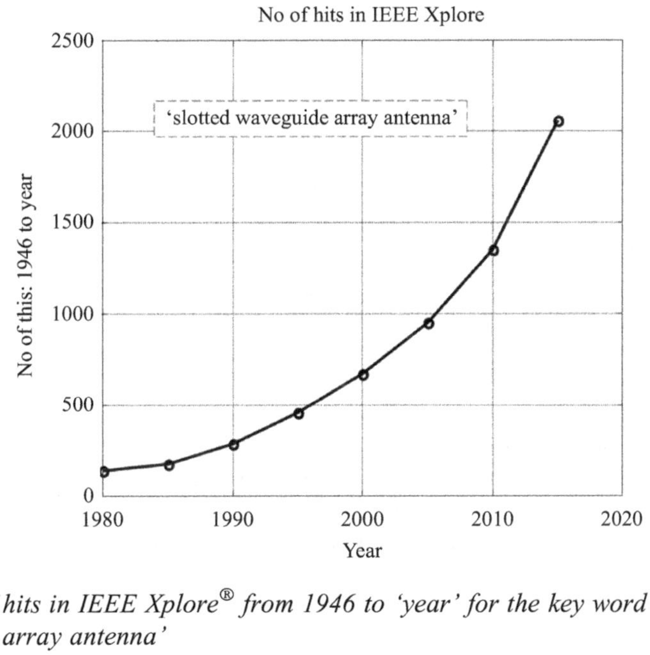

Number of hits in IEEE Xplore® from 1946 to 'year' for the key word 'slotted waveguide array antenna'

How do other antenna types such as microstrip patch array antennas and reflector antennas compare with slotted waveguide array antennas? Without going into details we try a simple comparison:

Antenna type	**Characteristics**
Microstrip patch arrays	– Transmission losses, surface waves
Reflector antennas	– Fixed beam, require depth
Slotted waveguide arrays	+ Low losses, high precision

There are of course both pros and cons for all types, depending on the application. Several books discuss microstrip antennas and/or reflector antennas. The present book is the first totally devoted to slotted waveguide array antennas.

With specific examples of waveguide array designs, accompanied by detailed illustrations and antenna characteristics, the book is a must-have reference for engineers involved in the design and development of slotted waveguide array antennas. An additional value is provided by a thorough overview of the related literature in the field and the history of these antennas given in several chapters. With this book system engineers and design and development engineers in industry and government will have a single source with a coherent treatment instead of turning to numerous papers in the literature.

The book goes well beyond some of the commonly discussed topics on slotted waveguide array design, into areas that include

- higher-order mode coupling and edge effects
- performance optimisation in terms of bandwidth and pattern performance
- special slot excitation methods
- applications such as monopulse and phase steering
- manufacturing techniques and tolerances, etc.

The book is organised as follows:

The first few pages contain a list of abbreviations and acronyms that are common when dealing with slotted waveguide array antennas and related applications. There is also a list of symbols with definitions, representing mathematical and physical quantities that appear in the text. Finally, a table providing definitions of waveguide bands and designations, with corresponding rectangular waveguide inner dimensions is included. The following chapters all start with some introductory material such as the basic concepts that are essential to get an understanding of the more advanced concepts to follow.

Chapter 1, 'Introduction', is a short overview of the radiating mechanism of slots in waveguides carrying the fundamental TE_{10} mode. There is also a short discussion of equivalent circuits for various types of resonant slots: longitudinal, transverse and inclined slots.

Chapter 2, 'Review of electromagnetic theory, starts from Maxwell's equations. Important concepts such as boundary conditions, reciprocity theorems, field equivalence and Green's functions and other definitions are included. The purpose is to serve as a reference to be consulted if the need arises when reading the theoretical chapters that follow.

Chapter 3, 'History', is an account of the history of early slotted waveguide array antenna developments from the Second World War and onwards. A few descriptions of more recent designs are also presented, demonstrating the wide range of applications of the technology (radar, remote sensing, communication, etc.).

Chapter 4, 'The slot antenna', starts with an analysis of a general aperture in an infinite ground plane, the near and far fields, and the aperture admittance. It continues with resonant slots and their radiation conductance. General expressions for waveguide modes are derived, to be followed by modes in rectangular waveguides. The case with longitudinal slots is analysed in more detail including an analysis of mutual coupling among pairs of rectangular slots in the E-plane and in the H-plane. The *array coupling* is also calculated, that is, the coupling between two slots in the presence of other slots in the array. The impact on coupling of edge diffraction effects in finite ground planes is also analysed.

Chapter 5, 'Slot Models', theoretical models for characterising slots in waveguides are introduced. An integral equation formulation is presented and expressions for a solution by the Method of Moments are presented. Computed results for longitudinal and transverse slots are discussed, including the effect of waveguide wall thickness, frequency dependence, waveguide dimensions, etc.

Chapter 6, 'The linear slotted waveguide array antenna'. In this chapter algorithms for designing a linear array of slots in a single waveguide are derived. The influence of mutual coupling is included and typical results are presented. Some examples of the design of standing wave and travelling wave arrays are discussed.

Chapter 7, 'Design of planar slotted waveguide array antennas'. Here several slotted waveguides form a planar array. Mutual coupling plays a large role and a coupling network is added to feed the waveguides. Examples of planar array design are presented. Other design methods in the literature are discussed.

Chapter 8, 'Concepts and models for advanced designs'. The influence of higher-order modes is treated in more detail, including coupling slots. Designs with edge wall slots, compound slots, iris-excited longitudinal slots, slots in ridge waveguides and slots with dielectric cover are treated as well as optimisation methods.

Chapter 9, 'Antenna systems and special requirements'. This chapter treats system aspects such as slotted array antennas with phase and/or frequency beam steering, monopulse for angular tracking, conformal arrays and polarisation control. Bandwidth limitations are also addressed, as well as omnidirectional antennas, spurious lobe generation and elimination, and scattering cross-sectional analysis. A short section on slotted arrays using stripline technology is also included. The chapter can be read separately as an introduction into the many system applications of slotted waveguide array antennas. Several references to the literature on system applications are included.

Chapter 10, 'Slot arrays in special waveguide technologies'. In this chapter we discuss slot arrays in different waveguide technologies such as two types of parallel plate waveguides propagating the TEM mode, the substrate integrated waveguides (SIW) and the gap waveguide propagating quasi TEM mode. In the circular parallel plate waveguide an electromagnetic wave propagates in the radial direction inward or outward, while slots cut in one of the parallel plates radiate. Such an array is convenient for large apertures requiring high gain and mass production. The second type of parallel plate waveguide has a rectangular shape propagating TEM mode waveguide with a planar wave front. Slot arrays in SIW have great potential for easy integration with monolithic microwave integrated circuit devices. Finally slot arrays in gap waveguides, exhibiting wider bandwidth at millimetre-wave frequencies and beyond, are presented.

Chapter 11, 'Manufacturing aspects'. This is a wide subject so we concentrate on critical parameters such as electrical and mechanical tolerances that relate to electrical performance. We also discuss joining methods for metal materials, for example, dip brazing of aluminium. Applications at high frequencies (millimetre waves) in particular call for very high precision and special methods. An important area is the use of slotted carbon fibre-reinforced plastic waveguides that offer light weight and thermal stability, typically required for large antenna systems for space applications. The technology is also used in some ground and airborne radar systems. The exciting developments in low-cost fabrication using plastic materials and metallisation techniques are discussed. This area is related to the microelectronic area, particularly for high frequencies and highly integrated antenna/microwave assemblies.

Chapter 12, 'Outlook for the future'. In this last chapter we try to summarise the state of the art and discuss expected new applications of slotted waveguide array antennas. Important areas include integration of antenna and circuit technology, the trend towards higher frequencies, spaceborne systems, design tools, the impact of advanced software, new materials, cost reduction and many more. However, in this final overview it is not possible to mention all the details of the evolving field; the reader is referred to the respective chapters and the references for more information.

Extensive literature references are included in each chapter. A Subject Index is included at the back.

Acknowledgements

Much pioneering work in the field of slotted waveguide array antennas was done at the Radiation Laboratory at the Massachusetts Institute of Technology (MIT) during the years 1940–45. The Radiation Laboratory Series of publications (28 volumes) remain a valuable source of reference even today.

The book draws on material from several short courses on 'Slotted Waveguide Array Antenna Technology' by the authors. One such course was presented together with Dr M Ando and Dr J Hirokawa from the Tokyo Institute of Technology.

Numerous colleagues and friends have supported our work with their advice, encouragement and contributions. In particular we would like to mention Dr Hans Steyskal and Dr Bob Mailloux for their valuable comments.

We also remember with gratitude Professor Bob Elliott of UCLA, a mentor and friend, who introduced us to many exciting aspects of slotted waveguide arrays.

We are grateful for the permission to use results from several studies on slotted waveguide arrays. This includes Ericsson AB, Saab AB, RUAG Space AB, in Sweden and many more.

One of us (SR) would like to acknowledge his research sponsors including Hughes Aircraft, University of California, Saab Ericsson Space, JPL, Rantec, Lockheed Martin, DSO Singapore and CSUN Foundation, and his many students and colleagues who have contributed to his work.

Last but not least, we express our appreciation and gratitude to our families for their encouragement, understanding and patience during the writing of this book.

Chapter 1
Introduction

The first successful slotted waveguide array antennas were developed in Canada during the Second World War. The immediate applications were in military ground and airborne radar systems for target detection and tracking. Later applications include remote sensing from aircraft and space vehicles and microwave communication links. Spaceborne synthetic aperture radar (SAR) with slotted waveguide arrays are used for weather forecasting, environmental monitoring, climate change studies, etc. The slotted array antenna is also considered in automobile collision avoidance systems.

For the interested reader not too familiar with slotted waveguide array antennas a few simple concepts are introduced in the following text.

Slots in *rectangular* waveguides are the most commonly used. The waveguide can be standard dimension, or half height or even quarter height, in order to save space and weight. Many other realisations are possible, however, as will be discussed in later chapters. The fields inside the waveguide are related to the currents in the waveguide walls. For the fundamental TE_{10} mode we have the current distribution in the waveguide wall as shown in Figure 1.1. Typical slot positions are illustrated in Figure 1.2. By comparing the two figures we realise that slot positions *c* and *e* are not very useful since these slots do not interrupt any wall current and hence do not couple to the waveguide mode. Slots *a* (transverse

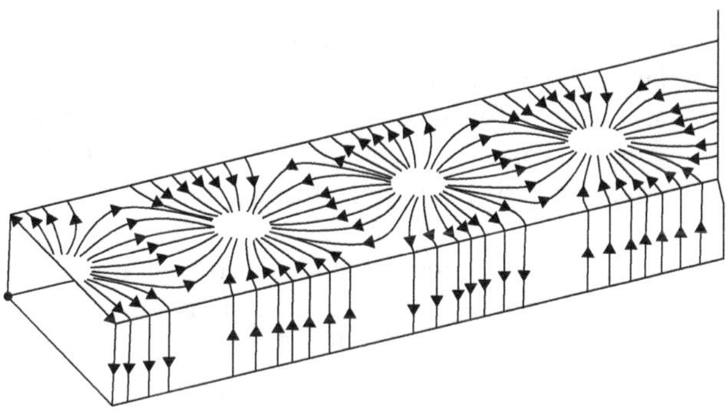

Figure 1.1 Surface currents on the waveguide wall for the TE_{10} mode in a rectangular waveguide

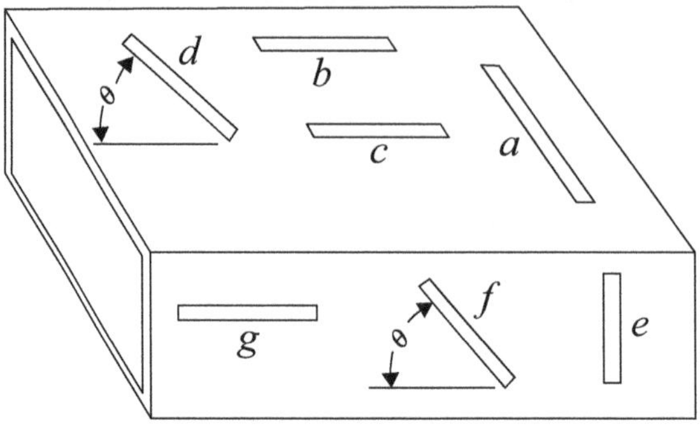

Figure 1.2 Possible slot positions in a rectangular waveguide. Adapted from [1]

broad wall slot) and *g*, on the other hand, couple strongly, and slot *f* (inclined sidewall slot) couples to a degree depending on the inclination angle θ. Slot *b* (longitudinal broad wall slot) also couples strongly when displaced far from the waveguide centreline.

The rectangular slot in a large conducting ground plane radiates approximately as a dipole of same length and width. Thus, the slot in a waveguide has a broad antenna pattern approximately equal to the corresponding dipole, only that E- and H-planes are interchanged. The corresponding antenna impedances differ considerably, however (Babinet's principle) [2, p. 336].

A. F. Stevenson at the National Research Council in Canada [3] found suitable equivalent networks for the most common slots (at resonance) as shown in Figure 1.3.

A more detailed analysis of waveguide modes and slot characteristics will be presented in subsequent chapters. Also array design principles and applications will be discussed.

New fabrication techniques for slotted arrays such as electroplating on dielectric materials have demonstrated good performance at frequencies as high as $\geq$100 GHz. Thus, highly directive antennas with small dimensions can be made, and a very large signal bandwidth becomes possible. Applications at such high frequencies are mainly for short ranges: covert communication, imaging radar, collision avoidance, planetary landing, etc. An important step forward is the possibility to integrate active microelectronics and circuits with the antenna structure.

Important progress over the recent years has also been made in the design and optimisation of slotted waveguide array antennas. It is now possible to fabricate planar arrays in a cost-effective and simplified way in several layers using new techniques; there are new ways to match and optimise the electrical design, increase bandwidth, etc.

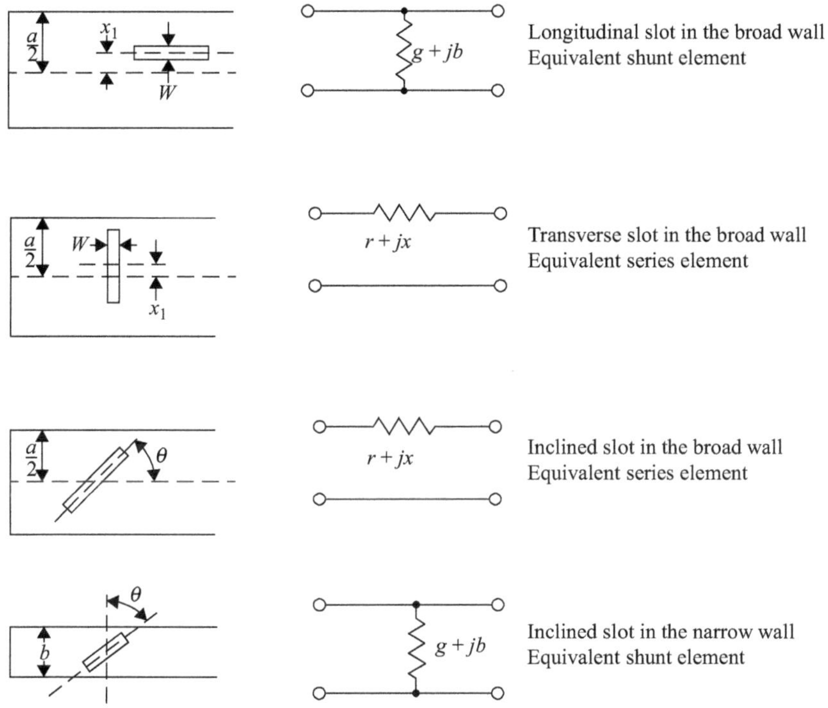

Figure 1.3 Approximate equivalent networks for resonant slots. Adapted from [1]

References

[1] Silver S. ed. *Microwave Antenna Theory and Design*, MIT Rad. Lab. Series, Vol. 12, McGraw-Hill, New York, 1949.

[2] Elliott R. S. *Antenna Theory and Design.* Prentice-Hall, Englewood Cliffs, NJ, 1981.

[3] Stevenson A. F. *Series of Slots in Rectangular Waveguides*, National Research Council of Canada, Radio Reports 12 and 13, 1944.

Chapter 2
Review of electromagnetic theory

In this chapter, we review the electromagnetic theory and concepts used in later chapters. We will start with Maxwell's equations in time harmonic form, followed by boundary conditions. Expressions for energy and power are derived. We then discuss the reciprocity theorem. Vector and scalar potentials are derived. Image principle and the field equivalence principle are presented. Green's functions are discussed with a presentation of dyadic Green's functions for the magnetic current in a waveguide.

2.1 Maxwell's equations

Steady-state electromagnetic fields under sinusoidal excitations are expressed in phasor form. Time dependence in the form of exp($j\,\omega t$), where ω is the angular frequency and $j = \sqrt{-1}$, is assumed and suppressed in Maxwell's equations (2.1)–(2.4).

$$\nabla \times \vec{E} = -j\omega\mu\vec{H} \tag{2.1}$$

$$\nabla \times \vec{H} = j\omega\varepsilon\vec{E} + \vec{J} \tag{2.2}$$

$$\nabla \bullet \vec{D} = \rho_{\rm v} \tag{2.3}$$

$$\nabla \bullet \vec{B} = 0 \tag{2.4}$$

In these equations $\vec{E}$, $\vec{H}$, $\vec{D}$ and $\vec{B}$ are the electric field, magnetic field, electric displacement density and magnetic flux density, respectively. ε and μ are the permittivity and permeability, respectively. In lossy media $\varepsilon = \varepsilon' - j\varepsilon''$ and $\mu = \mu' - j\mu''$. The sources $\vec{J}$ and $\rho_{\rm v}$ are the electric current density and volume charge density, respectively. The current density term $\vec{J}$ may consist of both induced currents and impressed currents.

The constitutive relations are given next, where σ is the conductivity of the medium.

$$\vec{D} = \varepsilon\vec{E} \tag{2.5}$$

$$\vec{B} = \mu\vec{H} \tag{2.6}$$

$$\vec{J} = \sigma\vec{E} \tag{2.7}$$

$$\text{In lossy dielectric media, } \sigma = \omega\varepsilon'' \tag{2.8}$$

where ε'' is the negative imaginary part of the permittivity.

We will be working mostly with linear, homogeneous and isotropic media in this book. So the constitutive parameters are scalar quantities and may depend on frequency. In some equivalent problems there may be an impressed magnetic current density term, $\vec{M}_{\text{i}}$, on the right side of (2.1). Such a magnetic current density results from an impressed electric field, $\vec{E}_{\text{i}}$, given by (2.9).

$$\vec{M}_{\text{i}} = -\nabla \times \vec{E}_{\text{i}} \tag{2.9}$$

2.2 Boundary conditions

Let us consider two regions labelled 1 and 2 with $\hat{n}$, a unit vector at the boundary surface directed towards region 1 as shown in Figure 2.1. At the boundary surface the tangential components should satisfy the conditions given by (2.10) and (2.11), where the subscripts 1 and 2 refer to media 1 and 2, respectively.

$$\hat{n} \times \left(\vec{H}_1 - \vec{H}_2\right) = \vec{J}_{\text{s}} \tag{2.10}$$

$$-\hat{n} \times \left(\vec{E}_1 - \vec{E}_2\right) = \vec{M}_{\text{s}} \tag{2.11}$$

where $\vec{J}_{\text{s}}$ and $\vec{M}_{\text{s}}$ are the electric and magnetic surface current densities, respectively, at the boundary. The magnetic surface current density term on the right of (2.11) is non-zero only in aperture problems where equivalent magnetic surface currents exist. The boundary conditions for the normal components satisfy (2.12) and (2.13), where ρ_{s} is the electric surface charge density.

$$\hat{n} \bullet \left(\vec{D}_1 - \vec{D}_2\right) = \rho_{\text{s}} \tag{2.12}$$

$$\hat{n} \bullet \left(\vec{B}_1 - \vec{B}_2\right) = 0 \tag{2.13}$$

Special cases

If medium 2 is a perfect electric conductor (PEC),

$$\hat{n} \times \vec{H}_1 = \vec{J}_{\text{s}} \tag{2.14}$$

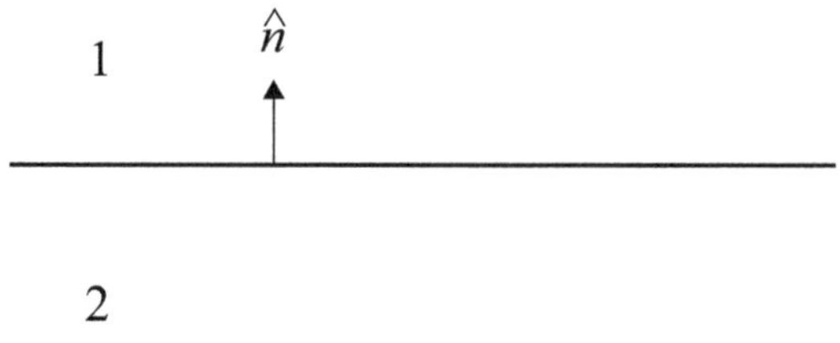

Figure 2.1 Two regions with a boundary surface separating them

and

$$\hat{n} \bullet \vec{D}_1 = \rho_s \tag{2.15}$$

If medium 2 is a perfect magnetic conductor (PMC), introduced in some equivalent field problems,

$$-\hat{n} \times \vec{E}_1 = \vec{M}_s \tag{2.16}$$

and

$$\hat{n} \bullet \vec{B}_1 = \rho_{ms} \tag{2.17}$$

where ρ_{ms} is the equivalent surface magnetic charge.

2.3 Energy and power

Maxwell's equations (2.1) and (2.2) are rewritten with the impressed magnetic current density $\vec{M}_i$, and the electric conduction and impressed current density terms $\vec{J}_c$ and $\vec{J}_i$, respectively, in (2.18) and (2.19).

$$\nabla \times \vec{E} = -j\omega\mu\vec{H} - \vec{M}_i \tag{2.18}$$

$$\nabla \times \vec{H} = j\omega\varepsilon\vec{E} + \vec{J}_c + \vec{J}_i \tag{2.19}$$

Let us consider the expression

$$\nabla \bullet \left(\vec{E} \times \vec{H}^*\right) = \vec{H}^* \bullet \left(\nabla \times \vec{E}\right) - \vec{E} \bullet \left(\nabla \times \vec{H}^*\right) \tag{2.20}$$

where the superscript * denotes the complex conjugate.

Both sides of (2.20) are integrated over a volume v, and the divergence theorem is applied to obtain

$$\oiint_s \left(\vec{E} \times \vec{H}^*\right) \bullet d\vec{s} = \iiint_v \left[\vec{H}^* \bullet \left(\nabla \times \vec{E}\right) - \vec{E} \bullet \left(\nabla \times \vec{H}^*\right)\right] dv \tag{2.21}$$

where s is the surface enclosing the volume v.

The use of (2.18), (2.19), (2.5), (2.6) and (2.7) in (2.21) yields

$$\begin{aligned} -\oiint_s \left(\vec{E} \times \vec{H}^*\right) \bullet d\vec{s} &= j\omega\mu \iiint_v |\vec{H}|^2 dv - j\omega\varepsilon \iiint_v |\vec{E}|^2 dv \\ &\quad + \iiint_v \left[\vec{M}_i \bullet \vec{H}^* + \vec{E} \bullet \vec{J}_i^*\right] dv + \sigma \iiint_v |\vec{E}|^2 dv \end{aligned} \tag{2.22}$$

Equation (2.22) is a statement of conservation in terms of complex power. It may be expressed as

$$-S = j2\omega(W_m - W_e) + P_d - P_s \tag{2.23}$$

where S is the complex power leaving the volume, W_m is the magnetic energy stored in the volume, W_e is the electric energy stored in the volume, P_d is the dissipated power and P_s is the complex power delivered by the sources. If μ and ε are complex, additional dissipation terms will be present.

2.4 Reciprocity theorem

Consider a region in space containing two sets of sources. Each set of fields is related to its sources via Maxwell's equations as given next.

$$\nabla \times \vec{E}_a = -j\omega\mu\vec{H}_a - \vec{M}_a \tag{2.24}$$

$$\nabla \times \vec{H}_a = j\omega\varepsilon\vec{E}_a + \vec{J}_a \tag{2.25}$$

$$\nabla \times \vec{E}_b = -j\omega\mu\vec{H}_b - \vec{M}_b \tag{2.26}$$

$$\nabla \times \vec{H}_b = j\omega\varepsilon\vec{E}_b + \vec{J}_b \tag{2.27}$$

Using vector identities, (2.28) and (2.29) are obtained.

$$\begin{aligned} -\nabla \bullet \left(\vec{E}_a \times \vec{H}_b\right) &= \vec{E}_a \bullet \left(\nabla \times \vec{H}_b\right) - \vec{H}_b \bullet \left(\nabla \times \vec{E}_a\right) \\ &= j\omega\varepsilon\vec{E}_a \bullet \vec{E}_b + \vec{E}_a \bullet \vec{J}_b + j\omega\mu\vec{H}_a \bullet \vec{H}_b + \vec{H}_b \bullet \vec{M}_a \end{aligned} \tag{2.28}$$

$$\begin{aligned} -\nabla \bullet \left(\vec{E}_b \times \vec{H}_a\right) &= \vec{E}_b \bullet \left(\nabla \times \vec{H}_a\right) - \vec{H}_a \bullet \left(\nabla \times \vec{E}_b\right) \\ &= j\omega\varepsilon\vec{E}_a \bullet \vec{E}_b + \vec{E}_b \bullet \vec{J}_a + j\omega\mu\vec{H}_a \bullet \vec{H}_b + \vec{H}_a \bullet \vec{M}_b \end{aligned} \tag{2.29}$$

Subtract (2.29) from (2.28) to obtain

$$-\nabla \bullet \left\{\vec{E}_a \times \vec{H}_b - \vec{E}_b \times \vec{H}_a\right\} = \vec{E}_a \bullet \vec{J}_b - \vec{H}_a \bullet \vec{M}_b - \vec{E}_b \bullet \vec{J}_a + \vec{H}_b \bullet \vec{M}_a \tag{2.30}$$

Integrating (2.30) over a volume containing all sources and by using the divergence theorem we obtain

$$-\oiint_s \left\{\vec{E}_a \times \vec{H}_b - \vec{E}_b \times \vec{H}_a\right\} \bullet d\vec{s} = \iiint_v \left\{\vec{E}_a \bullet \vec{J}_b - \vec{H}_a \bullet \vec{M}_b - \vec{E}_b \bullet \vec{J}_a + \vec{H}_b \bullet \vec{M}_a\right\} dv \tag{2.31}$$

Equation (2.31) is the most general form of reciprocity theorem.

2.4.1 *Source-free region*

If there are no sources in the volume of interest, terms on the right side of (2.31) are zero.

$$\therefore \oiint_s \left\{\vec{E}_a \times \vec{H}_b - \vec{E}_b \times \vec{H}_a\right\} \bullet d\vec{s} = 0 \tag{2.32}$$

Equation (2.32) is known as the Lorentz reciprocity theorem [1].

2.4.2 Infinite region or a region bounded by perfect conductors

If we consider a volume bounded by a surface S at infinity, the integrand on the left of (2.32) is zero everywhere. This is easy to prove using the Sommerfeld radiation condition stated in (2.33) for both sets of fields as

$$r\{\vec{E} - \eta_0 \vec{H} \times \hat{r}\} \to 0 \text{ as } \quad r \to \infty \tag{2.33}$$

where $\hat{r}$ is a unit vector in the r direction and η_0 is the intrinsic impedance of free space.

Left of (2.31) becomes zero also for the case of S being a perfect conductor. Then (2.31) becomes

$$\iiint\limits_{v} \{\vec{E}_a \bullet \vec{J}_b - \vec{H}_a \bullet \vec{M}_b\}\mathrm{d}v = \iiint\limits_{v} \{\vec{E}_b \bullet \vec{J}_a - \vec{H}_b \bullet \vec{M}_a\}\mathrm{d}v \tag{2.34}$$

Using the reaction concept [2] the above equation may be stated as

$$\langle a, b\rangle = \langle b, a\rangle \tag{2.35}$$

The term $\langle a,b\rangle$ is called the reaction of 'a' fields with 'b' sources. Reciprocity theorem has been widely applied in microwave circuits, antennas and scattering problems.

2.5 Vector and scalar potentials

Let us consider Maxwell's equations (2.1) and (2.2) in a region of space.

$$\nabla \times \vec{E} = -j\omega\mu\vec{H} \tag{2.36}$$

$$\nabla \times \vec{H} = j\omega\varepsilon\vec{E} + \vec{J} \tag{2.37}$$

Since $\nabla \bullet \vec{B} = 0$ everywhere $\vec{B}$ can be expressed as a curl of a vector.

$$\text{Let } \vec{B} = \mu\vec{H} = \nabla \times \vec{A} \tag{2.38}$$

where $\vec{A}$ is the magnetic vector potential [3]. From (2.36) and (2.38) we obtain

$$\nabla \times \left(\vec{E} + j\omega\vec{A}\right) = 0 \tag{2.39}$$

Since the curl of a gradient of any scalar function is zero, we express the electric field in terms of $\vec{A}$ and an electric scalar potential ϕ, as given next.

$$\vec{E} + j\omega\vec{A} = -\nabla\phi \quad \text{or} \quad \vec{E} = -j\omega\vec{A} - \nabla\phi \tag{2.40}$$

By eliminating $\vec{E}$ and $\vec{H}$ in (2.37) using (2.38) and (2.40) we obtain

$$\nabla \times \frac{1}{\mu}\nabla \times \vec{A} = j\omega\varepsilon\left[-j\omega\vec{A} - \nabla\phi\right] + \vec{J} \tag{2.41}$$

Assuming that the medium is homogeneous the above equation may be rewritten as

$$\nabla \times \nabla \times \vec{A} = k^2\vec{A} - j\omega\mu\varepsilon\nabla\phi + \mu\vec{J} \tag{2.42}$$

The use of the vector identity $\nabla \times \nabla \times \vec{A} = \nabla\nabla \bullet \vec{A} - \nabla^2\vec{A}$, and Lorentz gauge $\nabla \bullet \vec{A} = -j\omega\mu\varepsilon\phi$, yields the vector Helmholtz equation (2.43) for $\vec{A}$. The term $\nabla^2\vec{A}$ is the vector Laplacian of $\vec{A}$.

$$\nabla^2\vec{A} + k^2\vec{A} = -\mu\vec{J} \tag{2.43}$$

By performing the divergence operation on (2.40) one can derive the scalar wave equation

$$\nabla^2\phi + k^2\phi = -\frac{\rho}{\varepsilon} \tag{2.44}$$

One can solve (2.43) for the given source distribution and then determine $\vec{H}$ from (2.38) and E from

$$\vec{E} = -j\omega\vec{A} + \nabla\frac{\nabla \bullet \vec{A}}{j\omega\mu\varepsilon} \tag{2.45}$$

2.5.1 Electric vector potential

The dual problem of the electric vector potential produced by a magnetic current is solved in the charge-free region specified by $\nabla \bullet \vec{D} = 0$. Maxwell's equations in this case are

$$\nabla \times \vec{E} = -j\omega\mu\vec{H} - \vec{M} \tag{2.46}$$

$$\nabla \times \vec{H} = j\omega\varepsilon\vec{E} \tag{2.47}$$

The electric vector potential F is given by

$$\vec{D} = \varepsilon\vec{E} = -\nabla \times \vec{F}$$

Using duality [1] we obtain

$$\overline{H} = -j\omega\vec{F} - \nabla\varphi_{\mathrm{m}} \tag{2.48}$$

where φ_{m} is the scalar magnetic potential

$$\nabla^2\vec{F} + k^2\vec{F} = -\varepsilon\vec{M} \tag{2.49}$$

and

$$\vec{H} = -j\omega\vec{F} + \nabla\frac{\nabla \bullet \vec{F}}{j\omega\mu\varepsilon} \tag{2.50}$$

2.6 The image principle

The image principle allows one to work with the electric and/or magnetic current sources above an infinite planar PEC as shown in Figure 2.2(a). In order to create an equivalent problem for the half space above the PEC, we place image currents below the PEC and remove the PEC. Each image current and the corresponding original current have the same strength and shape, and they are equidistant from the plane of the PEC as shown in Figure 2.2(b). The original currents and the image currents make the total tangential electric field zero in the plane where the PEC was located originally, thus satisfying the uniqueness conditions for the electromagnetic fields for the half space above the PEC plane [1]. Thus the image principle works to produce the right equivalence for the upper half space.

In certain equivalent problems one uses an infinite planar PMC. If such a PMC exists in place of PEC in Figure 2.2, all the image currents will be reversed.

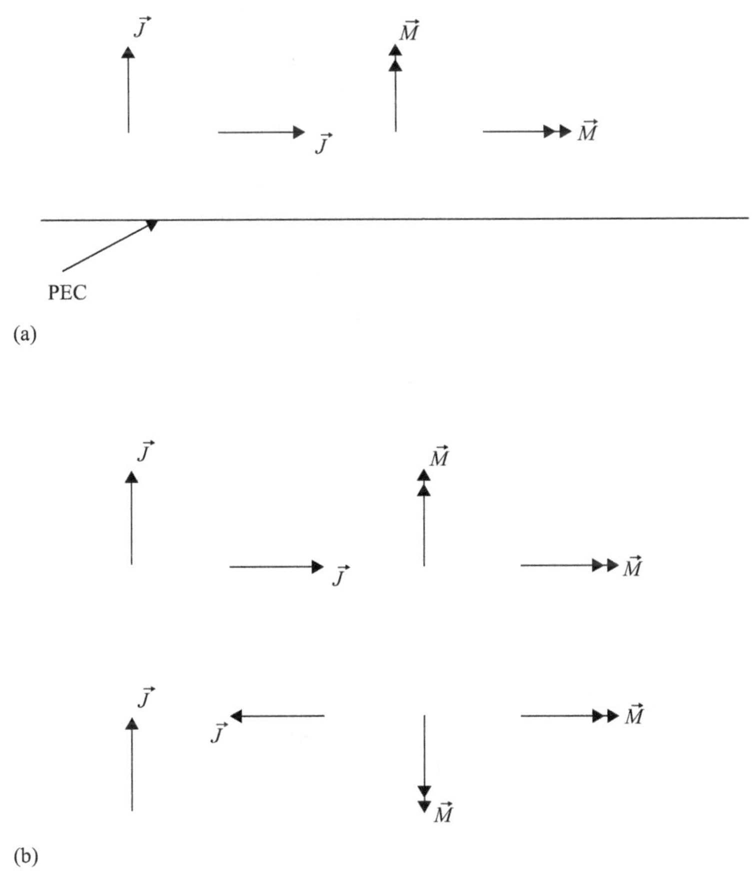

Figure 2.2 Application of the image principle: (a) electric and magnetic current sources above an infinite planar PEC; (b) equivalent problem for the upper half space

2.7 The field equivalence principle

Let us consider a set of sources inside a volume called region I, bounded by a surface S shown in Figure 2.3(a). The medium inside the volume has permittivity ε and permeability μ. The exterior region called region II is free space with permeability μ_0 and permittivity ε_0. The electric and magnetic fields produced by the sources are $\vec{E}$ and $\vec{H}$, respectively. Love's [4] equivalence principle allows us to create an equivalent problem that is valid for the region outside the volume, as shown in Figure 2.3(b). In the equivalent problem, the region I consists of no sources and null fields. At the boundary surface equivalent electric and magnetic currents given by (2.51) and (2.52) are set up.

$$\vec{J}_s = \hat{n} \times \vec{H} \tag{2.51}$$

$$\vec{M}_s = -\hat{n} \times \vec{E} \tag{2.52}$$

The unit vector $\hat{n}$ is normal to the boundary surface S and is directed into region II.

One could fill up region I with any material in Figure 2.3(b) since it consists of null fields and no sources. Therefore choosing the region I material same as that of region II makes it easy to compute the fields of the equivalent currents since they radiate in a homogeneous medium. One could also fill up region I with PEC or PMC [5]. These choices will short the electric and magnetic surface currents, respectively. Consequently, the number of unknowns on S will be reduced. However, the computation of the fields will be substantially complicated in those cases. A tutorial on the field equivalence principle with examples of planar cases has been presented by Rengarajan and Rahmat-Samii [6,7].

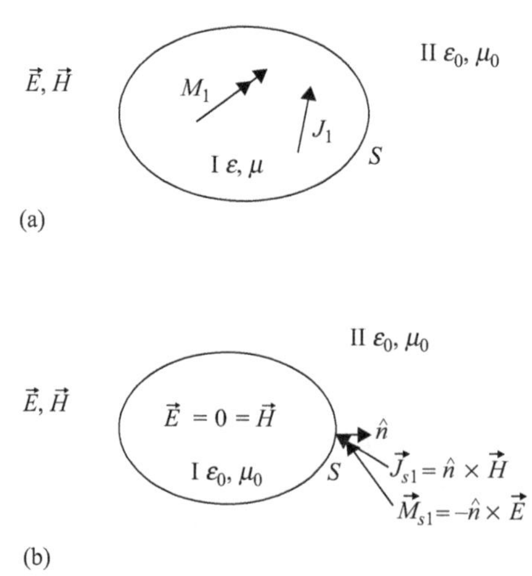

Figure 2.3 (a) Geometry of the original problem. (b) Love's equivalence for region II. (a, b) © IEEE 2000, reprinted with permission from [6]

2.8 Green's functions

Green's functions are solutions to the wave equations subject to boundary conditions, with point source excitations.

2.8.1 Free-space Green's function

Let us consider the Helmholtz equation given by

$$\nabla^2 \vec{G}_0(\vec{R},\vec{R}') + k^2 \vec{G}_0(\vec{R},\vec{R}') = -\hat{z}\ \delta(\vec{R}-\vec{R}') \tag{2.53}$$

where ∇^2 is the vector Laplacian operator, $\vec{G}_0$ is the free-space vector Green's function for the vector potential produced by the z-directed point source located at $\vec{R}'$. $\vec{R}$ and $\vec{R}'$ are position vectors of the field point and the source point, respectively. The solution should vanish at $R=\infty$. Thus $\vec{G}_0(\vec{R},\vec{R}') = \hat{z}\frac{e^{-jk|\vec{R}-\vec{R}'|}}{4\pi|\vec{R}-\vec{R}'|}$ [1]. The magnetic vector potential produced by an electric current distribution, $\vec{J}(\vec{R}')$, is given by

$$\vec{A}(\vec{R}) = \mu \iiint_V \vec{J}(\vec{R}') G_0(\vec{R},\vec{R}')\ \mathrm{d}v' \tag{2.54}$$

It is noted that the vector sign in the Green's function is left out since the direction of the vector potential is the same as that of the current density vector. The volume integral is carried out over the entire region of the current distribution. Alternatively we can express the Green's function as a dyad, noting that only the diagonal terms in the dyad will contribute to the free space dyadic Green's function, that is,

$$\overline{\overline{G}}_0(\vec{R},\vec{R}') = \frac{e^{-jk|\vec{R}-\vec{R}'|}}{4\pi|\vec{R}-\vec{R}'|}\overline{\overline{I}} \tag{2.55}$$

satisfies

$$\nabla^2 \overline{\overline{G}}_0(\vec{R},\vec{R}') + k^2 \overline{\overline{G}}_0(\vec{R},\vec{R}') = -\overline{\overline{I}}\ \delta(\vec{R}-\vec{R}') \tag{2.56}$$

where the unit dyad is given by

$$\overline{\overline{I}} = \begin{bmatrix} \hat{x}\hat{x} & 0 & 0 \\ 0 & \hat{y}\hat{y} & 0 \\ 0 & 0 & \hat{z}\hat{z} \end{bmatrix} \tag{2.57}$$

Then (2.54) may be written in terms of the dyadic Green's function as

$$\vec{A}(\vec{R}) = \mu \iiint_V \overline{\overline{G}}_0(\vec{R},\vec{R}') \bullet \vec{J}(\vec{R}')\mathrm{d}v' \tag{2.58}$$

2.8.2 Dyadic Green's function for the magnetic field of a magnetic current in a rectangular waveguide

The field scattered by a slot cut in a rectangular waveguide or coupling between two waveguides by a slot may be determined with the knowledge of the Green's functions in the rectangular waveguide. The slot aperture is shorted out and a magnetic equivalent current equal to $\vec{M} = \vec{E} \times \hat{n}$ is placed in the location of the slot. Here $\vec{E}$ is the electric field in the slot aperture and $\hat{n}$ is a unit normal vector directed into the waveguide. The magnetic field produced by $\vec{M}$ may be obtained from the integral

$$\vec{H}(\vec{R}) = \iiint_V \overline{\overline{G}}_1(\vec{R}, \vec{R}') \bullet \vec{M}(\vec{R}') \mathrm{d}v' \tag{2.59}$$

The dyadic Green's function satisfies the following differential equation:

$$\nabla \times \nabla \times \overline{\overline{G}}_1(\vec{R}, \vec{R}') - k^2 \overline{\overline{G}}_1(\vec{R}, \vec{R}') = -j\omega\varepsilon \overline{\overline{I}}\, \delta(\vec{R} - \vec{R}') \tag{2.60}$$

and the boundary conditions

$$\hat{n} \times \nabla \times \overline{\overline{G}}_1(\vec{R}, \vec{R}') = 0 \quad \text{on waveguide walls } x = 0, x = a, y = 0 \text{ and } y = b. \tag{2.61}$$

The dyadic Green's function is readily available [8] and is given next for the time harmonic field with time dependence in the form $\exp(j\omega t)$.

$$\begin{aligned}\overline{\overline{G}}_1(\vec{R}, \vec{R}') = &-\frac{\omega\varepsilon}{ab} \sum_m \sum_n \frac{(2-\delta_0)}{\beta_{mn}(k^2 - \beta_{mn}^2)} \exp[\mp j\beta_{mn}(z - z')] \\ &\cdot \left[\frac{1}{k^2} \left\{ \pm j\beta_{mn} \frac{m\pi}{a} sx\, cy\, \hat{x} \pm j\beta_{mn} \frac{n\pi}{b} cx\, sy\, \hat{y} + (k^2 - \beta_{mn}^2) cx\, cy\, \hat{z} \right\} \right. \\ &\cdot \left\{ \mp j\beta_{mn} \frac{m\pi}{a} sx'\, cy'\, \hat{x} \mp j\beta_{mn} \frac{n\pi}{b} cx'\, sy'\, \hat{y} + (k^2 - \beta_{mn}^2) cx'\, cy'\, \hat{z} \right\} \\ &+ \left. \left\{ \frac{n\pi}{b} sx\, cy\, \hat{x} - \frac{m\pi}{a} cx\, sy\, \hat{y} \right\} \left\{ \frac{n\pi}{b} sx'\, cy'\, \hat{x} - \frac{m\pi}{a} cx'\, sy'\, \hat{y} \right\} \right] \\ &+ \frac{j\omega\varepsilon}{k^2} \hat{z}\hat{z} \delta(\vec{R} - \vec{R}')\end{aligned} \tag{2.62}$$

where m and n vary from 0 to ∞ in the double summation except for $m = 0$ and $n = 0$, $sx = \sin(m\pi\, x/a)$, $sx' = \sin(m\pi\, x'/a)$, $cx = \cos(m\pi\, x/a)$, $cx' = \cos(m\pi\, x'/a)$, $sy = \sin(n\pi\, y/b)$, $sy' = \sin(n\pi\, y'/b)$, $cy = \cos(n\pi\, y/b)$, $cy' = \cos(n\pi\, y'/b)$, in $\pm$ and $\mp$ the upper sign is used if $z > z'$ while the lower sign is appropriate for $z < z'$. $\delta_0 = 1$

if $m = 0$ or $n = 0$ and it is 0 for $m \neq 0$ and $n \neq 0$. The singular term containing the delta function is zero if $\vec{R} \neq \vec{R}'$.

$$\beta_{mn} = \sqrt{k^2 - (m\pi/a)^2 - (n\pi/b)^2} \text{ for propagating modes and}$$

$$\beta_{mn} = -j\sqrt{(m\pi/a)^2 + (n\pi/b)^2 - k^2} \text{ for evanescent modes.}$$

For the case of a slot in the broad wall ($y = 0$), $cy = 1$ and $sy = 0$, whereas if $y = b$, $cy = (-1)^n$ and $sy = 0$. If $y = y' = 0$ or b, the Green's function reduces to the following simple form:

$$\overline{\overline{G}}_1\left(\vec{R},\vec{R}'\right) = -\frac{\omega\varepsilon}{k^2 ab}\sum_m\sum_n (2-\delta_0)\exp[\mp j\beta_{mn}(z-z')][g_{xx}\hat{x}\hat{x} + g_{xz}\hat{x}\hat{z} + g_{zx}\hat{z}\hat{x} + g_{zz}\hat{z}\hat{z}]$$
$$+\frac{j\omega\varepsilon}{k^2}\hat{z}\hat{z}\delta\left(\vec{R}-\vec{R}'\right) \tag{2.63}$$

where

$$\begin{aligned} g_{xx} &= \frac{\left[k^2 - (m\pi/a)^2\right]}{\beta_{mn}}\sin(m\pi x/a)\sin(m\pi x'/a), \\ g_{xz} &= \pm j(m\pi/a)\sin(m\pi x/a)\cos(m\pi x'/a), \\ g_{zx} &= \mp j(m\pi/a)\cos(m\pi x/a)\sin(m\pi x'/a) \quad \text{and} \\ g_{zz} &= \frac{\left[k^2 - \beta_{mn}^2\right]}{\beta_{mn}}\cos(m\pi x/a)\cos(m\pi x'/a). \end{aligned} \tag{2.64}$$

We have listed the tangential components of interest only here. These results have also been derived using Green's theorem by Elliott [9]. For sidewall slot problems the required Green's function components can be obtained easily from (2.62).

References

[1] Harrington R. F. *Time Harmonic Electromagnetic Fields.* McGraw-Hill, New York, 1961.

[2] Rumsey V. H. 'The reaction concept in electromagnetic theory'. *Physical Review.* 1954;**94**(6):1483–1491.

[3] Stutzman W., Thiele GA. *Antenna Theory and Design.* John Wiley and Sons, Hoboken, NJ, 2013.

[4] Love A. E. H. 'The integration of equations of propagation of electric wave'. *Philosophical Transactions of the Royal Society London, Ser A.*1901;**197**:1–45.

[5] Schelkunoff S. A. 'Some equivalent theorems of electromagnetics and their application to radiation problems'. *Bell System Technical Journal.* 1936;**15**: 99–112.

[6] Rengarajan S. R., Rahmat-Samii Y. 'The field equivalence principle: illustration of the establishment of the non-intuitive null fields'. *IEEE Antennas and Propagation Magazine.* 2000;**42**(4):122–128.

[7] Rengarajan S. R. 'Errata'. *IEEE Antennas and Propagation Magazine.* 2001; **43**(1):134.

[8] Tai C. T. *Dyadic Green functions in Electromagnetic Theory.* IEEE Press, New York, 1994.

[9] Elliott R. S. *An Introduction to Guided Waves and Microwave Circuits.* Prentice-Hall, Englewood Cliffs, NJ, 1993, p. 712.

Chapter 3
History

The most important technical advance in the 1930s was, in the opinion of the writer, the invention of the resonant slot. This was a device which was both an aperture radiator and a resonant structure. Its novelty was major. Nothing as important had appeared since Hertz invented the dipole and the loop and Lodge and Bose experimented with open-ended waveguide radiators.

The words are from Ramsay [1]. He mentions in particular Alan D. Blumlein at EMI Central Research Laboratories in the UK as the inventor of the resonant slot antenna (patent no. GB 515684, 1938). Blumlein also proposed linear arrays of slots.

3.1 The early years

Antennas for transmitting and receiving electromagnetic waves were developed already in the nineteenth century. In 1885 Edison patented a wireless communication system using vertical wire structures as antennas. Hertz and Marconi also used wire antennas. Much of the early work was made at microwave frequencies. Hertz demonstrated directional beams using a parabolic reflector at 66 cm wavelength in 1888. A waveguide radiator was used in 1894 by Lodge. Later the interest shifted to much longer wavelengths for long-distance communication, and various types of vertical top-loaded dipole antennas and antenna arrays came into use. In the 1930s and 1940s vacuum tubes for higher frequencies were developed and the interest in high frequencies arose again. Microwave directional antennas were developed for applications such as communication, radar and navigation. The technology advanced in particular during the Second World War. It was not until this time that the slotted waveguide array antenna emerged on the scene [2,3].

3.2 The golden years

The period 1935–45 has been named *the golden years* for advances in microwave technology. The reason was an urgent need to develop a radar for the detection of military targets, and both the United States and Great Britain were heavily involved in this effort. Radar development was independently carried out also in France and Germany in the 1930s, while Japan and Russia entered the field later.

In 1940 the United States and Great Britain started to exchange information about radar developments. The British brought with them an extremely important invention, the resonant-cavity magnetron, which could produce power in kilowatts at centimetre wavelengths [4]. At this time the Radiation Laboratory at MIT (1940–45) was established for the development of microwave radar and navigational equipment for military purposes. A major legacy even today is the Radiation Laboratory Series of 28 volumes published after the war by McGraw-Hill, describing the results of the research at Rad Lab in areas related to microwaves and radar. Worth mentioning in the present context are in particular the Volumes 10 and 12 edited by Nathan Marcuvitz (*Waveguide Handbook*) and Samuel Silver (*Microwave Antenna Theory and Design*), respectively [5–8].

3.3 Waveguide slot antennas

It was in Canada that the first successful waveguide slot antennas were developed. Watson and his research group at McGill University in Montreal demonstrated the usefulness of several slot configurations in rectangular waveguides: broad wall longitudinal and transverse slots and inclined edge wall slots [9]. An example of experimental data is shown in Figure 3.1, illustrating the equivalent slot resistance versus slot offset for longitudinal slots [6].

Stevenson [10] studied the electromagnetic boundary problem resulting from matching the (external) slot field, assuming resonance and the field in the waveguide. From power conservation he could predict the longitudinal slot conductance and show how it varied with the slot offset from the waveguide centreline. See also Chapter 4 in this book. Stegen [11] provided a set of carefully measured slot characteristics that could be normalised and used in array designs.

Further progress on slot modelling includes the variational solution by Oliner [12] and the moment method solution by Vu Khac [13] using pulse expansion functions, and many more. The goal was to predict the frequency dependence and the variations due to slot offset and other physical parameters such as waveguide dimensions and wall thickness. This subject will be discussed in detail in Chapter 5.

For the efficient design of slotted waveguide array antennas one has to combine slot models, mutual coupling among slots and models for the waveguide feeding system. Such an array design model was developed by Elliott [14] in 1983. Design methods are treated in Chapters 6–8. An overview of the progress in slotted waveguide array antennas up to 1999 has been presented by Rengarajan *et al.* [15].

3.4 The many shapes of slotted waveguide array antennas

An interesting mechanically scanned array antenna developed at the Radiation Lab is the long slotted coaxial line array shown in Figure 3.2. The inner conductor has a

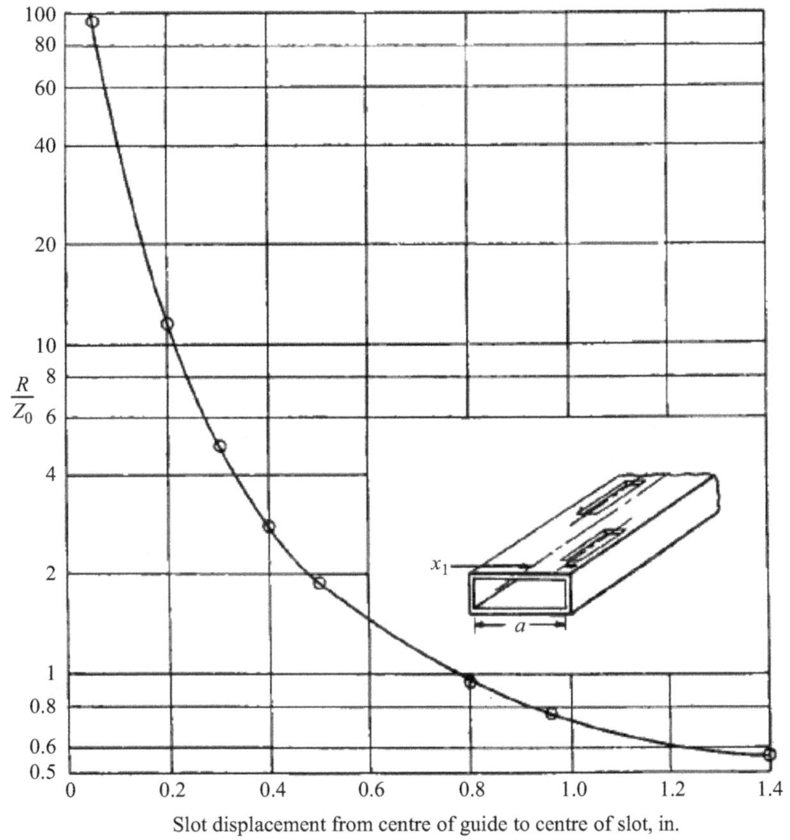

Figure 3.1 Measured longitudinal slot resistance versus offset from centreline [6]. From J. W. Dodds, E. W. Guptill and W. H. Watson, National Research Council of Canada

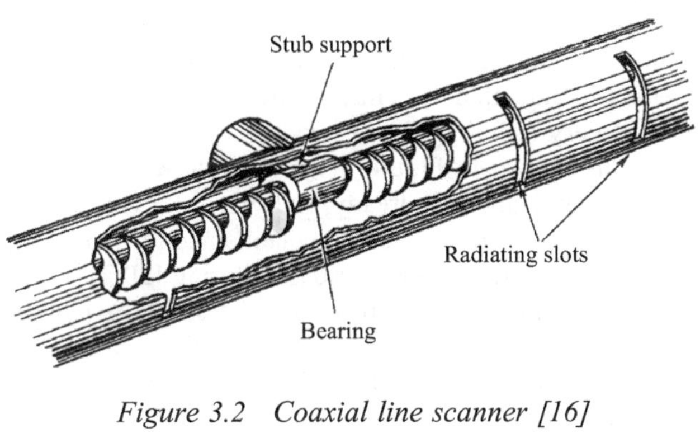

Figure 3.2 Coaxial line scanner [16]

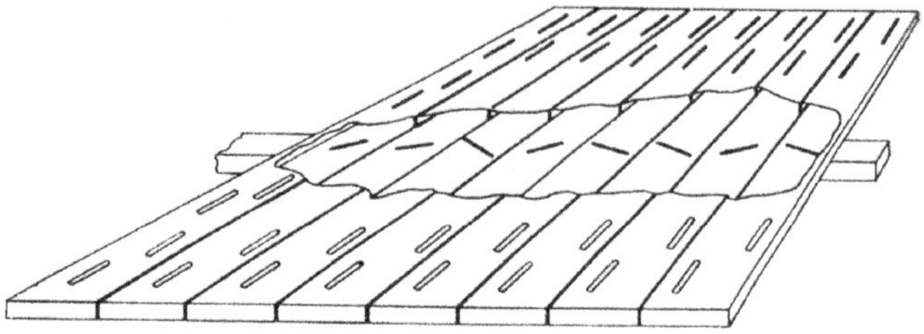

Figure 3.3 The principal layout of a slotted waveguide array antenna. © 1988 IEEE. Reprinted from [18], with permission

series of eccentric disks, displaced from the centre of the outer slotted conductor. As the inner conductor rotates, the wavelength in the line changes, causing the beam to scan.

Other scanners were also developed, for example, the Eagle Scanner, of which more than 2,000 units were produced before the end of the war. In the Eagle Scanner radiating dipoles were probe-fed from a rectangular waveguide. By mechanically changing the width of the waveguide the relative phases of the dipoles were changed, resembling frequency scanning. $\pm 30^\circ$ of scan could be realised [17]. Slots in circular waveguides are further discussed in Section 9.10.

A typical layout of a planar slotted waveguide array antenna, also called a flat plate antenna, is shown in Figure 3.3. The waveguides act both as a feeding structure and as structural support. The radiating elements are the openings (slots) in the front face. The antenna has a flat radiating surface and low thickness. It is suitable for applications in the range of about 1–100 GHz.

In Figure 3.3 we see eight waveguides with radiating slots forming a rectangular antenna aperture. The cut-out reveals the coupling from a feeding waveguide with inclined slots in the common wall. With slot dimensions and positions carefully optimised excellent radiation performance and very low losses are realised; more about this in the following chapters. Although the slotted waveguide array antenna was first used in military radar systems, it is also found in many non-military applications such as remote sensing, weather radar, navigation, radio links and several other communication systems. Figure 3.4(a) and (b) shows two slotted waveguide arrays of very different dimensions.

The SAR antenna shown in Figure 3.4(b) was used on the Earth Resources Satellite ERS-1 that was launched already in 1991, followed by ERS-2 in 1995. Several more similar systems have been launched since providing high-resolution information for land, ocean and atmospheric monitoring [20,21].

There are obviously numerous applications of slotted waveguide array antennas. A single slotted waveguide can act as a linear feed for a cylindrical reflector antenna. With a non-resonant (series) feed the antenna beam can be scanned by means of frequency variations. An example of an advanced phase/frequency

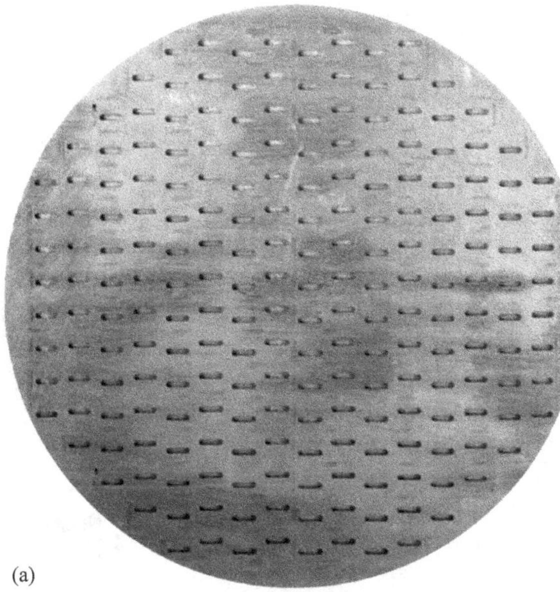

(a)

(b)

Figure 3.4 (a) Slotted array antenna. © 2017 Rantec Microwave Systems Inc. Reproduced from [19]. (b) The 10-m-long Satellite SAR antenna (C-band) for remote sensing undergoing near-field tests in an anechoic chamber. Courtesy of RUAG Space AB

scanned planar array is shown in Figure 3.5(a). It is used in a C-band radar system (Figure 3.5(b)) for pinpointing the location of hostile artillery. As seen in Figure 3.5(a) numerous vertical slotted waveguides make up the aperture. The antenna beam is scanned in azimuth by phase steering and in elevation by frequency variation. This is

(a) (b)

Figure 3.5 (a) Phase/frequency scanned slotted waveguide array; (b) radar unit. Courtesy of Saab AB

a much more cost-effective solution than two-dimensional phased arrays (active electronically steered arrays (AESA)) with thousands of phase-controlled antenna transmit/receive modules (TRMs).

The slotted waveguide array antenna has an attractive form factor: it is thin and flat. In many environments the flat shape is more acceptable than parabolic dish antennas. Other flat antennas can of course also be considered, such as microstrip patch arrays which have an equally attractive form factor. However, they have higher losses and require perhaps twice the area of the waveguide array to reach comparable performance. As higher frequencies are used the difference between the two technologies becomes even more apparent. See, for example, [22,23] for a discussion of communication slotted waveguide arrays at $\geq$20 GHz (Figure 3.6(a) and (b)).

Today slotted waveguide arrays are also proposed for very high frequencies (>100 GHz), such as in landing radar systems for space probes sent to distant planets where atmospheric attenuation is absent. This necessitates of course that the required high precision can be handled. The continued progress is also due to advances in electromagnetic modelling techniques for the analysis and design, improved fabrication techniques and new material technologies.

(a)

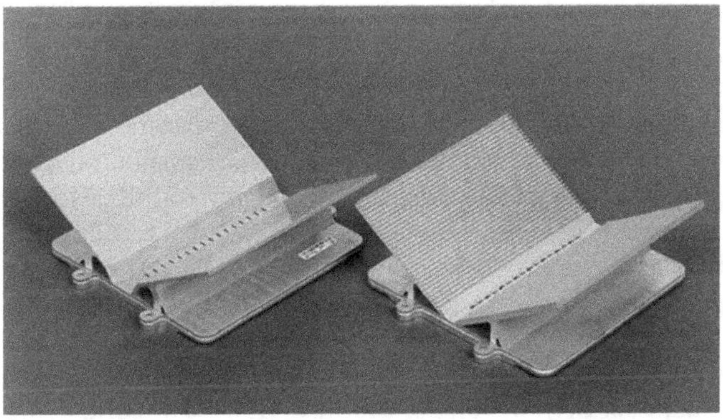

(b)

Figure 3.6 Radio link antennas: (a) antenna with narrow beam for point-to-point systems; (b) two antennas (E- and H-plane) with sector beam for point-to-multipoint systems. Courtesy of Ericsson AB

References

[1] Ramsay J. 'Highlights of antenna history'. *IEEE Antennas and Propagation Newsletter.* December 1981. pp. 8–20.

[2] Van Atta L. C. 'A history of early microwave antenna development'. *IEEE Microwave Theory and Techniques Newsletter.* October 1981. pp. 10–14.

[3] Sobol H., Tomiyasu K. 'Milestones of microwaves'. *IEEE Transactions on Microwave Theory and Techniques.* 2002;**50**(3):594–611.

[4] Skolnik M. I. *Introduction to Radar Systems.* McGraw-Hill, New York, 1962.

[5] Burns R. W., ed. 'The background to the development of the cavity magnetron. In *Radar Development to 1945.* Peter Peregrinus, London, 1988.

[6] Silver S., ed. *Microwave Antenna Theory and Design.* MIT Rad. Lab. Series, Vol. 12, McGraw-Hill, New York, 1949.
[7] Marcuvitz N., ed. *Waveguide Handbook.* MIT Rad. Lab. Series, Vol. 10, McGraw-Hill, New York, 1951.
[8] Flock W. L. 'The radiation laboratory, fifty years later'. *IEEE Antennas and Propagation Magazine*. 1991;**33**(5):43–48.
[9] Watson W. H. 'Resonant slots'. *Journal of the Institution of Electrical Engineers.* 1946;**93**(4):747–777.
[10] Stevenson A. F. 'Theory of slots in rectangular waveguides'. *Journal of Applied Physics.* 1948;**19**(1):24–38.
[11] Stegen R. J. 'Slot radiators and arrays at X-band'. *IRE Transactions on Antennas and Propagation.* 1952;**1**(1):62–84.
[12] Oliner A. A. 'The impedance properties of narrow radiating slots in the broad face of rectangular waveguide'. *IRE Transactions on Antennas and Propagation.* 1957;**5**(1):4–20.
[13] Vu Khac T. *A Study of Some Slot Discontinuities in Rectangular Waveguides*. Ph.D. Dissertation, Monash University, Australia. November 1974.
[14] Elliott R. S. 'An improved design procedure for small arrays of shunt slots'. *IEEE Transactions on Antennas and Propagation.* 1983;**31**(1):48–53.
[15] Rengarajan S. R., Josefsson L. G., Elliott R. S. 'Waveguide-fed slot antennas and arrays: a review'. *Electromagnetics*. 1999;**19**(1):3–22.
[16] Ridenour L. N. *Radar System Engineering.* MIT Rad. Lab. Series, Vol. 1, McGraw-Hill, New York, 1947.
[17] Cady W. M., Karelitz M. B., Turner L. A. *Radar Scanners and Radomes.* MIT Rad. Lab. Series, Vol. 26, McGraw-Hill, New York, 1948.
[18] Petersson R., Ingvarson P. 'The planar array antennas for the European remote sensing satellite ERS-1'. *Proceedings of the European Microwave Conference*. 1988. pp. 289–294.
[19] http://www.rantecantennas.com, retrieved 2017-04-28.
[20] van't Klooster K. 'ERS-1, European remote satellite was launched 20 years ago'. *21st International Crimean Conference on Microwave and Telecommunication Technology* (CriMiCo'2011), September 2011, Sevastopol, Crimea, Ukraine. pp. 117–118.
[21] Moreira A. 'A golden age for spaceborne SAR systems'. *20th International Conference on Microwaves, Radar and Wireless Communication* (MIKON). 2014. pp. 1–4.
[22] Svensson B. 'Dual use of slotted waveguide array antennas'. *IEEE Conference on Antennas and Propagation for Wireless Communication*, November 2000. pp. 149–152.
[23] Svensson B., Manholm L., Wikgren E. 'A waveguide sector antenna for point-to-multipoint systems'. *IEEE Antennas and Propagation Society Symposium.* 2003. pp. 1185–1188.

Chapter 4
The slot antenna

In this chapter, we will consider the single-slot antenna. We will first discuss an arbitrarily shaped aperture in an infinite conducting ground plane. Then we specialise to a rectangular resonant slot antenna and derive its radiation conductance. We continue with the special but important case of a longitudinal slot in the broad wall of a rectangular waveguide. The normalised modal functions for describing fields in waveguides are introduced. The equivalent slot conductance as seen from the feeding waveguide is then calculated. Mutual coupling to other slots and ground plane edge effects is also discussed.

4.1 An aperture in an infinite ground plane

4.1.1 Problem formulation

An aperture is located in the plane $z=0$, as illustrated by the shaded area in Figure 4.1. The plane $z=0$ is assumed to be perfectly conducting, except for the aperture, and of infinite (or very large) extent. The figure to the right presents a side view, indicating the electric field in the aperture $\overline{E}_a(x,y)$. The source of this field is not specified, but one could think of a waveguide feed from below, ending in the aperture plane as suggested by the figure.

The field in the upper half space (i.e. for $z>0$) can be uniquely determined from equivalent sources on the plane S (i.e. for $z=0$). This follows from Schelkunoff's field equivalence principles for a plane surface [1, p. 37]. The equivalent sources are electric and magnetic surface currents which replace the original sources (at $z<0$). Since we are only interested in the external fields in the upper region we can choose to impose a null field for $z<0$. Let us then place a thin electric conductor to cover all of the plane $z=0$. The electric surface current will be shorted and we are left with the magnetic current on the (shorted) aperture as the only remaining source. Over the rest of the ground plane the magnetic current is zero since the tangential electric field is zero. See also Section 2.7 on the field equivalence principle.

The last step is to combine the magnetic surface current and its image and remove the infinite ground plane; cf. [2, p. 581]. What remains is the equivalent source, a magnetic current $\overline{M}_s$ radiating in free space:

$$\overline{M}_s = -2\hat{n} \times \overline{E}_a \tag{4.1}$$

Here $\hat{n} = \hat{z}$ is the direction normal to the x/y plane.

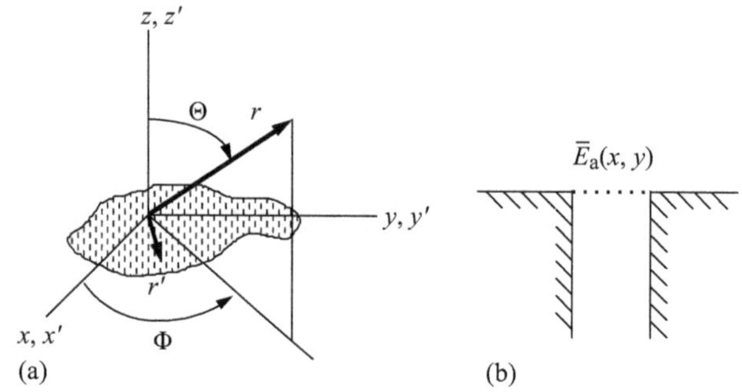

Figure 4.1 (a) A general aperture in a ground plane; (b) cross section

4.1.2 Far field

The radiated far field from the magnetic current distribution is easy to find via the electric vector potential $\overline{F}$ [2, p. 120]:

$$\overline{F} = \frac{\varepsilon}{4\pi} \iint_S \overline{M}_s \frac{e^{-jkR}}{R} dS \tag{4.2}$$

In this general expression the integration is carried out over the magnetic surface current distribution. R is the distance from any point in the source to the observation point and ε is the permittivity of the medium.[1]

In the far-field region we can write $R = r - \hat{r} \cdot \overline{r}'$ (see Figure 4.1) which gives

$$\overline{F} = \frac{\varepsilon e^{-jkr}}{4\pi r} \iint_S \overline{M}_s \, e^{j\left(k_x x' + k_y y'\right)} \, dx' dy' \tag{4.3}$$

where k_x and k_y are the directional cosines of the wave vector, that is, the far-field direction. From $\overline{F}$ we get the electric far field

$$\begin{aligned} E_\Theta &= -j\omega\eta F_\Phi \\ E_\Phi &= j\omega\eta F_\Theta \end{aligned} \tag{4.4}$$

where $E_r = 0$ in the far field, η is the free space impedance.

By inserting (4.1) in (4.3) we obtain the vector potential in the following form:

$$\begin{pmatrix} F_x \\ F_y \end{pmatrix} = \frac{\varepsilon\, e^{-jkr}}{2\pi r} \iint_S \begin{pmatrix} E_{ay} \\ -E_{ax} \end{pmatrix} e^{j\left(k_x x' + k_y y'\right)} \, dx' dy' \tag{4.5}$$

Equation (4.5) resembles a Fourier transform, except for the integration limits. However, let us extend the integration limits to infinity (makes no difference, since

[1] We have chosen the most common definition of the potential $\overline{F}$ from the relation $\varepsilon\overline{E} = -\nabla \times \overline{F}$. However, in [3,4] the relation $\overline{E} = -\nabla \times \overline{F}$ is used.

the aperture field is zero outside the aperture). Hence, we introduce the following Fourier transform:

$$\overline{f}_t(k_x, k_y) = \int_{-\infty}^{\infty}\int_{-\infty}^{\infty} \overline{E}_{\mathrm{a}}(x, y)e^{j(k_x x + k_y y)}\, dxdy \tag{4.6}$$

where $\overline{f}_t$ is a vector with the components (f_x, f_y). Thus, we can write

$$\begin{pmatrix} F_x \\ F_y \end{pmatrix} = \frac{\varepsilon\, e^{-jkr}}{2\pi\, r} \cdot \begin{pmatrix} f_y \\ -f_x \end{pmatrix} \tag{4.7}$$

From the x- and y-components of $\overline{F}$ we get the θ- and ϕ-components:

$$\begin{aligned} F_\Theta &= \left(F_x \cos\phi + F_y \sin\phi\right)\cos\theta \\ F_\Phi &= -F_x \sin\phi + F_y \cos\phi \end{aligned} \tag{4.8}$$

Combining (4.4), (4.7) and (4.8) yields

$$\begin{cases} E_\Theta = \dfrac{j\omega\eta\varepsilon\, e^{-jkr}}{2\pi r}\left(f_y \sin\phi + f_x \cos\phi\right) \\ E_\Phi = \dfrac{j\omega\eta\varepsilon\, e^{-jkr}}{2\pi r}\cos\theta\left(f_y \cos\phi - f_x \sin\phi\right) \end{cases} \tag{4.9}$$

This result shows the electric far field as a function of the Fourier transform of the aperture field. The inverse transform, cf. (4.6), is

$$\overline{E}_{\mathrm{a}}(x, y) = \frac{1}{4\pi^2}\iint \overline{f}_t\left(k_x, k_y\right)e^{-j(k_x x + k_y y)}dk_x dk_y \tag{4.10}$$

4.1.3 General field expressions

So far we have only discussed the far field. We will now derive a general solution for the electric field in all of the source-free half-space above the aperture. The solution will have to satisfy Helmholtz wave equation in this entire region, both near and far fields:

$$(\nabla^2 + k^2)\overline{E} = 0 \tag{4.11}$$

We introduce the following Fourier transform pair

$$\tilde{\overline{E}}\left(k_x, k_y, z\right) = \iint \overline{E}(x, y, z)e^{j(k_x x + k_y y)}\, dxdy \tag{4.12}$$

$$\overline{E}(x, y, z) = \frac{1}{4\pi^2}\iint \tilde{\overline{E}}\left(k_x, k_y, z\right)e^{-j(k_x x + k_y y)}\, dk_x dk_y \tag{4.13}$$

and put (4.13) into the wave equation, yielding

$$\left(\frac{\partial^2}{\partial z^2} + k_z{}^2\right)\tilde{\bar{E}} = 0 \tag{4.14}$$

The solution to this is

$$\tilde{\bar{E}}\left(k_x, k_y, z\right) = \bar{f}\left(k_x, k_y\right) \cdot e^{-jk_z z} \tag{4.15}$$

and thus

$$\bar{E}(x, y, z) = \frac{1}{4\pi^2}\iint \bar{f}\left(k_x, k_y\right)e^{-j\bar{k}\cdot\bar{r}}dk_x dk_y \tag{4.16}$$

This solution describes the electric field as a superposition of plane waves [5]. It is valid in the half plane above the aperture, both in the near field and in the far field.

The integrals in (4.13) and (4.16) extend over the full $k_x k_y$ plane. k_z is implicitly included since

$$k^2 = k_x{}^2 + k_y{}^2 + k_z{}^2 \tag{4.17}$$

That is,

$$\begin{cases} k_z = +\sqrt{k^2 - k_x^2 - k_y^2} \text{ when } k_x^2 + k_y^2 < k^2 \\ k_z = -j\sqrt{k_x^2 + k_y^2 - k^2} \text{ when } k_x^2 + k_y^2 > k^2 \end{cases} \tag{4.18}$$

k_z is negative imaginary in the so-called invisible space, characterised by attenuated waves (see Figure 4.2).

While $\bar{f}(k_x, k_y)$ in (4.16) describes the electric field as a spectrum of plane waves, the corresponding plane wave spectrum for the magnetic field is analogously given by

$$\bar{h}\left(k_x, k_y\right) = \frac{1}{k\eta}\bar{k} \times \bar{f}\left(k_x, k_y\right) \tag{4.19}$$

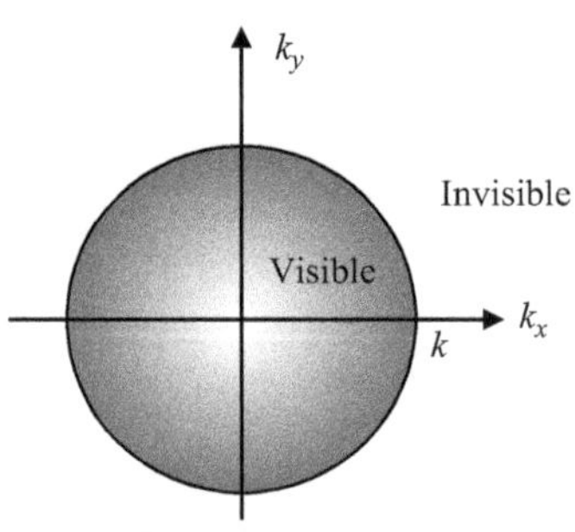

Figure 4.2 Visible and invisible space

4.1.4 Radiated complex power

For simplicity, let us assume that the aperture electric field is x-directed, that is, $\overline{E}_a(x,y) = E_x(x,y)\hat{x}$. We have therefore

$$E_x(x,y) = \frac{1}{4\pi^2}\iint f_x(k_x,k_y)e^{-j(k_xx+k_yy)}dk_xdk_y \tag{4.20}$$

From (4.19) we get the magnetic field in the aperture as

$$H_y(x,y) = \frac{1}{4\pi^2 k\eta}\iint \frac{k^2 - k_y{}^2}{k_z} f_x(k_x,k_y)e^{-j(k_xx+k_yy)}dk_xdk_y \tag{4.21}$$

since $\overline{f}\cdot\overline{k} = 0$, that is, the plane wave fields are perpendicular to the direction of propagation, see also [2, p. 624].

Knowing the aperture fields we can now formulate an expression for the complex power flow out of the aperture:

$$P = P_r + jP_i = \iint_S \frac{1}{2}\left(E_xH_y^*\right)dS = \frac{1}{8\pi^2 k\eta}\int_{-\infty}^{+\infty}\int_{-\infty}^{+\infty}\frac{k^2 - k_y^2}{k_z^*}|f_x(k_x,k_y)|^2 dk_xdk_y \tag{4.22}$$

Inserting E_x and H_y from (4.20) and (4.21) into the expression for the complex power integral given earlier leads to a sextuple integral. However, it can be reduced to this form (4.22, right member) using the relation

$$\int_{-\infty}^{+\infty} e^{-j(k_1-k_2)x}dx = 2\pi\delta(k_1 - k_2) \tag{4.23}$$

An excellent illustration of the complex power from a narrow slot aperture has been given by Rhodes [6] (see Figure 4.3). The volume within the visible region $k_x^2 + k_y^2 < k^2$ corresponds to the radiated power. The reactive power is outside this region. Note that the reactive, non-radiating (stored) power goes both negative and positive (electric and magnetic energy, respectively). When they balance each other the aperture is resonant.

4.1.5 The aperture admittance

Assume that a suitable port with excitation voltage V_0 and current I_0 is defined. Then the aperture admittance Y can be defined from the complex power P according to

$$P = P_r + jP_i = \frac{V_0 I_0^*}{2} = \frac{|V_0^2|}{2}Y^* \tag{4.24}$$

We have so far only discussed the characteristics of a general aperture. In practice an aperture is backed by some feeding arrangement, cavity or waveguide,

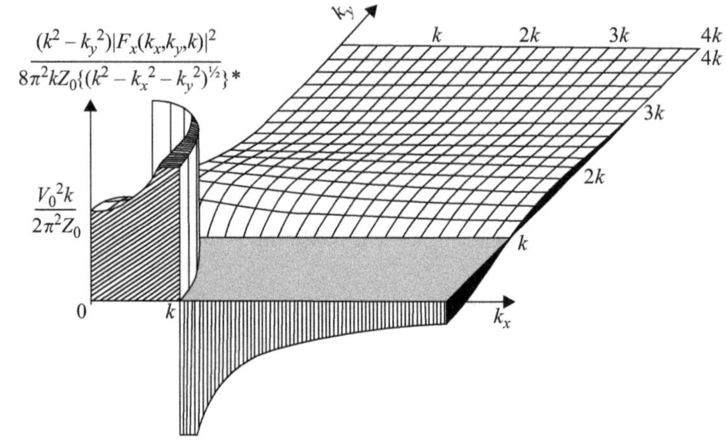

Figure 4.3 One quadrant of the complex power density from a thin half wave slot antenna with applied voltage V_0 at the centre. Reproduced from [6, Fig. 3.1, p. 50]. By permission of Oxford University Press

etc., and the total admittance is modified by these structures. Before going into more detail on this we will in the next section analyse the conductance of a narrow rectangular slot in an infinite ground plane.

4.2 The rectangular slot antenna

Consider a rectangular slot in a large ground plane coincident with the x/z-plane as shown in Figure 4.4. The feeding voltage is V_0 in the slot centre. The slot is narrow, z-directed, with width w and length $2L$.

As in (4.24) we can relate the slot conductance G_{slot} to the radiated real power P_r:

$$\mathrm{Re}(P) = P_r = \frac{1}{2} V_0^2 G_{slot} \tag{4.25}$$

We will find the radiated power by integrating the power pattern from the slot. Since a narrow slot and a thin dipole have identical radiation patterns (if E- and H-fields are interchanged) we can use known results for the electric dipole, the dual problem [3].

In the dipole case with a sinusoidal current I_z fed in the dipole centre

$$I_z(z) = I_0 \sin[k(L - |z|)]$$

the pattern is [4]

$$E_\theta = -\frac{j\eta I_0}{2\pi} \frac{e^{-jkr}}{r} \frac{\cos(kL\cos\theta) - \cos(kL)}{\sin\theta} \tag{4.26}$$

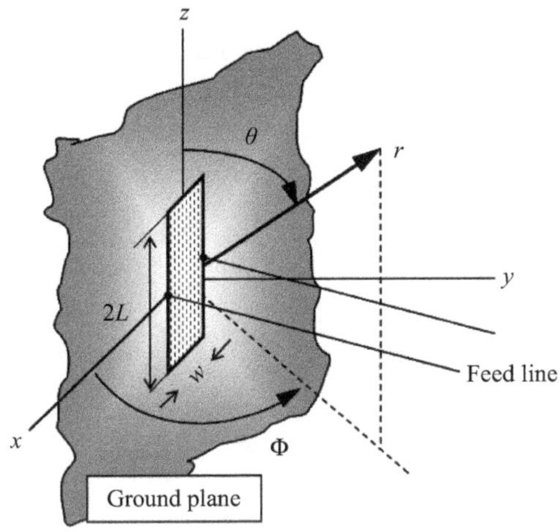

Figure 4.4 A rectangular slot antenna in a large ground plane

With a resonant length $2L = \lambda/2$ this becomes

$$E_\theta = -\frac{j\eta I_0}{2\pi}\frac{e^{-jkr}}{r}\frac{\cos\left(\frac{\pi}{2}\cos\theta\right)}{\sin\theta} \tag{4.27}$$

The corresponding resonant slot is assumed to be excited according to

$$E_{sx}(z) = \frac{V_0}{w}\cos kz \tag{4.28}$$

The equivalent magnetic current source is

$$\overline{M} = -2\overline{E} \times \hat{y}, \tag{4.29}$$

which is z-directed, hence

$$M_z(z) = -2\frac{V_0}{w}\cos kz \tag{4.30}$$

The resulting far field is

$$H_\theta = \frac{jV_0}{\eta\pi}\frac{e^{-jkr}}{r}\frac{\cos\left(\frac{\pi}{2}\cos\theta\right)}{\sin\theta} \tag{4.31}$$

The last expression is obtained directly from the dipole case (4.27), by replacing the electric current I_0 by the magnetic current $-2V_0/w$ and have the impedance of free space η replaced by $1/\eta$.

The radiated power is obtained by integrating the radiated average power density in the far field (here we assume that the slot radiates only over one-half space). The power density is

$$|S| = \frac{1}{2}|\overline{E} \times \overline{H}^*| = \frac{1}{2}\eta|H_\theta|^2 \tag{4.32}$$

thus

$$P_r = \eta\frac{1}{2}\int_0^\pi d\Phi \int_0^\pi r^2 \sin\theta |H_\theta|^2 d\theta \tag{4.33}$$

We get

$$P_r = \frac{V_0^2}{\eta\pi}I, \text{ where } I = \int_0^{\pi/2} \frac{\cos^2\left(\frac{\pi}{2}\cos\theta\right)}{\sin\theta} d\theta \tag{4.34}$$

This last integral is a bit complicated to evaluate. The solution can be formulated in terms of the cosine integral function *Ci*(*x*), see, for example, [7, p. 224], or be solved numerically.

Changing variables in (4.34) we obtain for the integral *I*

$$I = \frac{1}{4}\int_0^{2\pi} \frac{1 - \cos v}{v} dv \tag{4.35}$$

The cosine integral function *Ci*(*x*) can be written

$$Ci(x) = \gamma + \ln x - \int_0^x \frac{1 - \cos v}{v} dv \tag{4.36}$$

where $\gamma = 0.5772$ is Euler's constant. Thus

$$I = \frac{1}{4}[0.5772 + \ln(2\pi) - Ci(2\pi)] \tag{4.37}$$

Ci(*x*) is tabulated in Standard Mathematical Tables [8, p. 532] and we obtain finally

$$P_r = 0.609\frac{V_0^2}{\eta\pi} \tag{4.38}$$

We have already in (4.25)

$$P_r = \frac{1}{2}V_0^2 G_{\text{slot}} \qquad (4.39) = (4.25)$$

From the last two expressions we obtain $G_{\text{slot}} = \dfrac{0.609}{60\pi^2} \approx \dfrac{1}{1{,}000}$ siemens, thus about 1,000 Ω (or 500 Ω if radiating on both sides of the ground plane).

The impedance Z_d of a dipole and that of a slot Z_s (complementary antennas) are related as

$$Z_d Z_s = \eta^2/4 \tag{4.40}$$

(Booker's relation) [4,9] where η is the intrinsic impedance of free space ~120π Ω. The typical half-wave dipole has about 73 Ω impedance. From (4.40) we can expect the complementary slot to have approximately 35,000/73 = 479 Ω. The conclusion is that a slot antenna has a rather high impedance compared to common transmission line impedances with typically 50 or 70 Ω characteristic impedance. This might indicate a potential matching problem. This can be solved in actual slot antennas and slot arrays as we will see in subsequent chapters.

4.3 Waveguide modes

For the analysis of waveguide-fed slot antennas we need adequate solutions for the fields in the waveguide. A set of normalised expressions for the waveguide modes is presented here, starting with a general case and later specialising to TE_{10} modes.

4.3.1 Definitions

A general expression for the waveguide fields can be written as [1]

$$\boxed{\begin{aligned}\overline{E}(x,y,z) &= \sum_i A_i\left(\overline{e}_i^t \pm e_i^z\hat{z}\right)e^{\mp\gamma_i z}\\ \overline{H}(x,y,z) &= \sum_i A_i\left(\pm\overline{h}_i^t + h_i^z\hat{z}\right)e^{\mp\gamma_i z}\end{aligned}} \tag{4.41}$$

where A_i is the amplitude of mode i. The superscript t indicates transverse fields, that is, x- and y-components and z the direction of propagation (Figure 4.5).

4.3.1.1 Sign convention

The time dependence $e^{j\omega t}$ is assumed. The upper (lower) sign in (4.41) refers to propagation in the positive (negative) z direction. The functions $\overline{e}_i^t, \overline{h}_i^t, e_i^z, h_i^z$ are all independent of the direction of propagation. These conventions are identical to those in [1] and other places.

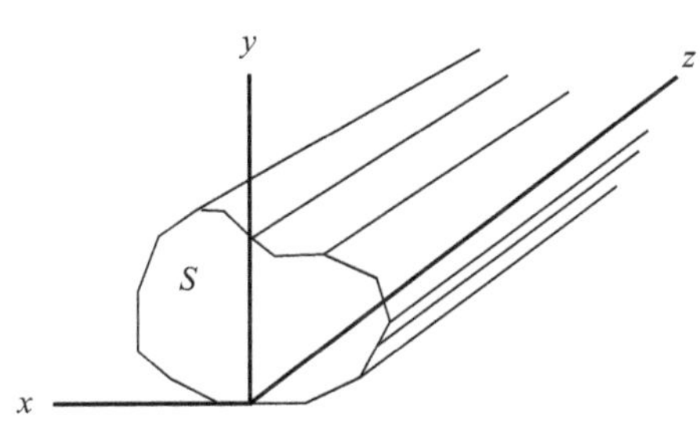

Figure 4.5 A waveguide with arbitrary cross section (S)

4.3.1.2 Modal indices

The mode number i can be understood as a reordering of the triple index pmn, where $p=1$ indicates a TE wave and $p=2$ indicates a TM wave. mn is the usual mode index for variations in the x and y directions, respectively.[2]

4.3.1.3 Modal admittance

$$\begin{aligned}&\text{For TE waves:} \quad Y_i = Y_{1mn} = \gamma_i/(j\omega\mu_0)\\&\text{For TM waves:} \quad Y_i = Y_{2mn} = j\omega\varepsilon_0/\gamma_i\end{aligned} \tag{4.42}$$

4.3.1.4 Normalisation

We choose

$$\iint_S \overline{e}_i^t \cdot \overline{e}_i^t dS = \delta_{ij} = \begin{cases} 0, & i \neq j \\ 1, & i = j \end{cases} \tag{4.43}$$

where the integration is over $S=$ the waveguide cross section.

Other criteria are sometimes used, for example, in [1,10]:

$$\iint_S \overline{e}_i^t \times \overline{h}_j^t \cdot \hat{z} dS = \begin{cases} 0, & i \neq j \\ 1, & i = j \end{cases}$$

which differs from our choice by a factor = the modal admittance. See also [11].

Our choice according to (4.43) results in

$$\iint_S \overline{e}_i^t \times \overline{h}_j^t \cdot \hat{z} dS = Y_i \cdot \delta_{ij} \tag{4.44}$$

4.3.2 *The rectangular waveguide*

The waveguide has a cross section $a \times b$ with z the direction of propagation (see Figure 4.6). The materials involved are assumed to be lossless.

Propagation constant γ_i

$$\gamma_i = \alpha_i + j\beta_i \tag{4.45}$$

$$\gamma_i^2 = {k_{c,mn}}^2 - k^2 \tag{4.46}$$

where

$${k_{c,mn}}^2 = \left(\frac{m\pi}{a}\right)^2 + \left(\frac{n\pi}{b}\right)^2 \tag{4.47}$$

and

$$k = \frac{2\pi}{\lambda}$$

[2] Note, however, that Collin [1] uses nm.

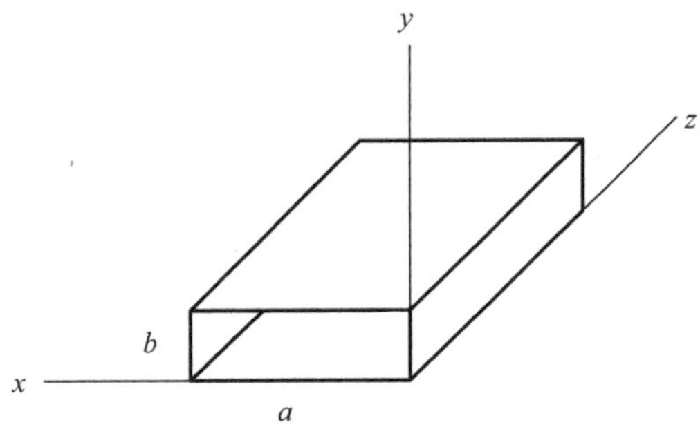

Figure 4.6 Rectangular waveguide

γ_i is positive real for non-propagating waves. It is positive imaginary for propagating waves.

4.3.2.1 Normalised mode expressions

TE case:

$$\bar{e}^t_{mn} = \frac{1}{k_{c,mn}}\sqrt{\frac{\varepsilon_m\varepsilon_n}{ab}}\left[\frac{n\pi}{b}\cos\frac{m\pi x}{a}\sin\frac{n\pi y}{b}\hat{x} - \frac{m\pi}{a}\sin\frac{m\pi x}{a}\cos\frac{n\pi y}{b}\hat{y}\right] \tag{4.48}$$

$$\bar{h}^t_{mn} = \frac{\gamma_{mn}}{j\omega\mu_0 k_{c,mn}}\sqrt{\frac{\varepsilon_m\varepsilon_n}{ab}}\left[\frac{m\pi}{a}\sin\frac{m\pi x}{a}\cos\frac{n\pi y}{b}\hat{x} + \frac{n\pi}{b}\cos\frac{m\pi x}{a}\sin\frac{n\pi y}{b}\hat{y}\right] \tag{4.49}$$

$$h^z_{mn} = \frac{k_{c,mn}}{j\omega\mu_0}\sqrt{\frac{\varepsilon_m\varepsilon_n}{ab}}\cos\frac{m\pi x}{a}\cos\frac{n\pi y}{b} \tag{4.50}$$

TM case:

$$\bar{e}^t_{mn} = \frac{-2}{k_{c,mn}\sqrt{ab}}\left[\frac{m\pi}{a}\cos\frac{m\pi x}{a}\sin\frac{n\pi y}{b}\hat{x} + \frac{n\pi}{b}\sin\frac{m\pi x}{a}\cos\frac{n\pi y}{b}\hat{y}\right] \tag{4.51}$$

$$e^z_{mn} = \frac{2k_{c,mn}}{\gamma_{mn}\sqrt{ab}}\sin\frac{m\pi x}{a}\sin\frac{n\pi y}{b} \tag{4.52}$$

$$\bar{h}^t_{mn} = \frac{2j\omega\varepsilon_0}{\gamma_{mn}k_{c,mn}\sqrt{ab}}\left[\frac{n\pi}{b}\sin\frac{m\pi x}{a}\cos\frac{n\pi y}{b}\hat{x} - \frac{m\pi}{a}\cos\frac{m\pi x}{a}\sin\frac{n\pi y}{b}\hat{y}\right] \tag{4.53}$$

In these expressions ε_m and ε_n are the Neumann factors defined as follows:

$$\varepsilon_m = \begin{cases} 1 & \text{for} \quad m = 0 \\ 2 & \text{for} \quad m > 0 \end{cases} \tag{4.54}$$

The special case with $m = n = 0$ is often neglected since it does not contribute anything to the power flow in the waveguide [12]. It contributes, however, to the stored (reactive) power and is sometimes essential in the analysis of slots in the wall of a waveguide [13].

4.3.3 The TE_{10} mode case

In this special case the following expressions are valid.

Wave admittance: $$Y_{10} = \frac{\gamma_{10}}{j\omega\mu_0} = \frac{\beta_{10}}{\omega\mu_0} = \frac{\beta_{10}}{k\eta} = \frac{\lambda}{\lambda_g\eta} \tag{4.55}$$

Propagation constant: $$\gamma_{10}^2 = k_{c,10}^2 - k^2 \tag{4.56}$$

$$k_{c,10} = \frac{\pi}{a} \tag{4.57}$$

$$\beta_{10} = \sqrt{k^2 - \left(\frac{\pi}{a}\right)^2} = \frac{2\pi}{\lambda_g} \tag{4.58}$$

where $\lambda_g = \lambda_{10}$ is the guide wavelength. The cut-off wavelength is $\lambda_c = 2a$.

Mode expressions: $$\overline{e}_{10}^t = -\sqrt{\frac{2}{ab}}\sin\frac{\pi x}{a}\hat{y} \tag{4.59}$$

$$h_{10}^z = \frac{\pi/a}{j\omega\mu_0}\sqrt{\frac{2}{ab}}\cos\frac{\pi x}{a} \tag{4.60}$$

where x is measured from the waveguide side wall (Figure 4.6).

For the transverse magnetic field we have in general $\overline{h}_i^t = Y_i\hat{z} \times \overline{e}_i^t$ and obtain for TE_{10}

$$\overline{h}_{10}^t = Y_{10}\hat{z} \times \overline{e}_{10}^t = Y_{10}e_{10}^t\hat{x} \tag{4.61}$$

As an example we have the transverse magnetic field of a mode with amplitude A_{10}

$$\overline{H}_{10}^t = \pm A_{10}Y_{10}e_{10}^t\hat{x} \tag{4.62}$$

where again the upper/lower sign indicates propagation in the $+/-$ z direction, respectively.

4.4 The longitudinal slot in a waveguide wall

4.4.1 Preliminaries

Figure 4.7 shows a general aperture in a general waveguide.

Suppose the aperture field $\overline{E}_s$ (see Figure 4.7) is known. This field will excite waves in the waveguide propagating backward (amplitudes A_i^-) and forward (amplitudes A_i^+) passing through the waveguide cross sections S_1 and S_2. The amplitudes $A_i^\pm$ can be found with the aid of the Reciprocity Theorem (see e.g. [14, p. 24]), applied to the volume enclosed by $S = S_1 + S_2 + S_3$. The method is outlined next; cf. also Section 2.4.

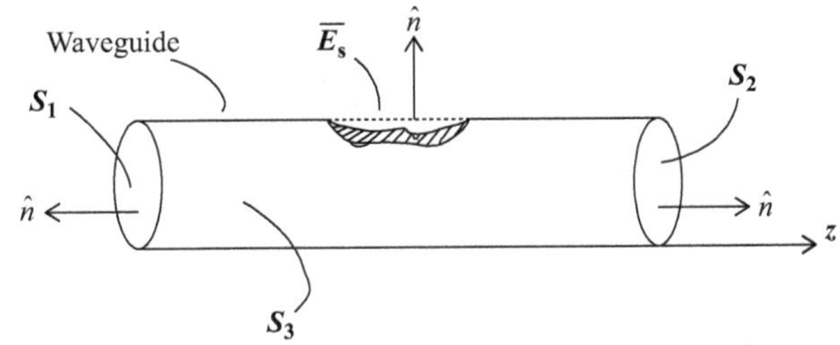

Figure 4.7 An aperture with aperture field $\overline{E}_s$ in a waveguide wall

For two field solutions $\overline{E}_1/\overline{H}_1$ and $\overline{E}_2/\overline{H}_2$ in a source-free region the following relation is true:

$$\int_S (\overline{E}_1 \times \overline{H}_2 - \overline{E}_2 \times \overline{H}_1) \cdot \hat{n}\ dS = 0 \tag{4.63}$$

We assume

$\overline{E}_1/\overline{H}_1$ = the field produced by the excited aperture
$\overline{E}_2/\overline{H}_2$ = a testing field (a normalised waveguide mode)

and rewrite (4.63) in the following form:

$$\int_S \{(\hat{n} \times \overline{E}_1) \cdot \overline{H}_2 - (\hat{n} \times \overline{E}_2) \cdot \overline{H}_1\} dS = 0 \tag{4.64}$$

Now

$\hat{n} \times \overline{E}_1 = 0$ on S_3 except in the aperture
$\hat{n} \times \overline{E}_2 = 0$ on all of S_3.

Thus

$$\int_{\text{aperture}} (\hat{n} \times \overline{E}_s) \cdot \overline{H}_2 dS + \int_{S_1} (\overline{E}_1 \times \overline{H}_2 - \overline{E}_2 \times \overline{H}_1) \cdot (-\hat{z}) dS + \int_{S_2} (\overline{E}_1 \times \overline{H}_2 - \overline{E}_2 \times \overline{H}_1) \cdot \hat{z} dS = 0 \tag{4.65}$$

Take two modes, i and j with amplitudes A_i and A_j, respectively. Then

$$\int_{\substack{\text{cross}\\ \text{section}}} \overline{E}_i \times \overline{H}_j \cdot \hat{z} dS = \pm A_i A_j Y_i \delta_{ij} \tag{4.66}$$

(plus sign if A_j propagates in the positive z direction).

Let $\overline{E}_1/\overline{H}_1$ be expressed as a summation of modes with amplitudes A_j.
Let $\overline{E}_2/\overline{H}_2$ be a normalised mode number i (amplitude = 1) propagating in the negative z direction.

This gives from (4.63) + (4.64)

$$\int_{\text{slot}} (\hat{n} \times \overline{E}_s) \cdot \overline{h}_i e^{\gamma_i z} dS + Y_i A_i - Y_i A_i - Y_i A_i - Y_i A_i = 0 \tag{4.67}$$

The result is

$$\boxed{A_i^{\pm} = \frac{1}{2Y_i} \int_{\text{aperture}} (\hat{n} \times \overline{E}_s) \cdot \overline{h}_i^{\mp} e^{\pm \gamma_i z} dS} \tag{4.68}$$

where $A_i^{\pm}$ is the ith mode amplitude (voltage) propagating in the positive/negative z-direction; Y_i is the modal admittance; $\hat{n}$ is the outward normal; $\overline{h}_i^{\mp} = \mp \overline{h}_t + h_i^z \hat{z}$ is the normalised modal function for the magnetic field, propagating in the negative/positive z-direction.

Thus, if we know the aperture field $(\overline{E}_s)$ we can find the amplitudes of all modes scattered forward and backward in the waveguide. The expression (4.68) is quite general; it is valid for many types of waveguides with circular, rectangular and even ridge-shaped cross sections.

4.4.2 The longitudinal rectangular slot

The aperture will now be specialised to a longitudinal slot with length $2L$ (see Figure 4.8). We assume the slot to be narrow enough so that only the transverse E-field (x-directed) is significant.[3] We will furthermore assume the slot to be resonant, that is, with a cosinusoidal field distribution along the z-direction.

We apply (4.68), noting that the outward normal in the slot region is y-directed, i.e.

$$(\hat{n} \times \overline{E}_s) \cdot \overline{h}_i = -E_{sx} \cdot h_i^z \tag{4.69}$$

We obtain

$$A_i^{\pm} = -\frac{1}{2Y_i} \int_{\text{slot}} E_{sx}(x, z) \cdot h_i^z(x) e^{\pm \gamma_i z} dS \tag{4.70}$$

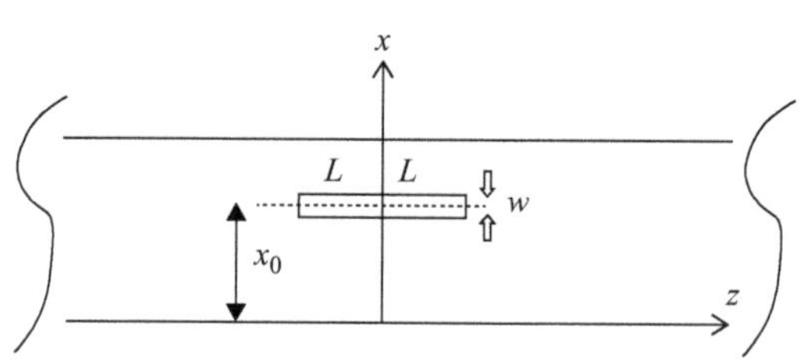

Figure 4.8 A longitudinal slot with coordinates

[3] How narrow is 'narrow enough'? From experience a possible criterion is $w/2L < 0.1$ to 0.2. For wider slots the longitudinal E-field may become significant.

The slot electric field distribution is assumed to be cosinusoidal:

$$E_{sx}(z) = \frac{V_0}{w}\cos\frac{\pi z}{2L} \tag{4.71}$$

where V_0 is the slot voltage. Inserting (4.71) in (4.70) and performing the z integration gives

$$\boxed{A_i^+ = A_i^- = \frac{-V_0}{k_0 w Y_i}\cdot\frac{\frac{\pi}{2k_0L}\cos\beta_i L}{\left(\frac{\pi}{2k_0L}\right)^2-\left(\frac{\beta_i}{k_0}\right)^2}\cdot\int_{x_0-w/2}^{x_0+w/2} h_i^z(x)dx} \tag{4.72}$$

where β_i is the propagation constant for mode i in the waveguide and x_0 is the slot position from the side wall. Note that $A_i^+ = A_i^-$ since the slot field was assumed to be symmetric.

This result is valid for longitudinal, narrow slots in arbitrary waveguides. As the next step we will study the special case with rectangular waveguides and calculate the TE_{10} scattering, see Figure 4.9.

4.4.3 *Rectangular waveguide and TE_{10} mode scattering*

For the fundamental TE_{10} mode we have the normalised mode function for the magnetic field (see Section 4.3.3):

$$h_i^z = h_{10}^z = \frac{\pi/a}{j\omega\mu_0}\sqrt{\frac{2}{ab}}\cos\frac{\pi x}{a} \tag{4.73}$$

Carrying out the integration in (4.72), neglecting variations across the narrow slot width, we obtain

$$A_{1,0}^+ = A_{1,0}^- = \frac{V_0 j\pi\sqrt{2}}{\beta_{1,0}a\sqrt{ab}}\cdot\frac{\frac{\pi}{2L}\cos\beta_{1,0}L}{\left(\frac{\pi}{2L}\right)^2-\beta_{1,0}{}^2}\cdot\cos\frac{\pi x_0}{a} \tag{4.74}$$

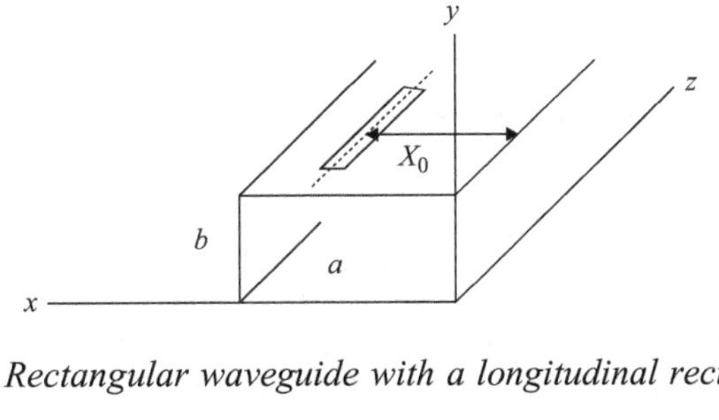

Figure 4.9 Rectangular waveguide with a longitudinal rectangular slot

Inserting the resonant length $2L = \lambda/2$ and $\beta_{1,0} = 2\pi/\lambda_g$, where λ_g is the guide wavelength, we get finally:

$$\boxed{A_{1,0}^{\pm} = V_0 \frac{j\sqrt{2}\lambda_g}{\pi\lambda}\sqrt{\frac{a}{b}}\cos\left(\frac{\pi\lambda}{2\lambda_g}\right)\cos\left(\frac{\pi x_0}{a}\right)} \tag{4.75}$$

To summarise we have now a relation between the slot voltage V_0, the scattered wave amplitude $A_{1,0}^{\pm}$ and the geometrical parameters. The simplifying assumptions leading to (4.75) were

- narrow rectangular longitudinal slot
- slot voltage V_0 given, cosinusoidal field distribution
- slot length $2L = \lambda/2$
- fundamental TE_{10} mode
- zero wall thickness

For non-zero wall thickness, V_0 applies to the inner slot field. The inner and outer fields are about the same for a practical wall thickness. Nevertheless, the thickness has an impact on the slot admittance and the slot resonant length as we will see later.

We still need to find the relation between the incident waveguide mode amplitude and the slot excitation.

4.4.4 The equivalent slot conductance

We take a circuit approach and consider the slot to be represented by a shunt admittance element (shunt conductance G_s if resonant) (see Figure 4.10). The shunt model is motivated by the symmetrical scattering from a slot with symmetrical field distribution as already discussed. The slot is excited from an incident waveguide mode with amplitude A_i^{inc}.

Only the fundamental mode is assumed to be propagating but higher-order modes will be excited near the slot. Although they do not propagate in the waveguide, they will contribute to the shunt admittance.

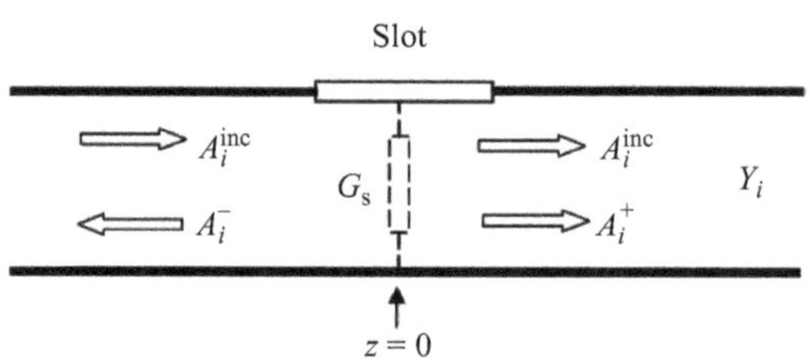

Figure 4.10 Circuit model with a shunt conductance G_s representing the slot loading of the waveguide. $A_i^{\pm}$ represent the scattered wave amplitudes (voltages) referenced to the slot position $z = 0$. Y_i is the characteristic admittance of the (waveguide) transmission line

The power radiated by the slot is represented by the power dissipated in the shunt conductance in our circuit model:

$$P_{\text{dis}} = \frac{1}{2}|V_n|^2 G_s \tag{4.76}$$

where the mode voltage $V_n = A_i^{\text{inc}} + A_i^-$.

This power must be equal to the radiated power already derived:

$$P_{\text{rad}} = 0.609\frac{V_0^2}{\eta\pi} \qquad (4.77) = (4.38)$$

The normalised shunt conductance G_s/Y_i is related to the reflection coefficient $\Gamma = A_i^-/A_i^{\text{inc}}$ and we get (assuming resonance)

$$G_s/Y_i = -2\Gamma/(1+\Gamma) = -2A_i^-/V_n \tag{4.78}$$

Furthermore, with the slot length $2L = \lambda/2$ we have the scattered TE_{10} mode amplitude

$$A_i^{\pm} = A_{1,0}^{\pm} = V_0\frac{j\sqrt{2}\lambda_g}{\pi\lambda}\sqrt{\frac{a}{b}}\cos\left(\frac{\pi\lambda}{2\lambda_g}\right)\cos\left(\frac{\pi x_0}{a}\right) \qquad (4.79) = (4.75)$$

Combining (4.78) and (4.79) and writing $Y_i = Y_{10} = G_0$ we get

$$G_s/G_0 = -2\frac{V_0}{V_n}\frac{j\sqrt{2}\lambda_g}{\pi\lambda}\sqrt{\frac{a}{b}}\cos\left(\frac{\pi\lambda}{2\lambda_g}\right)\cos\left(\frac{\pi x_0}{a}\right) \tag{4.80}$$

The two power expressions ($P_{\text{rad}} = P_{\text{dis}}$) give us

$$\left|\frac{V_0}{V_n}\right|^2 = \frac{\eta\pi}{0.609}G_s/2 \tag{4.81}$$

Squaring (4.80) leads to

$$\frac{G_s^2}{G_0^2} = \left|\frac{V_0}{V_n}\right|^2 \frac{8\lambda_g^2}{\pi^2\lambda^2}\frac{a}{b}\cos^2\left(\frac{\pi\lambda}{2\lambda_g}\right)\cos^2\left(\frac{\pi x_0}{a}\right) \tag{4.82}$$

Inserting (4.81) in (4.82) and noting $G_0 = Y_{10} = \dfrac{\beta_{10}}{k\eta} = \dfrac{\lambda}{\lambda_g 120\pi}$ gives the classical result

$$\boxed{\frac{G_s}{G_0} = 2.09\frac{a}{b}\frac{\lambda_g}{\lambda}\cos^2\left(\frac{\pi\lambda}{2\lambda_g}\right)\cos^2\left(\frac{\pi x_0}{a}\right)} \tag{4.83}$$

This important result, originally derived by Stevenson [15], shows how the conductance (i.e. the excitation) of a slot depends on the slot position. We reproduce the Stevenson result in Figure 4.11.

So far we have discussed the slot conductance for resonant slots. The assumption has been that the slot length $\lambda/2$ automatically results in slot resonance

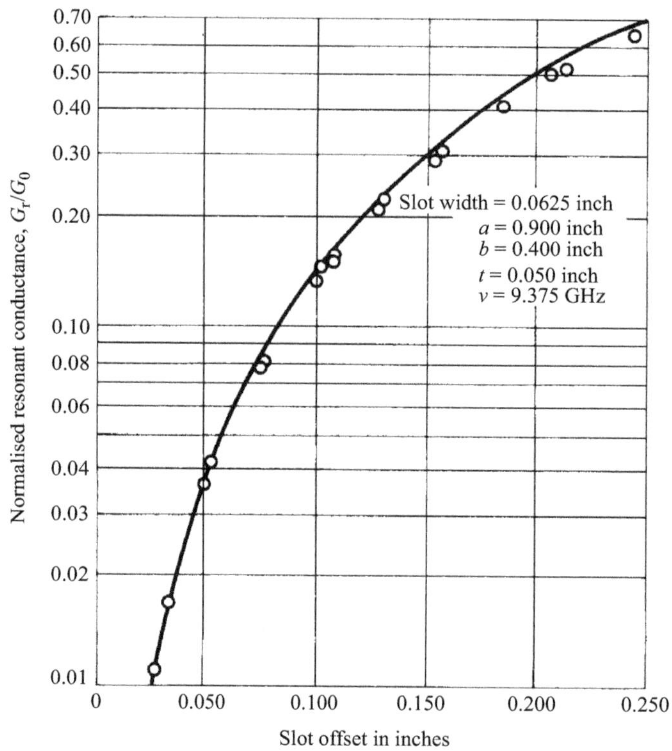

Figure 4.11 *Resonant conductance versus slot offset from the waveguide centreline for a longitudinal slot in standard X-band waveguide, calculated by Stevenson [15]. Points are experimental results according to Stegen [16]. © 1981 J. H. Elliott. Reprinted from [4], with permission*

and that the slot admittance is real. However, resonance also depends on the slot shape, the waveguide dimensions and the effects of mutual coupling among the slots in an array. We are also interested in the complete slot admittance and how it varies over a band of frequencies. A more detailed analysis is therefore necessary in order to control these parameters and optimise the slot (array) antenna for best performance including bandwidth.

Consequently it is not necessarily true that a slot length of $\lambda/2$ results in resonance. From the more detailed slot modelling examples discussed in Chapter 5, we will find that both shorter and longer slots (than $\lambda/2$) can be resonant, depending on several parameters such as slot offset, waveguide height and dielectric loading. In order to accurately calculate the slot admittance we need to include the reactive fields in the slot vicinity. For the external region we already have found one field representation that includes this: the plane wave spectrum [(4.20) and (4.21)]. For the internal waveguide region we have similarly the 'spectrum' of waveguide modes (4.68). Before going into detailed discussions on how these and other representations can be used in order to find the true slot field distribution and the slot admittance we will take a look at the mutual coupling effects.

4.5 Mutual coupling

4.5.1 Introduction

The mutual coupling between aperture-type antennas has been discussed by several authors [3,17]. For the analysis of waveguide slot arrays it has been assumed that an equivalent array of dipoles could be used as a model for the mutual coupling [18]. An improved model, especially for dielectric filled waveguides, was later developed by Elliott [19] in which the slot field is assumed to be half cosinusoidal. This agrees with assuming the slot to have the fundamental TE_{10} mode excitation. This is also our approach in our mutual coupling analysis in this chapter.

Consider the small array antenna with 52 slot elements in Figure 4.12. In this array each slot has its individual stripline feed with a connector and a matched load. The slot spacing is $d/\lambda = 0.625$ in a square lattice. In the diagram to the right are shown two measured E-plane patterns for one of the slots near the array centre, the *isolated pattern* and the *embedded pattern.* In the former case all other slots were covered with conducting tape, in the latter case all other slots were present and terminated in matched loads.

We see two important effects. The isolated E-plane element pattern would ideally be flat over $\pm 90°$, but exhibits a significant ripple. This is due to the diffraction from the ground plane edges. The second observation is the drastically modified embedded pattern due to mutual coupling among the slots. A significant narrowing and a dip occur at about $\pm 35°$, as expected from phased array theory

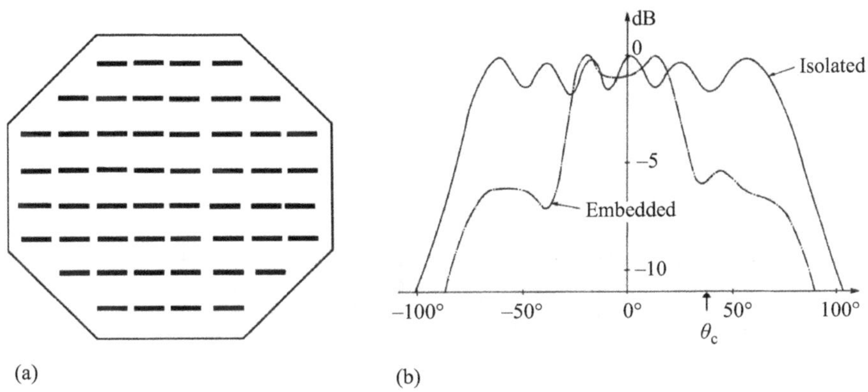

Figure 4.12 (a) A slotted stripline array; (b) measured isolated and embedded E-plane patterns for an element in the centre of the array. © IEEE 1974. Adapted from [20], with permission

[21]. See also Section 9.1. From the slot spacing the critical 'blind angle' θ_c is predicted to be at $\pm 37^\circ$ according to the grating lobe angle equation.

$$\sin\theta_c = \lambda/d - 1 \qquad (4.84)$$

The H-plane coupling for slots is much weaker than the E-plane coupling, and can sometimes be disregarded in non-scanned waveguide slotted array designs. However, our example demonstrates that the array excitation can be dramatically changed by mutual coupling and that we need to accommodate this effect in a general design procedure.

A final word on coupling definitions used here: *isolated coupling* stands for the case with only two elements present. *Array coupling* is again coupling between two elements, but now all other elements are also present and terminated in matched loads. *Active impedance/admittance* stands for the situation where all elements are fed. *Isolated (or self) impedance/admittance* is for one single element only. Typical *isolated* and *embedded patterns* are compared in Figure 4.12.

4.5.2 Mutual coupling calculations

4.5.2.1 Introduction

In the slotted waveguide array antenna the effect of the waveguide wall thickness can be modelled as short waveguides connecting the two apertures: the inner slot and the outer slot. We have discussed the equivalent slot conductance as seen from the feeding waveguide in Section 4.4.4 based on the inner slot field. We will now focus on the outer slot field for the mutual coupling calculation. The two regions will later be connected (Chapter 5).

Figure 4.13 shows a cross section of a number of waveguide openings in a large ground plane, suggesting a coupled signal from element m to element n. For the present discussion we assume that each aperture is characterised by one mode only. This mode will be the TE_{10} mode with an x-directed electric field (see Figure 4.13).

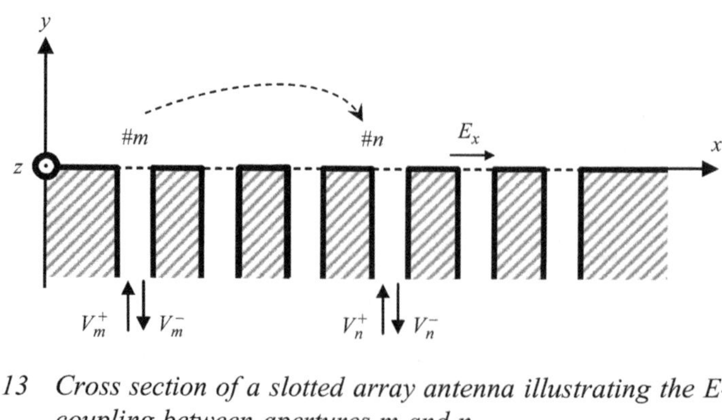

Figure 4.13 Cross section of a slotted array antenna illustrating the E-plane coupling between apertures m and n

4.5.2.2 Equivalent circuit approach

Since we assume a single-mode situation and also have reciprocity and linearity we can represent the slot array as seen from the feeding side by a simple circuit model (Figure 4.14).

For the circuit model the currents and voltages are related as follows:

$$\left.\begin{aligned} J_1 &= Y_{11}V_1 + Y_{12}V_2 + Y_{13}V_3 + \cdots \\ J_2 &= Y_{12}V_1 + Y_{22}V_2 + Y_{23}V_3 + \cdots \\ J_3 &= Y_{31}V_1 + Y_{32}V_2 + Y_{33}V_3 + \cdots \\ J_4 &= \cdots \end{aligned}\right\} \tag{4.85}$$

or in matrix form

$$[J] = [Y][V] \tag{4.86}$$

As an example we can find the matrix element $Y_{13} = J_1/V_3$ provided all the other voltages $V_1, V_2, V_4, \ldots = 0$, for example, by short circuiting the corresponding apertures. Repeating this procedure we can find the complete admittance matrix from the mutual admittances between all pairs of elements! In fact, for a regularly spaced array with identical elements, the mutual admittance values are only dependent on the relative positions between the elements (the matrix Y is a Toeplitz matrix).

We can think of the slot array as one where each slot has a feed port across the slot centre (cf. Figure 4.4). Shorting the port with a conducting element leaves two smaller apertures on each side of the short. The small apertures will be very far from resonance and do not radiate. In practice it would make no difference if the ground plane covered the slot completely. (A similar type of reasoning applies to arrays of resonant dipoles when calculating the impedance matrix. In that case the ports are left open circuited instead.)

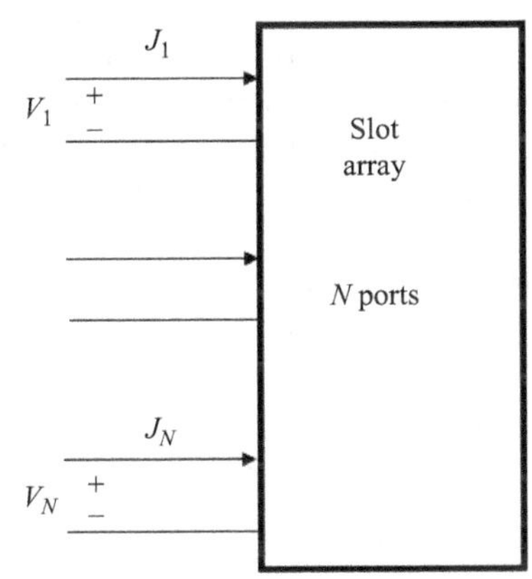

Figure 4.14 Circuit model of a slot array assuming single-mode apertures or minimum scattering antennas (MSA) [22]

4.5.2.3 Single-slot admittance

We will now analyse the self-admittance of a single slot (which we assign the number n) in a large ground plane. No other slots are present.

Let the upward and downward propagating modes in the feeding waveguide just below the slot aperture be $V_n^+ e_\mathrm{n}$ and $V_n^- e_\mathrm{n}$, respectively (cf. Figure 4.13), where e_n is the normalised (electric) modal function. Thus, the total (electric, x-directed) field at aperture n is $E^{\mathrm{int}} = (V_n^+ + V_n^-)e_\mathrm{n}$. The total ($z$-directed) magnetic field in the waveguide just below the aperture is then $H^{\mathrm{int}} = (V_n^+ - V_n^-)Y_0 e_\mathrm{n}$, where Y_0 is the modal admittance of the waveguide. The external magnetic field at the aperture can be written as $H^{\mathrm{ext}} = (V_n^+ + V_n^-)H_n^{\mathrm{ext}}(e_\mathrm{n})$, where $H_n^{\mathrm{ext}}(e_\mathrm{n})$ is a function describing the external magnetic field for a given normalised electric aperture field. We want to determine V_n^- while V_n^+ is assumed known.

We now enforce continuity of the magnetic field at the aperture, by putting $H^{\mathrm{int}} = H^{\mathrm{ext}}$, leading to $(V_n^+ - V_n^-)Y_0 e_\mathrm{n} = (V_n^+ + V_n^-)H_n^{\mathrm{ext}}(e_\mathrm{n})$. An approximate solution can be found according to the Method of Moments [23] by taking inner products with the normalised modal function e_n as testing function. We define this inner product (or projection) as follows:

$$\langle e_p, e_q \rangle = \iint_{\mathrm{slot}} e_p(x,z) \cdot e_q(x,z)\,dxdz = \delta_{p,q}, \tag{4.87}$$

and obtain

$$(V_n^+ - V_n^-)Y_0 = (V_n^+ + V_n^-)\langle H_n^{\mathrm{ext}}(e_\mathrm{n}), e_\mathrm{n} \rangle \tag{4.88}$$

The part inside the brackets $\langle\ \rangle$ is the aperture self-admittance: $Y_{nn} = \langle H_n^{\mathrm{ext}}(e_\mathrm{n}), e_\mathrm{n} \rangle$. The reflection coefficient Γ becomes

$$\Gamma = V_n^- / V_n^+ = \frac{Y_0 - Y_{nn}}{Y_0 + Y_{nn}} \tag{4.89}$$

An accurate evaluation of the self-admittance is rather complicated (cf. Section 6.4), partly because we need to find and integrate the field inside the source region. Furthermore, in order to improve the accuracy of the solution; it is often necessary to include a number of higher-order modes besides the fundamental mode in the aperture; more about this in Chapter 5.

4.5.2.4 Mutual admittance between two slot apertures

Let us now take a look at the mutual admittance between two slot apertures (isolated coupling).

The external magnetic field at aperture n due to coupling from the electric field at aperture m (the source) can be written as

$$(V_m^+ + V_m^-)H_n^{\mathrm{ext}}(e_m) + V_n^- H_n^{\mathrm{ext}}(e_\mathrm{n}) \tag{4.90}$$

The waveguide at the receiving aperture n will be excited with a wave with internal magnetic field $-V_n^- Y_0 e_\mathrm{n}$. A wave scattered into space will also appear. Note that $H_n^{\text{ext}}(e_m)$ is the external field at aperture n due to the source at aperture m.

Matching internal and external magnetic fields at aperture n gives

$$V_n^- H_n^{\text{ext}}(e_n) + (V_m^+ + V_m^-) H_n^{\text{ext}}(e_m) = -Y_0 V_n^- e_n \tag{4.91}$$

Let us also match the magnetic fields at the source aperture m, giving

$$(V_m^+ + V_m^-) H_m^{\text{ext}}(e_m) + V_n^- H_m^{\text{ext}}(e_n) = Y_0 (V_m^+ - V_m^-) e_m \tag{4.92}$$

Taking inner products with the modal functions e_m and e_n as before yields the two equations:

$$\left.\begin{aligned} V_n^- Y_{nn} + (V_m^+ + V_m^-) Y_{nm} &= -Y_0 V_n^- \\ (V_m^+ + V_m^-) Y_{mm} + V_n^- Y_{mn} &= Y_0 (V_m^+ - V_m^-) \end{aligned}\right\} \tag{4.93}$$

Note that $Y_{nm} = Y_{mn} = \langle H_n^{\text{ext}}(e_m), e_n \rangle$ is the mutual admittance between the two apertures. Eliminating V_m^- gives the mutual coupling in terms of the scattering parameter S_{mn}

$$\boxed{S_{mn} = \frac{V_n^-}{V_m^+} = \frac{2Y_0 Y_{mn}}{Y_{mn}^2 - (Y_0 + Y_{mm})^2}} \tag{4.94}$$

The coupling between just two elements (no other element present) is called the isolated coupling. For slot apertures it decays with distance approximately as $1/R$ in the E-plane. In the H-plane the decay is roughly $1/R^2$.

4.5.2.5 Array coupling

The scattering from a large array is most easily written using matrices, cf. (4.89):

$$[S] = ([Y_0] + [Y^e])^{-1}([Y_0] - [Y^e]) \tag{4.95}$$

where $[S]$ is the scattering matrix. $[Y^e]$ is the mutual admittance matrix (with the exponent e indicating the *external* aperture interface) ($[Y^i]$ is reserved for the *internal* aperture interface). $[Y_0]$ is the diagonal matrix with the (identical) modal admittances of the stub waveguides connected to the apertures:

$$[Y_0] = \begin{pmatrix} Y_0 & \cdots & 0 \\ \vdots & Y_0 & \vdots \\ 0 & \cdots & Y_0 \end{pmatrix} = Y_0[I], \tag{4.96}$$

and $[I]$ is the identity matrix.

The excitation of the aperture array would be represented by the vector $[V^+]$ if there were no coupling. With coupling we get

$$[V^+] + [V^-] = ([I] + [S])[V^+] = 2([Y_0] + [Y^e])^{-1}[Y_0][V^+] \tag{4.97}$$

4.5.2.6 Calculating the mutual coupling

We have defined the mutual admittance between two slots m and n as the inner product $Y_{nm} = Y_{mn} = \langle H_n^{\text{ext}}(e_m), e_n \rangle$. In general terms it can be expressed as the electromagnetic reaction of the two sources through the scalar product (suitably defined) of the field of one source on the current of the other [3, p. 340]. The mathematical formulation can be written as [17]

$$\boxed{Y_{mn} = \frac{-1}{V_m V_n} \iint_{\text{slot } n} \overline{E}_n \times \hat{y} \cdot \overline{H}_m dS} \tag{4.98}$$

where the normal direction is assumed to be y-directed. $\overline{E}_n$ is the field of aperture n and $\overline{H}_m$ is the magnetic field at aperture n due to the excitation of aperture m. The integration is carried out over aperture n, requiring in total the evaluation of a fourfold integral. The integral can be approximated by neglecting the field variations across the rectangular apertures in the E-plane dimension, which should work well for narrow rectangular slots.

Further simplifications aiming at reducing the cost and time of numerical evaluations have been developed; see Chapter 6 and [24].

We have already discussed the far-field radiation from aperture antennas (Section 4.1.2). We can take the same approach starting from the vector potential $\overline{F}$, although we now need to include the near-field region. We will first derive the magnetic field from a single slot. The electric vector potential is

$$\overline{F} = \frac{\varepsilon}{4\pi} \iint_S \overline{M}_s \frac{e^{-jkR}}{R} dS \qquad (4.99) = (4.2)$$

where $\overline{M}_s$ is the magnetic surface current $\overline{M}_s = -2\hat{n} \times \overline{E}_a$. The normal is y-directed (see Figure 4.15). The slot electric field is x-directed and hence the magnetic current is z-directed:

$$\overline{M}_s = 2E_a\hat{z} \tag{4.100}$$

The H-field is obtained from the electric vector potential $\overline{F}$ [1,3]:

$$\overline{H} = -\frac{1}{j\omega\mu} \nabla \times \overline{E} = \frac{1}{j\omega\mu\varepsilon} \nabla \times \nabla \times \overline{F} \tag{4.101}$$

That is,

$$\overline{H} = \frac{1}{4\pi j\omega\mu} \nabla \times \nabla \times \iint_S \overline{M}_s \frac{e^{-jkR}}{R} dS \tag{4.102}$$

We can expand

$$\nabla \times \nabla \times \overline{F} = \nabla(\nabla\overline{F}) - \nabla^2\overline{F} \tag{4.103}$$

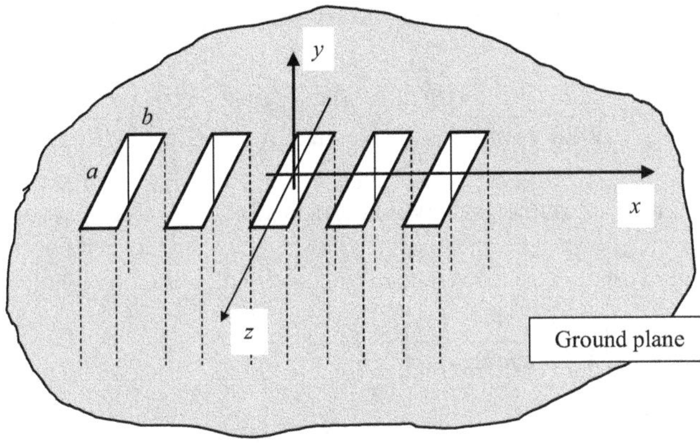

Figure 4.15 Slot apertures and major coordinate system

and, since $\overline{F}$ fulfils the wave equation,

$$\nabla^2 F + k^2 F = 0 \tag{4.104}$$

we obtain,

$$\nabla \times \nabla \times \overline{F} = \nabla(\nabla \overline{F}) + k^2 \overline{F} \tag{4.105}$$

$\overline{F}$ has only a z-component, so $\nabla(\nabla \overline{F}) = \dfrac{\partial^2}{\partial z^2} F_z$. Therefore,

$$H_z = \frac{1}{j\omega\mu\varepsilon}\left(\frac{\partial^2}{\partial z^2} + k^2\right) F_z \tag{4.106}$$

For this case it can be shown that derivation with respect to field coordinates is the same as derivation with respect to source coordinates. Thus,

$$H_z(x,z) = \frac{1}{2\pi j\omega\mu_0} \iint_{\text{slot}} E_{\text{a}}(x',z')\left(k^2 + \frac{\partial^2}{\partial z'^2}\right)\frac{e^{-jkR}}{R} dx'dz' \tag{4.107}$$

where $R = \sqrt{(x-x')^2 + (z-z')^2}$.

Neglecting variations in the aperture electric field across the slot (x-direction) we have

$$H_z(x,z) = \frac{b}{2\pi j\omega\mu_0} \int_{-a/2}^{a/2} E_{\text{a}}(z')\left(k^2 + \frac{\partial^2}{\partial z'^2}\right)\frac{e^{-jkR}}{R} dz' \tag{4.108}$$

The electric field variation along the slot is $E_{\text{a}}(z') = Ve_{10}(z')$, where $e_{10}(z')$ is the normalised modal function:

$$e_{10}(z') = -\sqrt{\frac{2}{ab}} \cos\frac{\pi z'}{a} \tag{4.109}$$

and V is the slot voltage.

Hence, the magnetic field in an arbitrary point $(x, 0, z)$ becomes

$$H_z(x,z) = -V\sqrt{\frac{b}{2a}}\frac{1}{\pi j\omega\mu_0}\int_{-a/2}^{a/2}\cos\left(\frac{\pi z'}{a}\right)\left(k^2+\frac{\partial^2}{\partial z'^2}\right)\frac{e^{-jkR}}{R}dz' \tag{4.110}$$

A form suited for numerical evaluation is obtained after partial integration, yielding

$$H_z(x,z) = -V\sqrt{\frac{b}{2a}}\frac{1}{\pi j\omega\mu_0}\left\{\frac{\pi}{a}\left[\frac{e^{-jkR_1}}{R_1}+\frac{e^{-jkR_2}}{R_2}\right]\right.$$
$$\left.+\left[k^2-\left(\frac{\pi}{a}\right)^2\right]\cdot\int_{-a/2}^{a/2}\cos\left(\frac{\pi z'}{a}\right)\frac{e^{-jkR}}{R}dz'\right\} \tag{4.111}$$

The distances R_1 and R_2 are shown in Figure 4.16.

Weighting the field in slot #n with the normalised modal function and assuming unit slot voltages (i.e. $V = 1$) now yields the mutual admittance:

$$Y_{nm} = \langle H_n^{\text{ext}}(e_{\text{m}}), e_{\text{n}}\rangle = -\iint_{\text{slot}\#n} H_z(x,z)\sqrt{\frac{2}{ab}}\cos\left(\frac{\pi z}{a}\right)dxdz \tag{4.112}$$

Again, neglecting field variations in the x-direction as before, we get

$$Y_{nm} = \langle H_n^{\text{ext}}(e_m), e_{\text{n}}\rangle = b\sqrt{\frac{2}{ab}}\int_{\text{slot}\#n} H_z(z)\cos\left(\frac{\pi z}{a}\right)dz \tag{4.113}$$

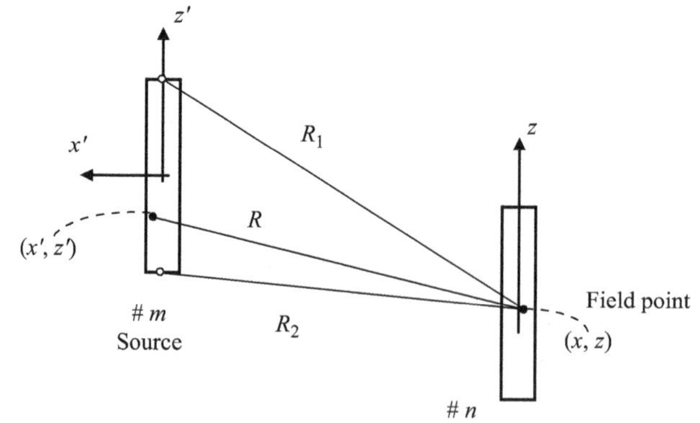

Figure 4.16 The two-slot geometry for calculating the magnetic field in slot #n due to coupling from the source slot #m. See also Section 6.4

and finally

$$\boxed{\begin{aligned} Y_{nm} = \frac{-jb}{\pi a \omega \mu_0} \int_{\#n} \cos\left(\frac{\pi z}{a}\right) \left\{ \frac{\pi}{a} \left[\frac{e^{-jkR_1}}{R_1} + \frac{e^{-jkR_2}}{R_2} \right] \right. \\ \left. + \left[k^2 - \left(\frac{\pi}{a}\right)^2 \right] \cdot \int_{\#m} \cos\left(\frac{\pi z'}{a}\right) \frac{e^{-jkR}}{R} dz' \right\} dz \end{aligned}} \tag{4.114}$$

In the literature about slotted waveguide array design, the dimension-less mutual coupling parameter g_{mn} is often introduced [19]. The mutual admittance Y_{mn} is related to g_{mn} as follows:

$$Y_{mn} = \frac{-jb}{\pi a \eta} g_{mn} \tag{4.115}$$

We need to evaluate the admittance for all unique *mn* combinations of an array in order to fill the complete admittance matrix. However, due to the Toeplitz characteristics, many relative positions repeat. For instance, in a linear array of N elements we only need to calculate the mutual admittances between element 1 and elements 2 to N. From this row we can fill the complete $N \times N$ admittance matrix. The scattering matrix is then given by

$$[S] = ([Y_0] + [Y^e])^{-1}([Y_0] - [Y^e]) \qquad (4.116) = (4.95)$$

The Toeplitz simplification is not valid if the array lattice is not exactly periodic and the slot dimensions are not identical. The consequence is that many more couplings will have to be calculated. However, the end result is still a scattering matrix $[S]$ that characterises the array.

It is usually necessary to include the mutual coupling in the design of slotted waveguide array antennas, but sometimes the coupling can be seen as a minor correction or be disregarded completely. More about this in Chapters 6 and 7. The following examples for slot aperture coupling in the E-plane and H-plane illustrate the magnitude of coupling for several cases. For these calculations periodic arrays of equal size apertures were assumed.

4.5.2.7 Examples

The mutual coupling between two elements, that is, with no other elements present, is defined as *isolated coupling*, just two slots and a common ground plane (Figure 4.17). In the full array, coupling between two slots will be influenced by the presence of the other slots, *array coupling*. One effect is that the other slots will absorb some power, thus reducing the coupling values. See the following examples showing isolated and array coupling for a linear array. The calculated results have been obtained with one aperture mode only, which appears to be sufficient in many cases. The variations across the slot width have not been included. However, a correction factor $\sim \text{sinc}^2(x)$ has been employed for the E-plane results [25]. This is not needed for more narrow slots (Figure 4.18).

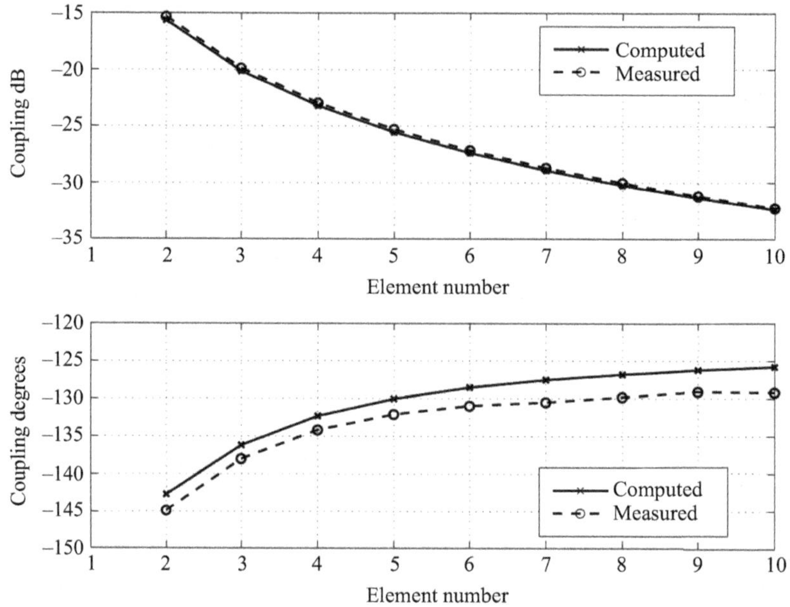

Figure 4.17 The magnitude and phase of the isolated coupling (two elements) for a 10 element E-plane array. The free space delay (phase) has been subtracted. Element no. 1 is fed. Aperture size: $0.75\lambda \times 0.20\lambda$

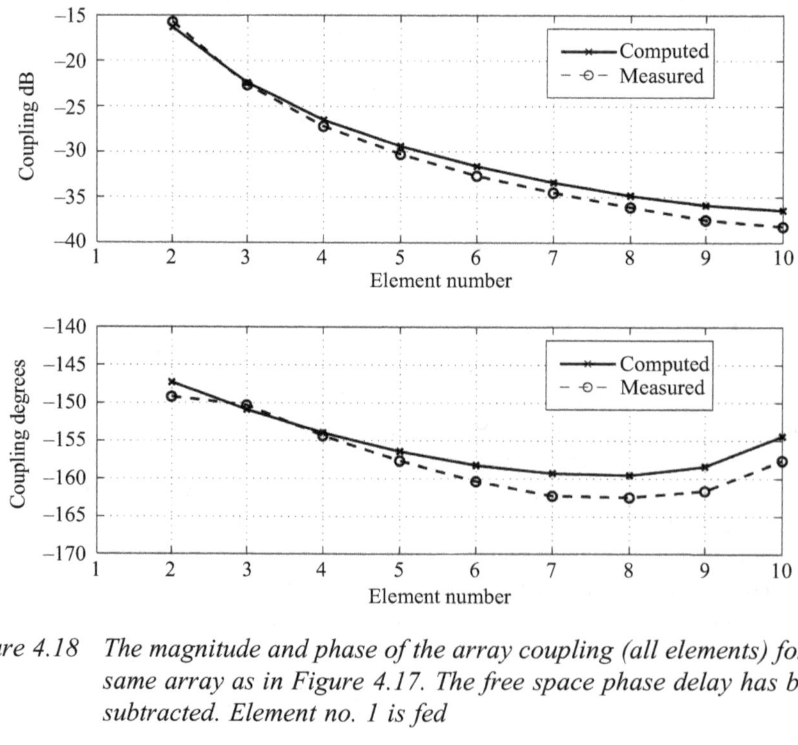

Figure 4.18 The magnitude and phase of the array coupling (all elements) for the same array as in Figure 4.17. The free space phase delay has been subtracted. Element no. 1 is fed

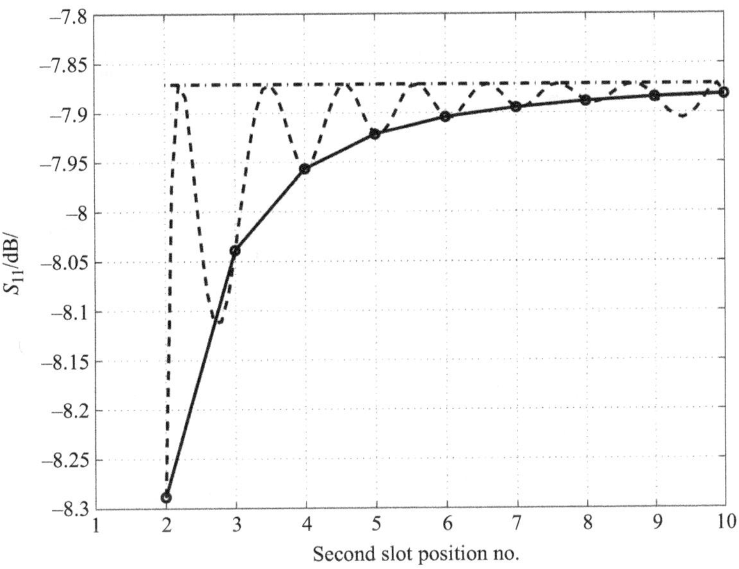

Figure 4.19 Reflection $|S_{11}|$ in dB of slot no. 1 with a single parasitic slot at increasing positions 2–10. Dashed line shows also intermediate positions. Dash-dotted (straight) line: the isolated reflection. E-plane

We find here that the scattering parameters, that is, elements of the scattering matrix, do not only depend on the apertures m and n, but also on the surrounding apertures. Even the self-reflection, that is, the S_{11} element in the scattering matrix, is shown in Figure 4.19 to depend on the distance to a second, parasitic slot (only two apertures present in this case). See also [26]. We have in this situation

$$S_{11} = \frac{Y_0^2 - Y_{11}^2 + Y_{12}^2}{(Y_0 + Y_{11})^2 - Y_{12}^2} \tag{4.117}$$

which becomes the conventional isolated (one element only) scattering if $Y_{12} \to 0$.

In terms of scattering parameters we can also write

$$S_{11} = \frac{S_{11}^0 - 1}{2} + \sqrt{\left(\frac{1 + S_{11}^0}{2}\right)^2 + S_{12}^2} \tag{4.118}$$

where S_{11}^0 stands for the scattering of the single (isolated) element.

Coupling in an *H*-plane linear slot array is shown in Figures 4.20 and 4.21.

In the *H*-plane example the phase agreement (measured vs. calculated) for array coupling is not as good as in the previous examples (*E*-plane). It has been found that higher-order modes are excited in the apertures and should be included if better agreement is sought. However, the coupled levels are very low so this effort is probably not needed in practice. As we can see from these results, the *H*-plane

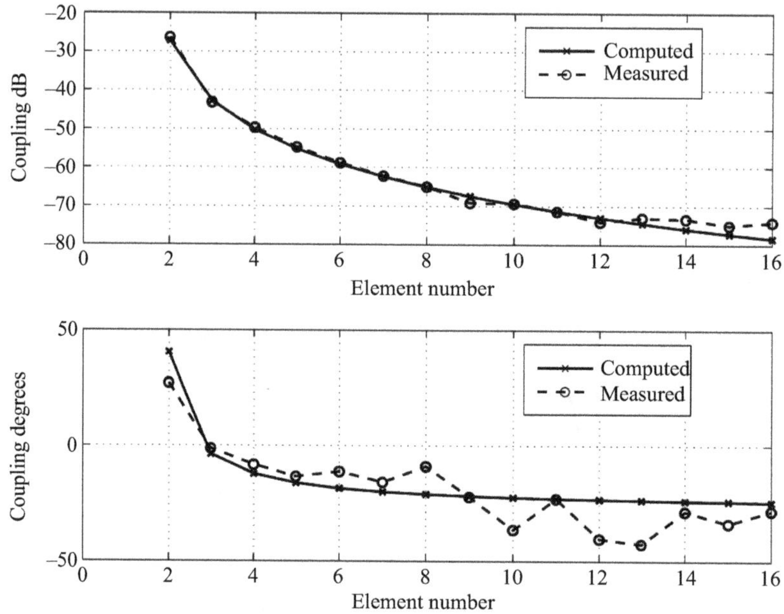

Figure 4.20 The magnitude and phase of the isolated coupling (two elements) for a 16-element H-plane array. Element no. 1 is fed. Aperture size: 0.73λ × 0.30λ

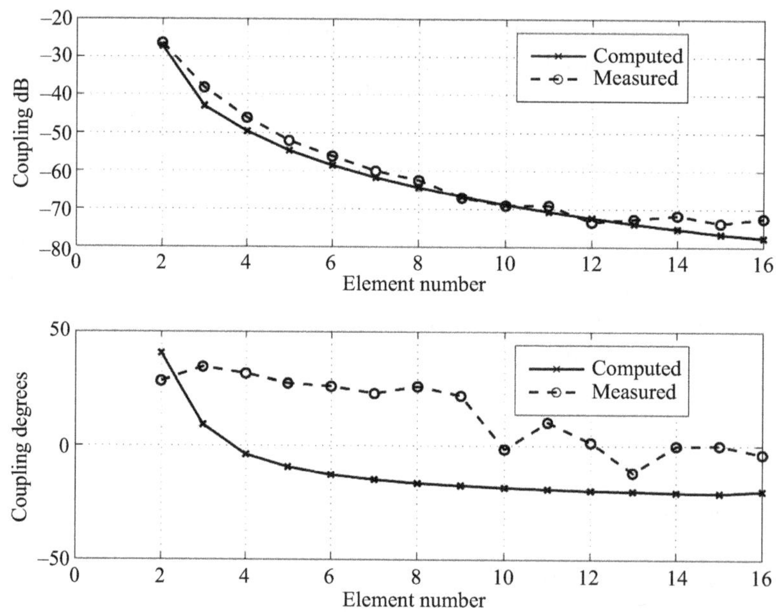

Figure 4.21 The H-plane array coupling for the same array as in Figure 4.20

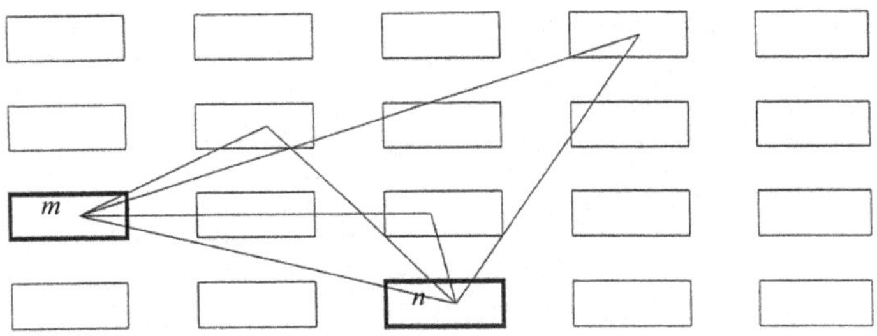

Figure 4.22 In a 2-D array mutual coupling is a multiple coupling phenomenon

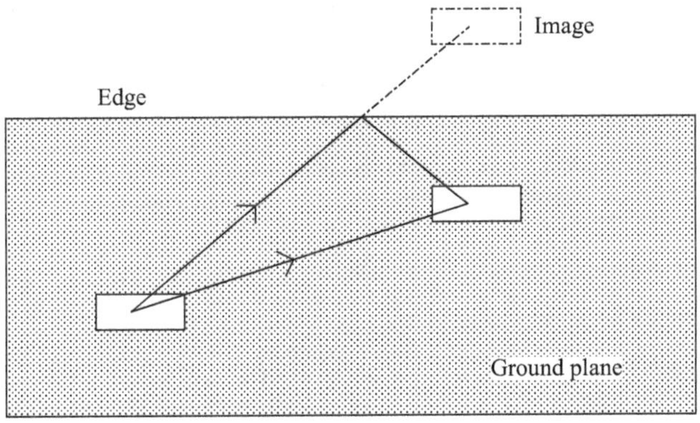

Figure 4.23 Edge reflection effect seen as a contribution from an image slot

coupling is very low for rectangular slots. This also leads to the interesting fact that isolated and array H-plane coupling (for linear arrays) are at about the same level. However, in two-dimensional arrays, the 'H-plane coupling' can seek other ways for the interaction. Figure 4.22 shows how elements m and n can interact via several of the elements in their surroundings.

4.5.3 Finite ground plane effects

The edges of a finite ground plane will cause diffraction effects as we have already seen in the radiation pattern (cf. Figure 4.12). The effect is of course strongest when the edge illumination is strong as is typically the case in the E-plane. There is also a scattering from the edges back to the slots in the ground plane, thus modifying the array excitation and mutual admittance to some extent. This has been studied by several authors [27–30].

In an experiment the coupling between two slots in a finite ground plane with edges was measured [29]. A comparison was made between the cases with the edges covered by absorbing material and with the edges exposed (not covered). The geometry and examples of measured and calculated isolated coupling are shown in Figures 4.23–4.25.

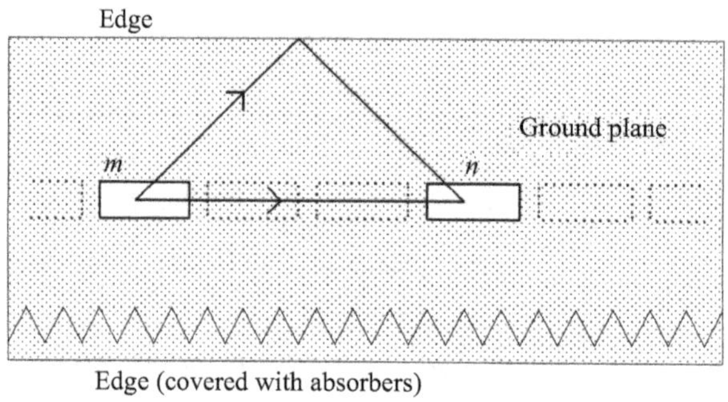

Figure 4.24 Set-up for measuring the edge effect on mutual coupling between two aperture elements m and n. The other apertures were covered with conducting tape. From [29]. Reproduced by permission of the Institution of Engineering & Technology

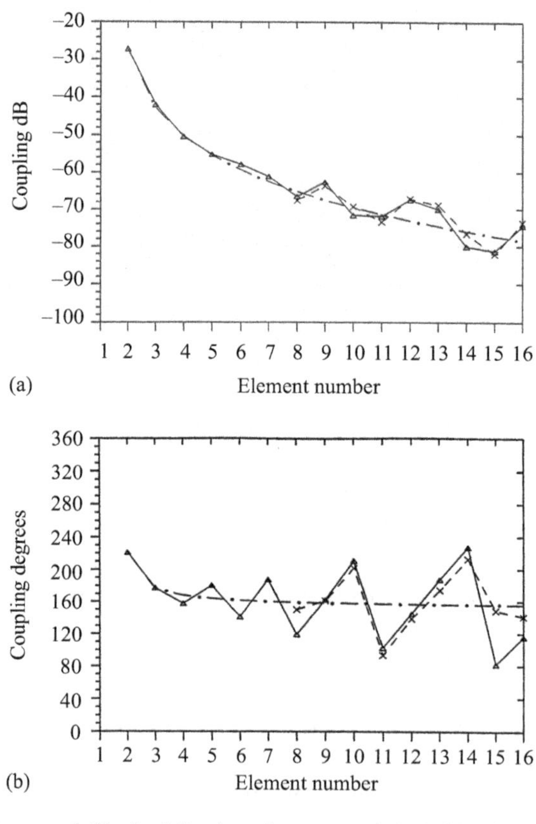

Figure 4.25 Measured (dashed line) and computed (solid line) mutual coupling including the effect of edge diffraction. The no edge case (ground plane edges covered with absorbing material) is shown as a dash-dotted line. (a) is magnitude and (b) is phase. From [29]. Reproduced by permission of the Institution of Engineering & Technology

The theoretical calculation is based on geometrical diffraction. Rays from the source slot are reflected by the exposed straight edge. In principle we introduce a mirror image of one of the slots. We get coupling contributions from the direct path and from the reflected path that add together. The reflected contribution is weighted by an edge diffraction coefficient [31]. In this case the direct path is in the *H*-plane and thus the edge influence is clearly visible in the results. More details are found in the references.

References

[1] Collin R. E. *Field Theory of Guided Waves*. IEEE Press, New York, 2nd ed., 1991.

[2] Balanis C. A. *Antenna Theory, Analysis and Design*. Wiley, New York, 1997.

[3] Harrington R. F. *Time-Harmonic Electromagnetic Fields*. IEEE Press, New York, 2001.

[4] Elliott R. S. *Antenna Theory and Design*. Prentice-Hall, Englewood Cliffs, NJ, 1981.

[5] Clemmow P. C. *The Plane Wave Spectrum Representation of Electromagnetic Fields*. Pergamon Press, Oxford, 1966.

[6] Rhodes D. R. *Synthesis of Planar Antenna Sources*. Clarendon Press, Oxford, 1974.

[7] Kraus J. D. *Antennas*. McGraw-Hill, New York, 2nd ed., 1988.

[8] Selby S. M. (ed.). *Standard Mathematical Tables*. The Chemical Rubber Co., Cleveland, OH, 1973.

[9] Booker H. G. 'Slot aerials and their relation to complementary wire aerials (Babinet's principle)'. *Journal of the Institution of Electrical Engineers*. Part. IIIA. 1946;**93**(4):620–626.

[10] Collin R. E. *Foundations for Microwave Engineering*. McGraw-Hill, New York, 1966.

[11] Pathak P. H. 'On the Eigenfunction expansion of electromagnetic dyadic Green's functions'. *IEEE Transactions on Antennas and Propagation*. 1983;**31**(6):837–846.

[12] Van Bladel J. 'Contribution of the Ψ=constant mode to the modal expansion in a waveguide'. *IEE Proceedings H – Microwaves, Antennas and Propagation*. 1981;**128**(5):247–251.

[13] Khac T. V., Carson C. T. 'm=0 n=0 mode and rectangular-waveguide slot discontinuity'. *Electronics Letters*. 1973;**9**(18):431–432.

[14] Collin R. E., Zucker F. J. *Antenna Theory, Part I*. McGraw-Hill, New York, 1969.

[15] Stevenson A. F. 'Theory of slots in rectangular waveguides'. *Journal of Applied Physics*. 1948;**19**(1):24–38.

[16] Stegen R. J. 'Slot radiators and arrays at X-band'. *IRE Transactions on Antennas and Propagation*. 1952;**1**(1):62–84.

[17] Borgiotti G. V. 'A novel expression for the mutual admittance of planar radiating elements'. *IEEE Transactions on Antennas and Propagation.* 1968;**16**(3):329–333.

[18] Elliott R. S. 'The design of small slot arrays'. *IEEE Transactions on Antennas and Propagation*. 1978;**26**(2):214–219.

[19] Elliott R. S. 'An improved design procedure for small arrays of shunt slots'. *IEEE Transactions on Antennas and Propagation*. 1983;**31**(1):48–53.

[20] Josefsson L., Moeschlin L., Svensson T. 'A stripline flat plate antenna with low sidelobes', *Proceedings of the IEEE Antennas and Propagation Symposium*, Atlanta, GA, USA, 1974, pp. 282–285.

[21] Amitay N., Galindo V., Vu C. P. *Theory and Analysis of Phased Array Antennas*. Wiley Interscience, New York, 1972.

[22] Wasylkiwskyj W., Kahn W. K. 'Theory of mutual coupling among minimum-scattering antennas'. *IEEE Transactions on Antennas and Propagation*. 1970;**18**(2):204–216.

[23] Harrington R. F. *Field Computation by Moment Methods*. The MacMillan Co., New York, 1968.

[24] Rengarajan S. R., Gabrelian E. 'Efficient and accurate evaluation of external mutual coupling between compound broad wall slots'. *IEEE Transactions on Antennas and Propagation*. 1992;**40**(6):733–737.

[25] Josefsson L. 'Mutual coupling effects on the performance of finite radar antenna arrays'. *Proceedings of the International Conference on Radar*. Paris, May 3–6, 1994, pp. 713–716.

[26] Mailloux R. J. 'Radiation and near-field coupling between two collinear open-ended waveguides'. *IEEE Transactions on Antennas and Propagation*. 1969;**17**(1):49–55.

[27] Bailey M. C. 'Mutual coupling between circular waveguide-fed apertures in a rectangular ground plane'. *IEEE Transactions on Antennas and Propagation*. 974;**22**(4):597–599.

[28] Commatas C. G., Sahalos J. N. 'Mutual coupling and radiation patterns of two slots asymmetrically located on a square plate'. *IEEE Transactions on Antennas and Propagation*. 1983;**31**(3):531–535.

[29] Josefsson L. 'Mutual coupling calculations including edge effects'. *Electronics Letters*. 1994;**30**(25):2087–2088.

[30] Mazzarella G., Panariello G. 'Evaluation of edge effects in slot arrays using the geometrical theory of diffraction'. *IEEE Transactions on Antennas and Propagation*. 1989;**37**(3):392–395.

[31] Kouyoumjian R. G. 'The geometrical theory of diffraction and its application', in R. Mittra (ed.). *Numerical and Asymptotic Techniques in Electromagnetics*. Springer-Verlag, Berlin Heidelberg, Chapter 6, 1975.

Chapter 5
Slot models

In this chapter we will discuss theoretical models for characterising slots in waveguides. In Chapter 4 we looked at basic theories for slot apertures and found expressions for the *conductance* of longitudinal slots in rectangular waveguides. We will now look at ways of finding the slot *susceptance*, the effect of slot offset on resonance conditions, influence of wall thickness, waveguide dimensions, etc. We will demonstrate how electromagnetic models can tell us about slot behaviour in greater detail. The longitudinal slot and the transverse slot in the broad wall of a rectangular waveguide will be treated in particular. We will thus establish a basis for an efficient design process of high-performance slotted waveguide arrays.

5.1 Modelling principles

5.1.1 Using experimental data

In Chapter 4 we derived the slot conductance for resonant slots. The assumption was that the slot length $\lambda/2$ automatically results in slot resonance and that the slot admittance is real. However, it has been found that resonance also depends on the slot shape, the waveguide dimensions and the mutual coupling among the slots in an array. We are also interested in the complete slot admittance and how it varies over a band of frequencies. A more detailed analysis is therefore necessary in order to control these parameters and optimise the slot (array) antenna for best performance including bandwidth. Although much theoretical work was performed on slot antennas in the years following the Second World War by several authors [1–5], the design of slotted waveguide array antennas relied in those days to a large degree on experimental data.

At Hughes Aircraft Co. in California, USA, R. J. Stegen made careful measurements on longitudinal (shunt) slots in rectangular waveguides [6]. He measured the slot admittance for various slot lengths and slot offsets and assembled the data in a set of normalised curves that can be used for array design. He introduced the normalised parameters:

Slot length normalised to the resonant length	$y = L/L_r$
Slot resonant length normalised to the free space wavelength	L_r/λ_0
Slot conductance normalised to the resonant conductance	$h_1 = G/G_0$
Slot susceptance normalised to the resonant conductance	$h_2 = B/G_0$

Stegen used the following waveguide and slot dimensions (standard *X*-band waveguide):

Slot width	0.0625 inch (1.59 mm)
Waveguide width	0.9 inch (22.86 mm)
Waveguide height	0.4 inch (10.16 mm)
Waveguide wall thickness	0.05 inch (1.27 mm)
Frequency	9.375 GHz

The normalised resonant conductance $G/G_0 = g(x)$ as a function of offset (x) was already introduced in Chapter 4 (Figure 4.11). Additionally, the resonant length as a function of offset is also needed. The data was presented in curves, and examples are shown in Figure 5.1(a) and (b).

Figure 5.1(a) $h_1(L/L_r)$ and $h_2(L/L_r)$ conductance and susceptance vs. slot length
Figure 5.1(b) L_r/λ_0 resonant length versus offset
Figure 4.11 $g(x)$ normalised conductance versus offset

The curves appear to be quite universal for the given waveguide dimensions. The data can easily be fitted to polynomials to be used in the array design iterations. However, new dimensions (half height waveguide, etc.) would require new measurements.

5.1.2 Equivalent circuit approach

In Chapter 4 the longitudinal slot, as seen from the feeding waveguide, was represented by an equivalent shunt admittance on a transmission line. This model was plausible under the assumption that the slot electric field was symmetrical (cosinusoidal distribution) with equal scattering forward and backward from the slot (Figure 5.2).

The reflection coefficient Γ is easily measured, and from Γ the shunt admittance is obtained:

$$Y/G_0 = -2\Gamma/(1+\Gamma) \tag{5.1}$$

In 1957 Oliner [4] presented a variational solution with equivalent circuits for shunt and series slots in rectangular waveguides. A correction for wall thickness was proposed by Yee [5]. The variational approach did not include the effect of slot offset from the centreline on the resonant length. However, the model was used successfully in combination with measured data. Additional parameters were introduced by Lagerlöf and Josefsson [8] (see Figure 5.3).

The parameters in Figure 5.3 represent

B_0	Residual susceptance for centred slot
n_d	Slot displacement coupling factor
Y_{sext}	Slot external admittance
B_{sint}	Slot internal susceptance
t	Waveguide wall thickness
$Z \rightleftarrows Y$	Impedance/admittance converter

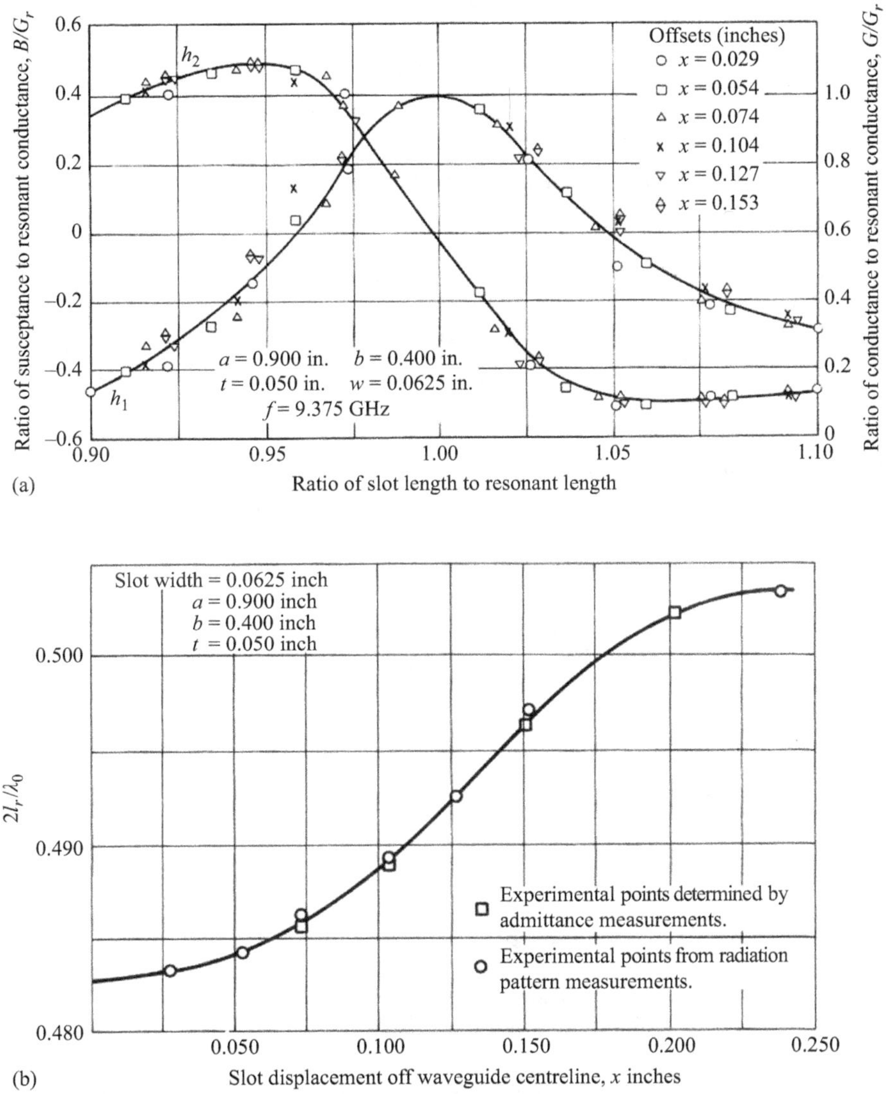

Figure 5.1 (a) Normalised admittance versus normalised slot length. (b) Resonant slot length versus slot offset from centreline. (a, b) © 1981 J. H. Elliott. Reprinted from [7] with permission

The outer slot admittance Y_{sext} could be determined theoretically [9]. The other parameters (B_{sint}, n_{d} and B_0) were determined from a set of experimental data on slot length and offset for several frequencies. Polynomials could then be derived from the experimental values and subsequently be used in array designs (within the range of the experimental data).

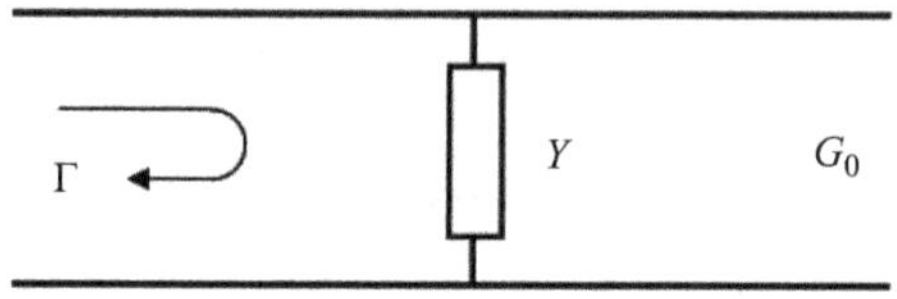

Figure 5.2 The equivalent shunt admittance

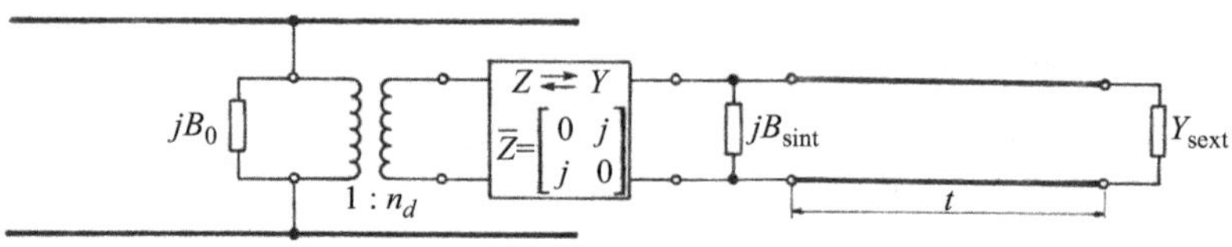

Figure 5.3 Equivalent circuit for a shunt slot in a rectangular waveguide. © 1990 Microwave Exhibitions and Publishers Ltd. Reproduced from [8] by permission of Microwave Exhibitions and Publishers Ltd

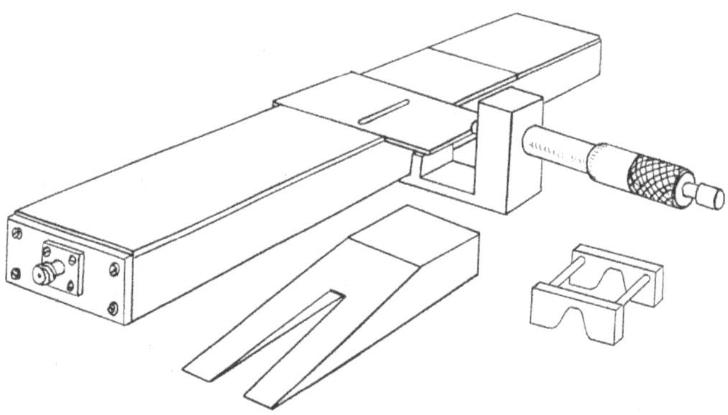

Figure 5.4 Test equipment for slot measurements. A wedge absorber and a waveguide short in the foreground

A test fixture for measuring slot characteristics is shown in Figure 5.4. The fixture is calibrated using a matched load (wedge absorber) and a moveable short. Various slot plates with different slot dimensions could be inserted and the offset carefully monitored with the micrometre shown. The example is from an evaluation of data for slots in a ridge loaded waveguide where theoretical models are somewhat complex [10].

5.1.3 Electromagnetic models

Theoretical calculations can in principle replace the tedious and costly collection of experimental data. From the theoretical results polynomials can be defined and used in the array design just as described above when equivalent circuits were postulated. However, the EM modelling can provide more information such as fields and currents on the physical structure. When this has been done we can find, for example, the electric field distribution in the slot aperture (e.g. by solving a boundary value problem, more about this later) instead of just *assuming* a particular distribution. For slot antennas, several authors have contributed in this area. Khac [11] solved the integral equation for the electric field by the Method of Moments using pulse expansion and point matching. Lyon and Sangster [12] used an entire basis with sinusoidal functions and concluded that as few as two basis functions were sufficient for 'power calculations'.

For the design of slotted waveguide arrays the resonant length must be known quite accurately. The slot is by itself narrow band with the bandwidth further reduced in the array by the mutual coupling among the slots [13,14]. An accuracy of about half a per cent or better is desired. This is achievable using moment method techniques as was demonstrated by Stern and Elliott [15]. They compared theoretical results to carefully obtained experimental data for slots in thin-walled waveguide with standard inner dimensions. Josefsson [16] presented Moment Method results and measurements for standard and reduced height waveguides including wall thickness.

5.1.3.1 The boundary problem

In Figure 5.5 four related boundary problems are illustrated. They can be solved using similar moment method techniques. No. 1 shows a radiation slot fed from a waveguide (bold arrow). The two thin arrows in the waveguide represent forward and backward scattering. In example no. 2 we have coupling between two waveguides. No. 3 is the slot antenna in a receive situation, and no. 4 shows a slot aperture in a metallic screen.

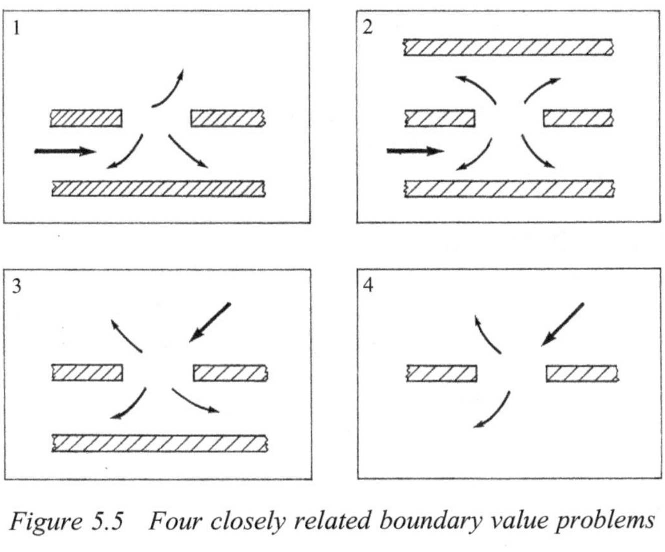

Figure 5.5 Four closely related boundary value problems

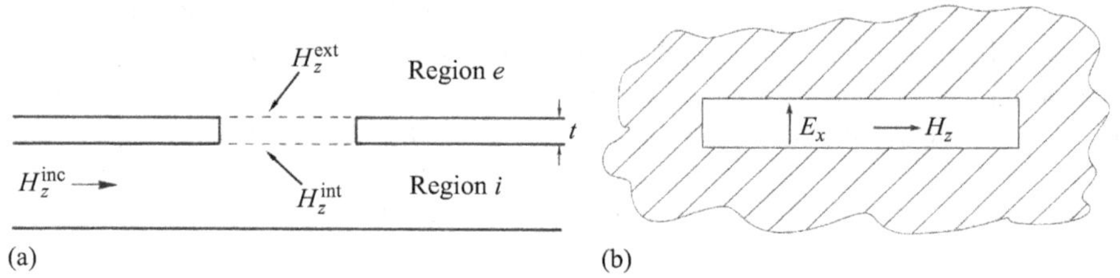

Figure 5.6 (a) The radiating slot problem. © IEEE 1987. Adapted from [16], with permission. (b) Major slot field components

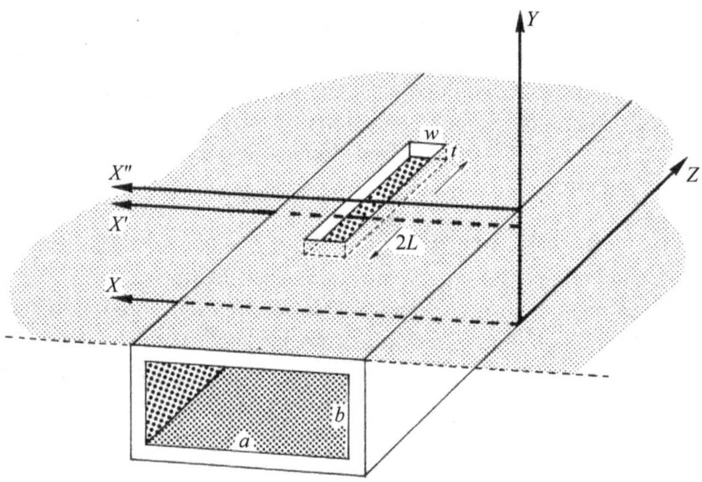

Figure 5.7 The slot geometry. 2L is the slot length and w the slot width. © URSI 1986. Reprinted from [33], with permission

The four problems in Figure 5.5 are all about a slot interface between two regions which we call the external (superscript e or ext) and the internal (superscript i or int); cf. Figure 5.6(a).

5.1.3.2 Zero wall thickness

Let us assume that the wall thickness t is very small and can be disregarded and that we only need to consider the major H_z and E_x slot field components, Figure 5.6(b). We will introduce finite wall thickness later. Problem no. 1 is first analysed with coordinates as shown in Figure 5.7.

With an incident field H_z^{inc} we have the boundary condition for the magnetic field:

$$H_z^e(E_x) - H_z^i(E_x) = H_z^{inc} \tag{5.2}$$

where E_x is the unknown transverse electric field in the slot, expanded in basis functions:

$$E_x(x', z') = \sum_{p=1}^{N} E_p e_p(z') \tag{5.3}$$

The solution to (5.2) is found in a projection sense according to the Moment Method [17], that is, we form inner products with testing functions $e_s(z)$ resulting in a system of equations:

$$\langle H_z^e(E_x), e_s \rangle - \langle H_z^i(E_x), e_s \rangle = \langle H_z^{\text{inc}}, e_s \rangle \tag{5.4}$$

or, in matrix form:

$$\left[[Y^e] - [Y^i] \right] [E] = [h] \tag{5.5}$$

The matrices $[Y^e]$, $[Y^i]$ have elements

$$\left. \begin{aligned} Y_{ps}^e &= \langle H_z^e(e_p), e_s \rangle \\ Y_{ps}^i &= \langle H_z^i(e_p), e_s \rangle \end{aligned} \right\} \tag{5.6}$$

The excitation vector is

$$[h] = \langle H_z^{\text{inc}}, e_s \rangle \tag{5.7}$$

$e_p(z)$ and $e_s(z)$ are the basis and testing functions, respectively. The solution to (5.2) is in the form of (5.3) with coefficients found in the vector

$$[E] = (E_1, E_2, \ldots, E_N) \tag{5.8}$$

obtained as

$$[E] = \left[[Y^e] - [Y^i] \right]^{-1} \cdot [h] \tag{5.9}$$

5.1.3.3 Choice of basis and testing functions

Simple functions such as pulse basis (expansion) and point testing have been used. Very complex functions can be modelled with these. However, at the slot ends the electric field must go to zero and this is not well represented by pulses. Khac [11] somewhat arbitrarily added one pulse of length 0.02 λ_0 to allow for the expected electric field variation down to zero at the slot ends. A sinusoidal entire basis may appear more attractive in this respect. In fact, just one term may sometimes be sufficient. No more than 10 terms have been found necessary even for accurate resonant length predictions (<0.5% error) [16]. More about the behaviour of fields near conducting edges (edge conditions) will be discussed in Section 5.3.

The results presented in this chapter have been obtained with entire basis and testing functions:

$$\left.\begin{aligned} &\text{basis function} \quad e_p(z) = \sin\frac{p\pi}{2L}(L+z) \\ &\text{testing function} \quad e_s(z) = \sin\frac{s\pi}{2L}(L+z) \end{aligned}\right\} \tag{5.10}$$

That is, the Galerkin case [17].

5.1.3.4 The other slot problems in Figure 5.5

No. 2: coupling between two waveguides.

Provided the waveguides are identical we have $H_z^{i1} = -H_z^{i2}$ and thus

$$[E] = -\frac{1}{2}[Y^i]^{-1} \cdot [h] \tag{5.11}$$

No. 3: waveguide slot with plane wave incident.

Only a sign change when the excitation has moved:

$$[E] = [[Y^i] - [Y^e]]^{-1} \cdot [h] \tag{5.12}$$

No. 4: slot in screen – no waveguide.

Symmetrical case like no. 2. $H_z^{e1} = -H_z^{e2}$, thus

$$[E] = -\frac{1}{2}[Y^e]^{-1} \cdot [h] \tag{5.13}$$

It is true that the excitation vector is not the same for waveguide mode excitation and plane wave excitation. However, the difference is not very large; more about this in Section 9.8.

5.1.4 *Finite wall thickness*

A rectangular slot in a waveguide wall of thickness t can be seen as a short rectangular waveguide by itself (Figure 5.8). The field inside the stub waveguide is described as a superposition of TE_{i0} modes. We neglect variations across the slot since the waveguide height (= slot width w) is assumed to be narrow. Our problem can be broken down into three partial boundary problems as illustrated in Figure 5.8(a)–(c).

We are concerned with matching the tangential fields at the upper (I/IIa) and lower (IIb/III) interfaces simultaneously. As in the case with zero wall thickness we only consider the E_x and the H_z field components at the two interfaces; cf. Figure 5.6(b). Our approach is based on solving the three partial problems formally and then couple the solutions via the mode expansion inside the stub waveguide.

The upper (external) interface is characterised by a scattering matrix $[R^e]$ relating the downward $[E^-]$ and upward $[E^+]$ propagating modes inside the waveguide:

$$[E^-] = [R^e][E^+] \tag{5.14}$$

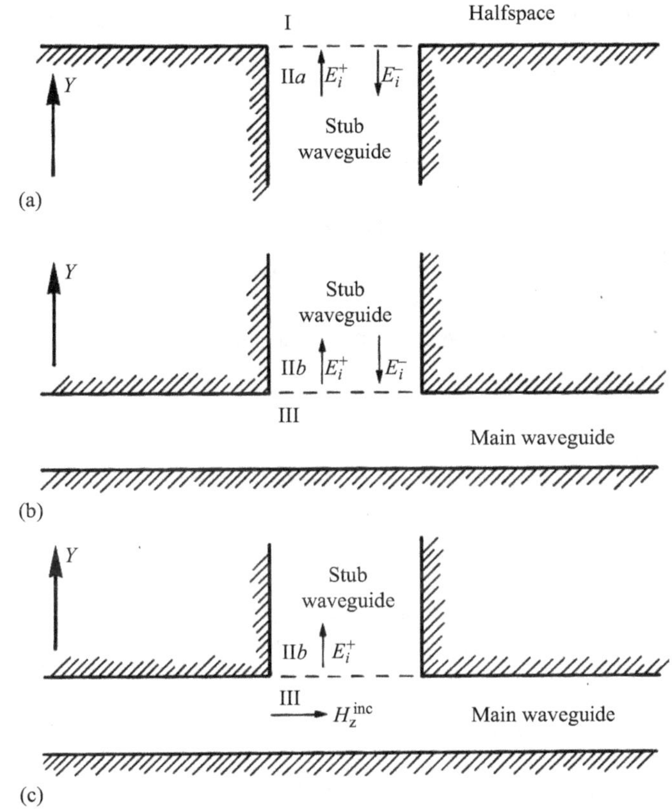

Figure 5.8 Slot in thick waveguide wall. The three partial problems (a)–(c). © URSI 1986. Reprinted from [33], with permission

Likewise, for the lower (internal) interface, we can write

$$[E^+] = [R^i][E^-] \tag{5.15}$$

These scattering matrices can be written out using appropriate quantities already defined for the zero thickness case. To summarise we obtain, for example, for the ith column of the matrix $[R^e]$

$$[R^e]_{i\text{:th column}} = [Y^{ew}]^{-1} \cdot [h_i^{ew}] \tag{5.16}$$

where the element number (p, s) of the matrix $[Y^{ew}]$ is

$$Y_{ps}^{ew} = Y_{ps}^e - Y_0 wL\delta_{ps} \tag{5.17}$$

$[Y^e]$ was defined in (5.6), Y_0 is the waveguide modal admittance and δ_{ps} is the Kronecker delta function. The vector $[h_i^{ew}]$ has elements

$$h_{is}^{ew} = -< H_z^{ext}(e_i), w_s > -Y_i wL\delta_{is} \tag{5.18}$$

The excitation by a unit incident mode in the main waveguide results in the formal solution

$$[E^+] = -[Y^{iw}]^{-1} \cdot [h^{\text{inc}}] \tag{5.19}$$

where the matrix $[Y^{iw}]$ has elements

$$Y^{iw}_{ps} = Y^{i}_{ps} + Y_0 wL\delta_{ps} \tag{5.20}$$

$[Y^i]$ was defined in (5.6) and $[h^{\text{inc}}]$ in (5.7). To this solution, (5.19), we will of course have to add those waves resulting from the E^- waves incident on the lower interface.

It remains to relate the waveguide modes at the upper interface to those at the lower interface:

$$[E^+]_{\text{upper interface}} = [B]\,[E^+]_{\text{lower interface}} \tag{5.21}$$

and similarly for the downward propagating modes. $[B]$ is a diagonal matrix with the elements

$$B_{ii} = \exp(-j\beta_i t) \tag{5.22}$$

where β_i is the propagation constant of the TE_{i0} mode and t is the wall thickness. Combining everything together we obtain the total tangential electric field at the lower interface

$$[E] = \{1 + [B][R^e][B]\}\{[R^i][B][R^e][B] - 1\}^{-1}.[Y^{iw}]^{-1}[h^{\text{inc}}] \tag{5.23}$$

This is the final solution for the field in the slot aperture (lower interface).

5.2 Integral equation solution

In the preceding section, we solved the slot admittance problem in a formal sense. We have now to derive the appropriate field expressions and compute the matrices required for the Moment Method solution. The external field is due to an equivalent magnetic current source in the upper slot aperture. Assuming the ground plane to be large we can describe this field by a free space Green's function; cf. Section 4.5.2. The internal field is a bit more complicated so we start with this.

5.2.1 *The internal field*

Field solutions for rectangular waveguides are traditionally expressed in terms of waveguide modes (Section 4.3.2). A complication occurs of course when there is an aperture (slot) in the waveguide wall. Stevenson [2] described the internal field including the slot region with the aid of a Green's function $G_2(P, P')$ [18].[1]

[1] Green's function = 'field of a point source' [32]. Georg Green was a British mathematician who in 1828 published 'An Essay on the Application of Mathematical Analysis to the Theories of Electricity and Magnetism' [18].

In our case we need the H_z field (TE mode) when there is an E_x field exciting the slot. Stevenson's expression for this case is

$$H_z^{\text{int}}(P) = \frac{-1}{j\omega\mu_0}\int_S E_x(P')\left(\frac{\partial^2}{\partial z'^2} + k^2\right)G_2(P,P')dS' \tag{5.24}$$

Complete expressions including other field components are found in, for example, [19]. See also Section 2.8.

In (5.24) S' denotes the slot region, P and P' refer to points in the slot, prime indicates source points, unprimed field points. The chosen Green's function is

$$G_2(P,P') = \sum_{m=0}^{\infty}\sum_{n=0}^{\infty}\psi_{mn}(x',y')\psi_{mn}(x,y)\frac{e^{-\gamma_{mn}|z-z'|}}{2\gamma_{mn}} \tag{5.25}$$

where $\psi_{mn}(x, y)$ is the TE modal function

$$\psi_{mn}(x,y) = \frac{\sqrt{\varepsilon_m\varepsilon_n}}{\sqrt{ab}}\cos\frac{m\pi x}{a}\cos\frac{n\pi y}{b} \tag{5.26}$$

Here $\varepsilon_m = 1$ if $m = 0$, $\varepsilon_m = 2$ if $m > 0$. Similarly for ε_n (the Neumann factor). γ_{mn} is the propagation constant:

$$\gamma_{mn}{}^2 = \left(\frac{m\pi}{a}\right)^2 + \left(\frac{n\pi}{b}\right)^2 - k^2 \tag{5.27}$$

Putting together (5.24) and (5.25) we have

$$H_z^{\text{int}}(P) = \frac{-1}{j\omega\mu_0}\sum_{m=0}^{\infty}\sum_{n=0}^{\infty}\frac{\psi_{mn}(x,y)}{2\gamma_{mn}}\int_S E_x(P')\psi_{mn}(x',y')\left(\frac{\partial^2}{\partial z'^2} + k^2\right)e^{-\gamma_{mn}|z-z'|}dS' \tag{5.28}$$

Note that the summation includes the TE_{00} mode. Normally, this mode is disregarded in waveguide problems since it does not propagate. However, in the source region it is essential. This has caught the attention of several authors [11,20–23].

We assume that the slot is in the upper broad wall, consequently $y = y' = b$. Furthermore, we can separate $E_x(P') = E_x(x', z') = E_1(x') \cdot E_2(z') = E_2(z')$; we assume that the electric field has no variation across the slot. We will discuss edge conditions in Section 5.3.

The derivative in 5.28 has to be handled with care when $z' = z$, that is, for points in the source region. Using the Dirac delta function concept we can write [21]

$$\frac{\partial^2}{\partial z'^2}e^{-\gamma|z'-z|} = \left[\gamma^2 - 2\gamma\delta(z'-z)\right]e^{-\gamma|z'-z|} \tag{5.29}$$

With this we get

$$H_z^{\text{int}}(P) = \frac{-1}{4j\omega\mu_0}\sum_{m=0}^{\infty}\sum_{n=0}^{\infty}\varepsilon_m\varepsilon_n F_m(x)\left\{\frac{k^2+\gamma_{mn}^2}{\gamma_{mn}}\int_S E_2(z')e^{-\gamma_{mn}|z-z'|}dz' - 2E_2(z)\right\} \tag{5.30}$$

where

$$F_m(x) = \cos\frac{m\pi x}{a}\int_{-w/2}^{w/2} E_1(x')\cos\frac{m\pi(x_0 + x')}{a}dx' \tag{5.31}$$

Here x_0 is the slot offset from the sidewall. With no variations across the slot we have

$$F_m(x) = w\cos\frac{m\pi x}{a}\cos\frac{m\pi x_0}{a}\,\mathrm{sinc}\frac{m\pi w}{2a} \tag{5.32}$$

Finally, with the expansion of the electric field variation along the slot; cf. (5.10):

$$E_2(z) = \sum_{p=1}^{N} E_p \sin\frac{p\pi}{2L}(L+z) \tag{5.33}$$

We arrive at the following expression for the internal H_z field:

$$\begin{aligned} H_z^{\text{int}}(x,z) = \frac{-1}{j\omega\mu_0 ab}\sum_{m=0}^{\infty}\sum_{n=0}^{\infty}\varepsilon_m\varepsilon_n F_m(x)\cdot \ldots \\ \cdot\sum_{p=1}^{N} E_p \left\{ \frac{\pi p}{4L\gamma_{mn}} \cdot \frac{\gamma_{mn}{}^2 + k^2}{\gamma_{mn}{}^2 + \left(\frac{\pi p}{2L}\right)^2} \cdot \left(e^{-\gamma_{mn}(L+z)} \pm e^{-\gamma_{mn}(L-z)}\right) \right. \\ \left. + \frac{k^2 - \left(\frac{\pi p}{2L}\right)^2}{\gamma_{mn}{}^2 + \left(\frac{\pi p}{2L}\right)^2}\sin\left(\frac{\pi p}{2L}(L+z)\right) \right\} \end{aligned} \tag{5.34}$$

The upper sign in (5.34) is valid for p odd, the lower for p even.

5.2.2 The external field

The magnetic field $H_z(x, z)$ due to the electric aperture field $E_a(x', z')$ for an aperture in an infinite ground plane was already derived in Chapter 4. We can start from (4.107):

$$H_z^{\text{ext}}(x,z) = \frac{1}{2\pi j\omega\mu_0}\iint_{\text{slot}} E_x(x',z')\left(k^2 + \frac{\partial^2}{\partial z'^2}\right)\frac{e^{-jkR}}{R}dx'dz' \qquad (5.35) = (4.107)$$

The distance $R = \sqrt{(x-x')^2 + (z-z')^2}$. The Green's function in (5.35) is a spherical wave, which can be transformed into a spectrum of plane waves [24,25]:

$$\frac{e^{-jkR}}{R} = \frac{1}{2\pi j}\int_{-\infty}^{\infty}\int_{-\infty}^{\infty}\frac{e^{-j[k_x(x-x')+k_z(z-z')+k_y y]}}{\sqrt{k^2 - k_x^2 - k_z^2}}dk_x dk_z \tag{5.36}$$

We insert this in (5.35), interchange the order of integration and evaluate the derivative with the following result:

$$H_z^{\text{ext}}(x,z) = \frac{-1}{2\pi\omega\mu_0}\int_{-\infty}^{\infty}\int_{-\infty}^{\infty}\frac{k^2-k_z^2}{k_y}\cdot g_x(k_x,k_z)e^{-j(k_xx+k_zz)}dk_xdk_z \tag{5.37}$$

where g_x (representing the radiation pattern):

$$g_x(k_x,k_z) = \frac{1}{2\pi}\iint_{\text{slot}} E_x(x',z')e^{j(k_xx'+k_zz')}dx'dz' \tag{5.38}$$

With the expansion of the electric slot field as before (5.33) with constant field across the slot we get finally

$$H_z^{\text{ext}}(x,z) = \sum_{p=1}^{NP}\frac{E_pwp}{\omega\mu_0 4\pi L}\int_{-\infty}^{\infty}\int_{-\infty}^{\infty}\frac{(k_z^2-k^2)\operatorname{sinc}(k_xw/2)}{k_y\left[k_z^2-\left(\frac{\pi p}{2L}\right)^2\right]}\cdot\begin{Bmatrix}-\cos k_zL\\ j\sin k_zL\end{Bmatrix}_{p\text{ even}}^{p\text{ odd}}$$

$$\cdot e^{-j(k_xx+k_zz)}dk_xdk_z \tag{5.39}$$

5.2.3 *Matrix elements*

The matrix elements are normalised as shown next

$$Y_{ps} = \langle H_z(E^p), e_s\rangle\frac{\omega\mu_0}{kE_p} \tag{5.40}$$

5.2.3.1 Internal field contribution

Forming the inner product we obtain the elements of the matrix $[Y^i]$

$$Y_{ps}^i = \frac{\lambda w^2}{\pi ab}\sum_{m=0}^{\infty}\sum_{n=0}^{\infty}\frac{\varepsilon_m\varepsilon_n}{4}\cdot W_m\cdot\ldots$$

$$\cdot\left\{\frac{\frac{j\pi p}{2L}}{\gamma_{mn}}\cdot\frac{\gamma_{mn}^2+k^2}{\gamma_{mn}^2+\left(\frac{\pi p}{2L}\right)^2}\cdot\frac{2\frac{\pi s}{2L}}{\gamma_{mn}^2+\left(\frac{\pi s}{2L}\right)^2}\cdot\left(e^{-2\gamma_{mn}L}\pm 1\right)\right.$$

$$\left.+2j\cdot\frac{k^2-\left(\frac{\pi p}{2L}\right)^2}{\gamma_{mn}^2+\left(\frac{\pi p}{2L}\right)^2}\cdot L\cdot\delta_{ps}\right\} \tag{5.41}$$

The function W_m is related to the weighting across the slot. With constant field we have

$$W_m = \cos^2\left(\frac{m\pi x_0}{a}\right)\operatorname{sinc}^2\left(\frac{m\pi w}{2a}\right) \tag{5.42}$$

where x_0 is the slot offset from the side wall as before.

In (5.41) the upper sign is valid for both p and s odd, the lower sign for both p and s even. When p is odd and s even or vice versa $Y^i_{ps} = 0$. $\delta_{ps} = 1$ for $p = s$, 0 otherwise.

5.2.3.2 External field contribution

We obtain the matrix element Y^e_{ps}

$$Y^e_{ps} = \frac{\lambda psw^2}{8\pi L^2}\int_{-\infty}^{\infty}\int_{-\infty}^{\infty}\frac{(k_z^2-k^2)\cdot \mathrm{sinc}^2\left(\frac{k_x w}{2}\right)\cdot\left\{\begin{matrix}\cos^2 k_z L\\ \sin^2 k_z L\end{matrix}\right\}\begin{matrix}p, s \text{ odd}\\ p, s \text{ even}\end{matrix}}{\sqrt{k^2-k_x^2-k_z^2}\left[k_z^2-\left(\frac{\pi p}{2L}\right)^2\right]\left[k_z^2-\left(\frac{\pi s}{2L}\right)^2\right]}\cdot dk_x dk_z \tag{5.43}$$

5.2.3.3 Excitation vector

The excitation vector becomes

$$h_s^{\mathrm{inc}} = \langle H_z^{\mathrm{inc}}, e_s\rangle = \frac{\frac{\pi s}{L}\cos\frac{\pi x_0}{a}\cdot\left\{\begin{matrix}\sin\beta_{10}L\\ -j\cos\beta_{10}L\end{matrix}\right\}\begin{matrix}s \text{ even}\\ s \text{ odd}\end{matrix}}{\beta_{10}^2-\left(\frac{\pi s}{2L}\right)^2} \tag{5.44}$$

where

$$\beta_{10} = \frac{\pi}{a}\sqrt{(2a/\lambda)^2-1} \tag{5.45}$$

5.3 Longitudinal slot characteristics

5.3.1 *Slot electric field distribution*

The integral equation solution has proved to be very useful for studies of the waveguide slot characteristics. A few results will be discussed here, starting with the electric field distribution in the slot. The results are naturally to a great degree dependent on the proper selection of basis and testing functions in the Moment Method. In Section 5.2 we made the following choice:

$$\left.\begin{matrix}\text{basis function } e_p(z) = \sin\frac{p\pi}{2L}(L+z)\\ \text{testing function } e_s(z) = \sin\frac{s\pi}{2L}(L+z)\end{matrix}\right\} \qquad (5.46) = (5.10)$$

Going back to the assumption of a shunt admittance model for the longitudinal slot, we have the relation between the reflection coefficient, $\Gamma = S_{11}$, that is, back scattering, and the slot admittance:

$$Y/G_0 = -2S_{11}/(1+S_{11}) \qquad (5.47) = (5.1)$$

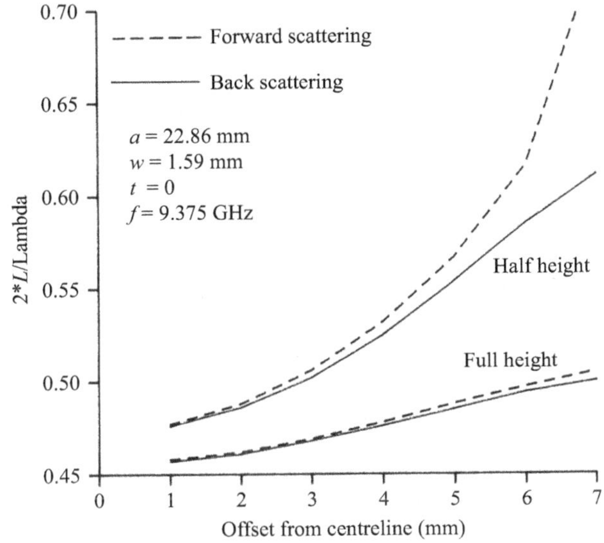

Figure 5.9 Computed resonant length based on the backscattered and forward scattered waves. Two waveguide heights are compared. Zero wall thickness. © IEEE 1987. Reprinted from [16], with permission

Based on the forward scattering (S_{12}) instead, still assuming a shunt representation, we would get

$$Y/G_0 = 2(1 - S_{12})/S_{12} \tag{5.48}$$

The two expressions give the same admittance for a pure shunt element since $S_{12} = 1 + S_{11}$ in that case. Once we have found the electric field distribution in the slot we can compute the backward and forward scattered wave amplitudes, for example, from (4.70), and from them calculate the assumed 'shunt' admittance values. Resonance is then defined from the admittance being pure real. An example is shown in Figure 5.9.

From this example it is clear that the scattering off the slot is not perfectly symmetrical. It has been found [11] that the electric field variation along the slot has an asymmetrical (odd) component as well as a symmetrical (even) component. The odd component gives rise to asymmetrical scattering. Examples of this have been given by Stern and Elliott [15] and others.

In Figure 5.9 the resonant length based on backscattering and on forward scattering is shown versus the slot displacement from the waveguide centreline. We see that the two results differ significantly when the offset is increased. The difference is even more pronounced for slots in reduced height waveguide. For slotted arrays with very stringent requirements it may be necessary to represent the slot by a more complex network (Tee or Pi network) replacing the simple shunt model. See also Compound Slots in Chapter 8.

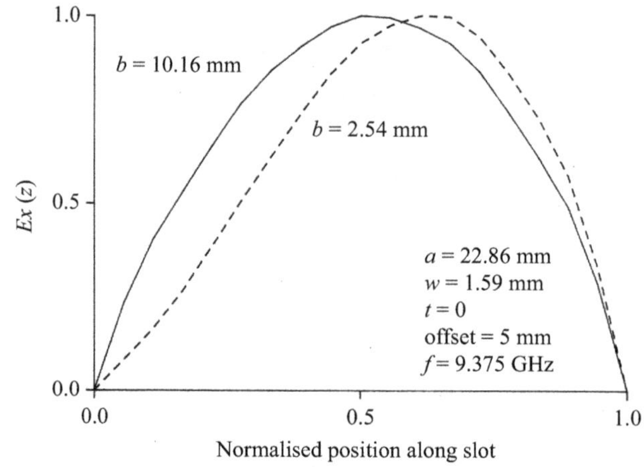

Figure 5.10 Slot electric field distribution for two waveguide heights. © IEEE 1987. Reprinted from [16], with permission

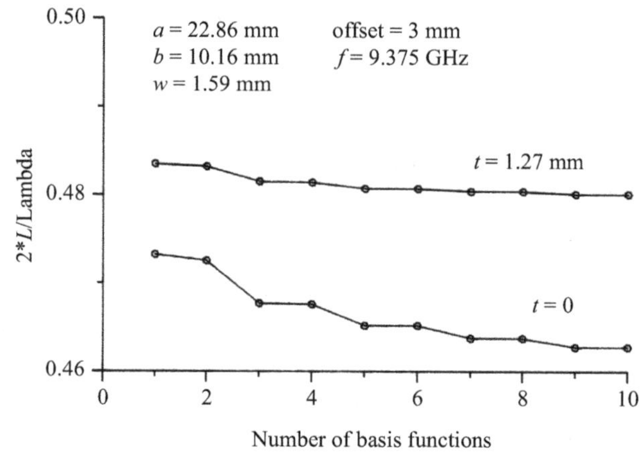

Figure 5.11 Convergence of resonant length versus number of basis functions. © URSI 1986. Reprinted from [33], with permission

The computed electric field distribution along a slot, $E_x(z')$, is shown in Figure 5.10 for two cases. The solid line is for a slot in full-height (standard) X-band guide, the broken line is for a slot in quarter-height guide. The latter case is characterised by a large asymmetrical component, indicating that the simple shunt model is a poor representation of the scattering from the slot.

For most of the results presented here nine or ten expansion functions for the slot field were used. The same number was also used in the stub waveguide to represent cases with finite wall thickness. Of course fewer terms can be used when less accurate results are acceptable to save computer time. Figure 5.11 shows for two wall thicknesses how the computed resonant length depends on the number of expansion functions chosen.

We can see that the zero wall thickness case ($t = 0$) converges more slowly than the case with finite wall thickness. It is also observed that the higher-order even terms in the expansion contribute very little to the computed result. The even terms represent the asymmetrical part of the slot field (which, however, becomes more important in reduced height waveguide; Figure 5.10). In general, the electric field distribution is slightly bowed out compared to the first sinusoidal term, resembling the shape of $\sqrt{\sin\,\theta}$.

The difference between the cases with $t = 0$ and $t = 1.27$ mm with respect to convergence can be understood when the edge condition at the slot ends is considered [26]. With zero wall thickness the electric field near the slot ends is expected to behave as $r^{1/2}$, r being the distance from the edge. This results in an expected decay rate for the coefficients $E_p \propto p^{-3/2}$. With finite thickness the right-angle edge gives the dependence $r^{2/3}$ and an expected decay rate $E_p \propto p^{-5/3}$. Thus, in the latter case, a slightly faster convergence is expected. See Figure 5.12, where the expansion coefficients are plotted on a logarithmic scale for two wall thicknesses. Slot data are the same as in Figure 5.11.

As discussed earlier one might consider adding a term with $\sim\sqrt{\sin\,\theta}$ shape in the expansion of the slot field (5.10). This term would have the proper variation at the edges $\sim r^{1/2}$ and might reduce the number of terms needed for a good field solution. A similar term would serve in the thick walled case [27, p. 99].

The entire basis results agree quite well with calculations using pulse expansion and point matching as demonstrated in Figure 5.13. It should be noted, however, that in the pulse expansion case one extra pulse of length 0.02 λ_0 was added somewhat arbitrarily to allow for the expected electric field variation down to zero amplitude at the slot ends. Including this, the agreement between the two solutions is very good, with a maximum difference of about 0.003 λ_0.

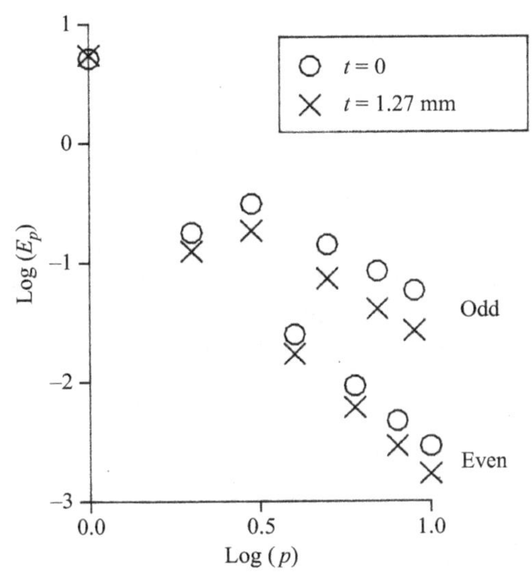

Figure 5.12 Decay of expansion coefficients E_p with increasing p. © IEEE 1987. Reprinted from [16], with permission

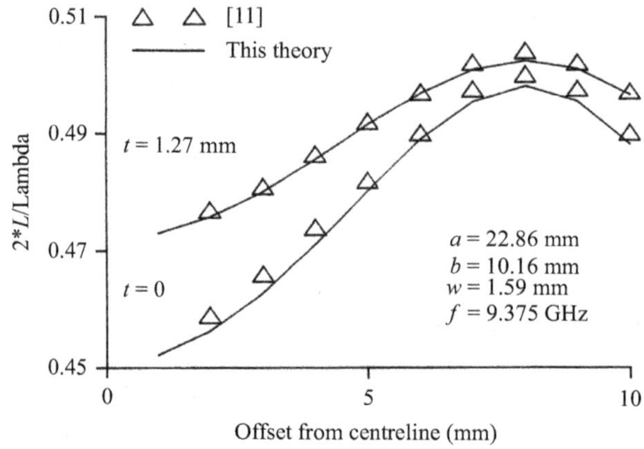

Figure 5.13 Computed resonant length versus slot offset. Entire basis (solid line) versus pulse basis (triangles). © URSI 1986. Modified from [33], with permission

In our expansion of the slot field we have assumed the electric field to be constant across the slot width. In fact, the expected physical behaviour [26] is a singularity such that the field E_x varies as $1/r^{1/2}$, where r is the distance to the edge (zero thickness case). A reasonable function modelling this behaviour is the electrostatic field variation in a narrow gap:

$$E_1(x') = \frac{w/\pi}{\sqrt{(w/2)^2 - x'^2}}, \ |x'| < w/2 \tag{5.49}$$

One could also add an odd component (and possibly even higher terms) with the same edge behaviour [28]:

$$E_1(x') = \frac{2x'/\pi}{\sqrt{(w/2)^2 - x'^2}}, \ |x'| < w/2 \tag{5.50}$$

Similar edge mode functions have been proposed for the dual case: scattering from a strip conductor [29]. For slotted waveguide array antennas with narrow slots it does not seem to improve the results very much to include these edge functions. However, for transverse slot problems using the Moment Method with pulse expansion it has been found that an edge mode may be needed in order to obtain convergence [30].

5.3.2 Slot magnetic field distribution

The purpose of the electromagnetic modelling was to find a solution to (5.2):

$$H_z^e(E_x) - H_z^i(E_x) = H_z^{\text{inc}} \qquad (5.51) = (5.2)$$

We have focused on the electric field in the slot, $E_x(z)$, and can claim that we have found good solutions with excellent agreement between measurements and calculations in terms of slot admittance, resonant length, etc. We expect therefore that (5.51) is fulfilled quite well.

The boundary conditions for the longitudinal H-field at the end points $z' = \pm L$ tell that the field shall behave as $1/r^{1/2}$ at the end points, that is, be singular. This is true both for the internal and the external fields. They are oppositely directed, so they would not cancel each other, but rather add in the zero wall thickness case; cf. (5.51). This is matched by the electric field going to zero with our choice of expansion functions, so the electromagnetic energy remains finite in the region close to the edge [26]. A test case illustrates the H-field singularities:

Standard X-band guide	$a \times b = 22.86 \times 10.16$ mm
Wall thickness	$t = 0$
Slot offset	$x_0 = 15.24$ mm
Slot width	$w = 1.587$ mm
Slot length	$2L = 15.53$ mm
Frequency	$f = 9.17$ GHz

The internal and external magnetic fields in the slot, $H_z^{\text{int}}(z')$ and $H_z^{\text{ext}}(z')$, were computed for a widely spaced infinite array ($D_x = D_z = 100$ mm), approximating the single slot case. The array situation was chosen in order to improve convergence for the external field. In total, 2,500 waveguide modes and 2,500 Floquet modes were included [14,31]. Nine expansion functions were used; see Figure 5.14.

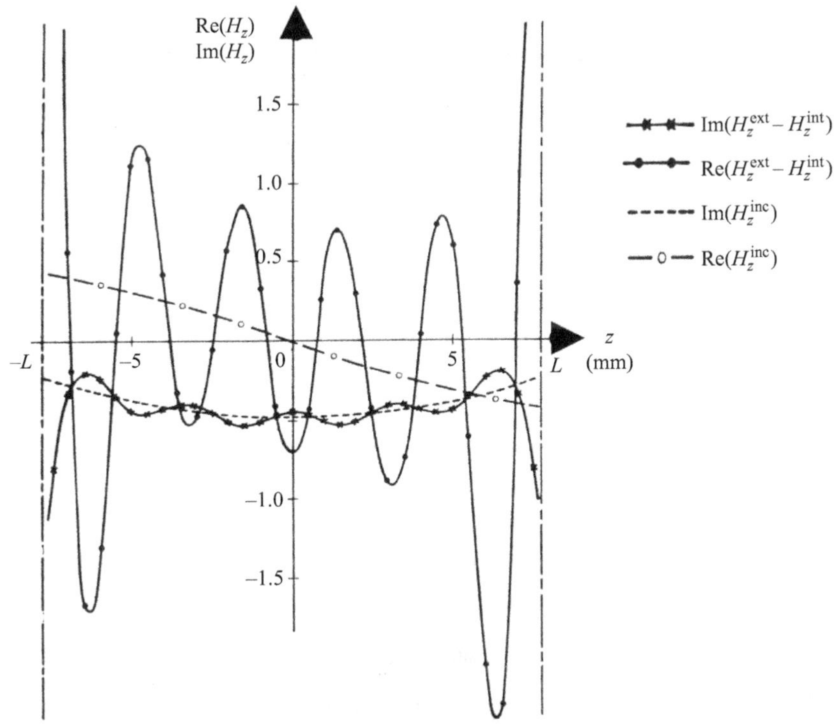

Figure 5.14 Magnetic field components in the slot [14]

Figure 5.14 shows a comparison based on the assumed identity:

$$H_z^e(E_x) - H_z^i(E_x) = H_z^{\text{inc}}$$

The right-hand side is

$$H_z^{\text{inc}} = j\cos(\pi x_0/a) \cdot e^{-j\beta_{10}z} \tag{5.52}$$

We get in our case

$$H_z^{\text{inc}} = -0.5\sin\beta_{10}z - j0.5\cos\beta_{10}z \tag{5.53}$$

In Figure 5.14 the real part and the imaginary part of H_z^{inc} are separately compared with the corresponding real and imaginary parts of $(H_z^{\text{ext}} - H_z^{\text{int}})$. We find that the imaginary parts agree rather well while the real part of $(H_z^{\text{ext}} - H_z^{\text{int}})$ demonstrates a strong oscillation centred on the real part of the incident field.

The imaginary part, which has a small oscillatory component, is in phase with the electric field in the slot. It is responsible for the real power radiated from the slot, that is, the conductance. The real part, in phase quadrature with the electric field, produces the reactive power. At resonance this is almost zero – the positive and negative peaks cancel each other. The more terms we use in the expansion of the electric field the higher will be the frequency of the ripple in the magnetic fields.

Figure 5.15 shows the magnitude $|H_z^{\text{ext}} - H_z^{\text{int}}|$ in the slot and also outside the slot. In the latter region the phase has stabilised and there is no ripple in the curve. The field magnitude here corresponds to a concentration of current density at the slot ends.

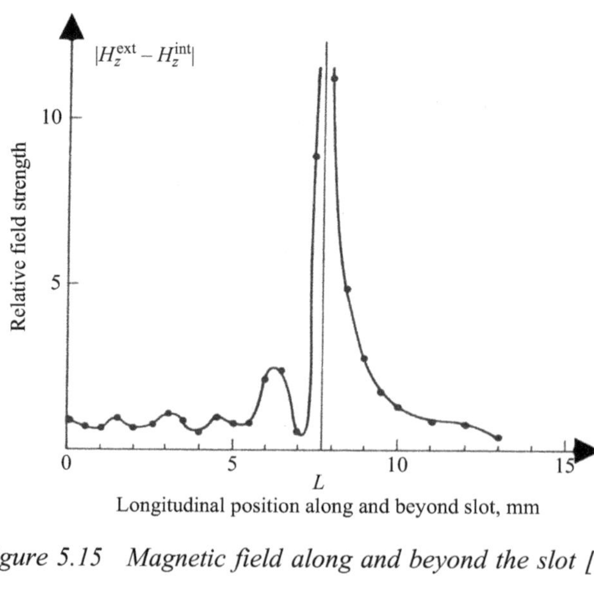

Figure 5.15 Magnetic field along and beyond the slot [14]

Now it might seem that we have not verified the boundary equation, from which we started:

$$H_z^e(E_x) - H_z^i(E_x) = H_z^{\text{inc}}$$

We solved this equation by projecting onto the space of the testing functions $e_s(z)$ by forming inner products $\langle H_z^{\text{int}}(E_x), e_s \rangle$ as already defined in (5.4). In this space our boundary condition is fulfilled. Provided that the errors in the assumed field quantities (expansion functions) are not too large, the errors in the secondary parameters such as admittance, resonance, reflection, etc., will be of second order [17,32].

5.3.3 Experimental verification

Figure 5.16 shows a comparison between measured and computed resonant lengths for rectangular slots with 0.38 mm wall thickness. Nine expansion terms were used in this computation. The difference between theory and experiment is maximum $0.002\lambda_0$, or about 0.4%, except for the largest offset where it is about twice that value.

The influence of different wall thicknesses was also studied experimentally. The resonant slot length versus slot offset was measured for 0.3 and 1.0 mm wall thickness. The slots had rounded ends, for which we still lack a theoretical model. However, the difference in resonant length for the two cases can be expected to be the same as with rectangular slots. As can be seen in Figure 5.17 the measured difference in resonant length agrees very well with the computed difference in resonant length.

In order to compensate for the rounded slot ends the following correction formula can be used:

$$2L_0 - 2L_\square = \Delta \cdot w \tag{5.54}$$

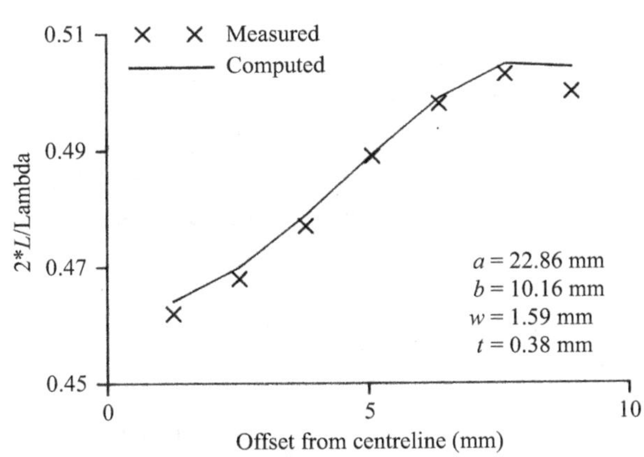

Figure 5.16 Computed and measured resonant lengths for X-band rectangular slots in standard waveguide. © IEEE 1987. Reprinted from [16], with permission

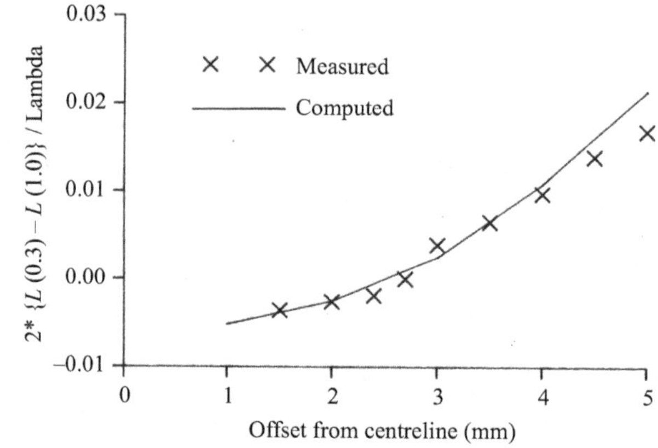

Figure 5.17 Difference in resonant length for two wall thicknesses, t = 0.3 mm and t = 1.0 mm, versus slot offset from waveguide centreline. Rounded slot ends. © IEEE 1987. Reprinted from [16], with permission

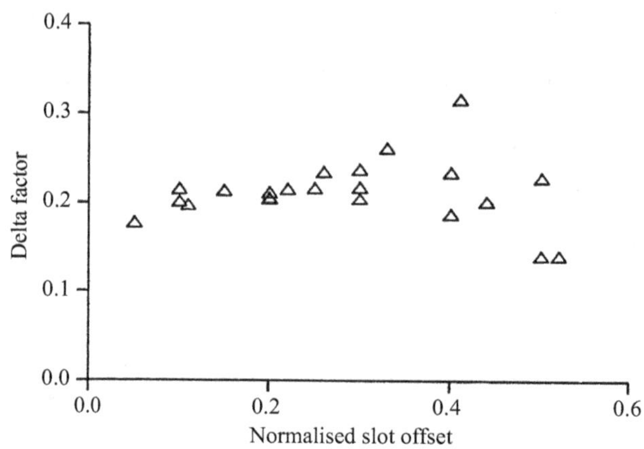

Figure 5.18 Rounded end correction factor from measured data versus slot offset normalised to half waveguide width. © URSI 1986. Reprinted from [33], with permission

Here L_0 and $L_\square$ are half the resonant slot lengths for rounded and rectangular slot ends, respectively, w is the slot width and Δ is a correction factor. It has been suggested that slots with equal area, alternatively slots with equal perimeter, should have equal resonant frequency. Yee [5] proposed an average between the two, which leads to $\Delta = 0.32$. More recently the correction factor was determined empirically from measured data for several cases with rounded slot ends and theoretical results for the corresponding rectangular slots, see Figure 5.18. The measurements were made at *C*- and *X*-band, and for wall thicknesses from 0.3 to 1.8 mm both in normal and half-height waveguide. Theoretical calculations using

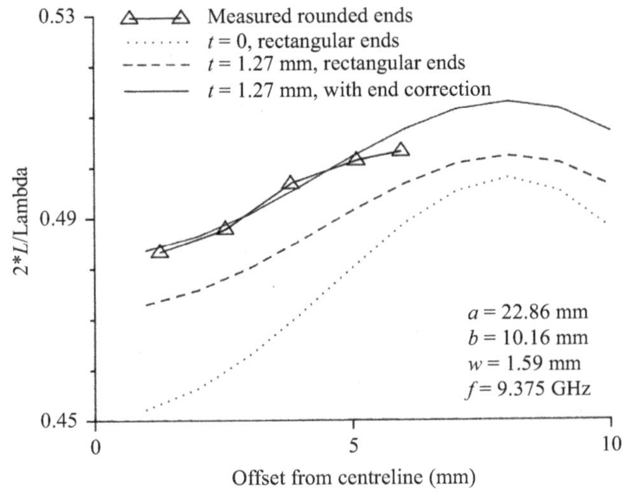

Figure 5.19 Computed resonant length versus slot offset for standard X-band guide compared with measured data from [35]. © IEEE 1987. Reprinted from [16], with permission

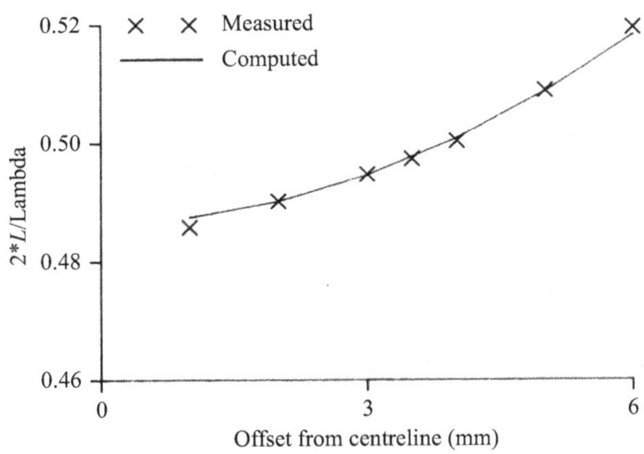

Figure 5.20 Computed and measured resonant length versus slot offset for C-band slot with rounded ends in half-height waveguide. $a = 39$ mm, $b = 10$ mm, $w = 2.8$ mm, $t = 0.6$ mm, $f = 5.3$ GHz. © URSI 1986. Reprinted from [33], with permission

MoM/FEM analysis have also been done [34]. Both studies support the assumption that slots with equal area have approximately equal resonant frequency, that is, $\Delta = 1 - \pi/4 = 0.215$. Assuming equal perimeter would lead to twice this correction.

Using this rounded end correction we obtain the results shown in Figure 5.19, where a comparison is made with Stegen's measured data [35]. Another example is given in Figure 5.20.

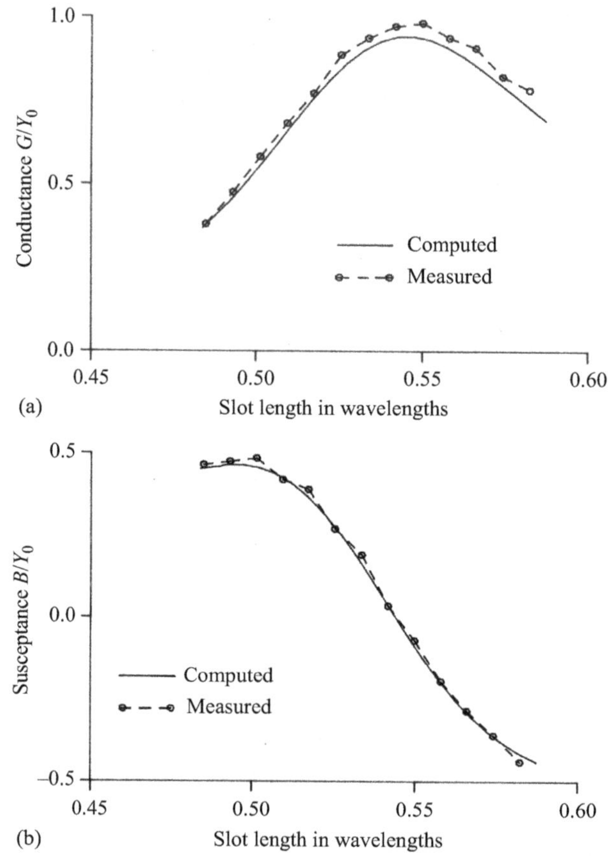

Figure 5.21 (a) Computed and measured slot conductance versus slot length for X-band slot in half-height waveguide. © IEEE 1987. Reprinted from [16], with permission. (b) Computed and measured slot susceptance versus slot length for X-band slot in half-height waveguide. © URSI 1986. Reprinted from [33], with permission

Figure 5.21 shows the computed slot conductance and susceptance versus slot length compared with measurements. The waveguide wall thickness was 0.3 mm. The slot had rounded ends and we applied the correction factor to account for this.

5.3.4 *Weakly excited slots*

A longitudinal slot with no offset, that is, positioned on the waveguide centreline, will not radiate. Very small offsets will result in weak radiation. This is a typical situation in long, travelling wave type slotted arrays. With this configuration a wide bandwidth is possible in spite of the long length, although the beam will squint with frequency – sometimes a desirable feature (see Section 9.2).

A phase/frequency scanned *C*-band array with long, non-resonant ridge waveguides has been described by Derneryd and Lorentzon [36]. High accuracy requirements combined with the small offsets led to an experimental approach in the slot modelling. Very low normalised conductance values had to be used in the design. In particular, the waveguide centreline to which slot offsets are referenced must be known accurately in order to avoid grating lobes [37].

With zero offset the slot will still cause a disturbance which is manifested as a small reactive loading of the waveguide. This was illustrated in the equivalent circuit in Figure 5.3 as a shunt element B_0, the residual susceptance for a centred slot. It can typically amount to $B_0/G_0 \approx -0.1$ or less and can be compensated for in the hardware with a small capacitive metallic post under the slot [38, p. 271]. Work on improved scattering models for weakly excited waveguide slots has been reported by, for example, Rengarajan and Derneryd [39]; see also Chapter 8.

5.4 Transverse slots

5.4.1 Introduction

Transverse slots in the *broad* wall of a rectangular waveguide have a potential application in low-thickness array antennas with the polarisation directed along the waveguide (Figure 5.22).

Transverse slots in the *narrow* wall of the guide require probes, irises or similar devices inside the waveguide in order to excite the slots [40,41]. Alternatively, one could use *inclined* narrow wall slots, which, however, radiate a cross-polarised field component. Watson [1] discussed the transverse waveguide slot, which he represented as a series element on an equivalent transmission line. Approximate expressions for the slot impedance were derived by Stevenson [2] and Oliner [4]. Das [42] reported on expressions for the slot impedance as a function of slot length, starting from an assumed sinusoidal electric field distribution in the slot.

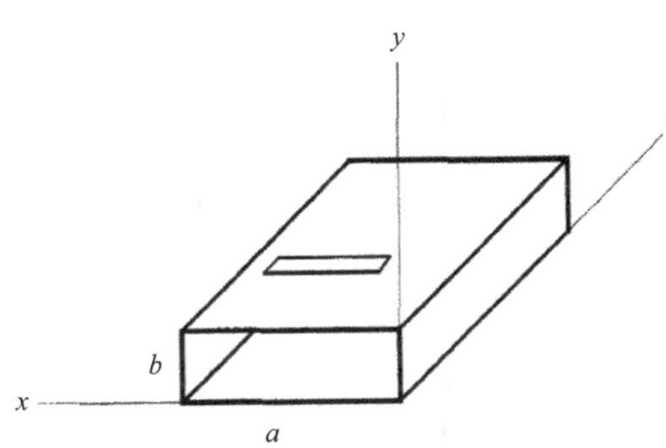

Figure 5.22 Transverse slot in a rectangular waveguide

We will here follow a similar procedure as we did when analysing the longitudinal slot characteristics in the previous sections. A main difference is that the slot field is z-directed leading to asymmetrical scattering in the waveguide. The slot can be represented by a series impedance on an equivalent transmission line.

5.4.2 Theoretical analysis

In order to analyse the slot impedance we make use of an integral equation for the slot electric field, which we arrive at by equating the magnetic fields on the two sides of the slot interface. The integral equation is solved by the Method of Moments. The wall thickness is easily and accurately included without increasing the number of unknowns. The procedure is the same as for longitudinal slots, the details are not repeated here [16,42].

The slot field is assumed to be of the form

$$E_z = \sum_{p=1}^{N} E_p \sin\frac{\pi p}{2L}(x - x_0 + L) \tag{5.55}$$

with no variation across the slot width, w (see Figure 5.23). The variable x_0 is the off-centre displacement, and L is the half length of the slot. Nine terms ($N = 9$) were used for the numerical results presented here. The E_x field along the slot is neglected as well as the magnetic field H_z across the slot.

The external magnetic field and the associated Moment Method matrix elements will be the same as in the longitudinal slot case (Section 5.2). The internal matrix elements, however, are different for the transverse slot.

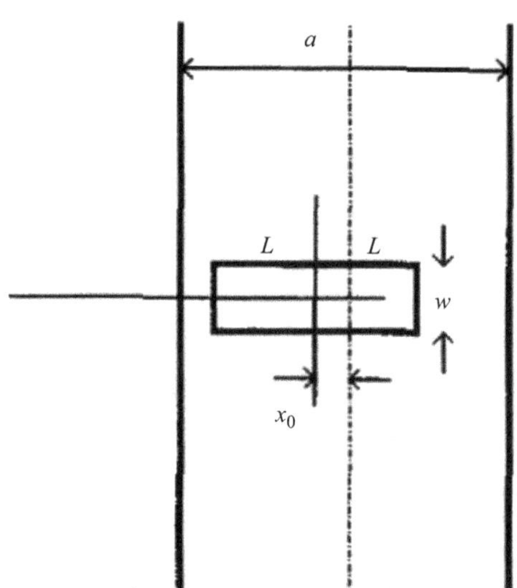

Figure 5.23 Definition of offset for a transverse slot

The appropriate interior Green function for the waveguide region is well known [2,19]. Limiting ourselves to the H_x component from the slot electric E_z field, we obtain both TE and TM contributions. They are summed to yield

$$H_x(x,z) = \sum_{p=1}^{N} E_p \sum_m \sum_n \frac{\varepsilon_m \varepsilon_n \frac{p\pi}{L}\left[\left(\frac{m\pi}{a}\right)^2 - k^2\right] \sin\frac{m\pi x}{a}}{j\omega\mu_0 ab\gamma_{nm}^2 \left[\left(\frac{p\pi}{2L}\right)^2 - \left(\frac{m\pi}{a}\right)^2\right]} \cdot \ldots$$

$$\cdot \begin{pmatrix} \sin\frac{m\pi x_0}{a}\cos\frac{m\pi L}{a} \\ -\cos\frac{m\pi x_0}{a}\sin\frac{m\pi L}{a} \end{pmatrix}_{p\ \text{even}}^{p\ \text{odd}} \cdot \left(1 - e^{-\gamma_{mn}w/2}\cosh\gamma_{mn}z\right) \tag{5.56}$$

Here

$$\gamma_{mn}^2 = \left(\frac{m\pi}{a}\right)^2 + \left(\frac{n\pi}{b}\right)^2 - k^2 \tag{5.57}$$

and $\varepsilon_m = 1$ for $m = 0$, while $\varepsilon_m = 2$ for $m > 0$.

For the Galerkin case we form inner products with the testing functions, w_s

$$w_s = \sin\frac{\pi s}{2L}(x - x_0 + L) \tag{5.58}$$

to yield the matrix element contributions Y_{ps}^{int}:

$$Y_{ps}^{\text{int}} = \sum_m \sum_n \frac{\varepsilon_m \varepsilon_n \frac{\pi p}{L}\frac{\pi s}{L}\left[\left(\frac{m\pi}{a}\right)^2 - k^2\right]\left[w - \frac{1 - e^{\gamma_{mn}w}}{\gamma_{mn}}\right]}{j\omega\mu_0 ab\gamma_{nm}^2\left[\left(\frac{p\pi}{2L}\right)^2 - \left(\frac{m\pi}{a}\right)^2\right]\left[\left(\frac{s\pi}{2L}\right)^2 - \left(\frac{m\pi}{a}\right)^2\right]} \cdot \ldots$$

$$\cdot \begin{pmatrix} \sin\frac{m\pi x_0}{a}\cos\frac{m\pi L}{a} \\ -\cos\frac{m\pi x_0}{a}\sin\frac{m\pi L}{a} \end{pmatrix}_{p\ \text{even}}^{p\ \text{odd}} \cdot \begin{pmatrix} \sin\frac{m\pi x_0}{a}\cos\frac{m\pi L}{a} \\ -\cos\frac{m\pi x_0}{a}\sin\frac{m\pi L}{a} \end{pmatrix}_{s\ \text{even}}^{s\ \text{odd}} \tag{5.59}$$

The driving vector, assuming a normalised incident TE_{10} field:

$$H_x^{\text{inc}} = \sqrt{\frac{2}{ab}}\sin\frac{\pi x}{a}e^{-j\beta_{10}z} \tag{5.60}$$

has elements h_s:

$$h_s = \langle H_x^{\text{inc}}, w_s\rangle = \frac{4\sqrt{2}}{\sqrt{ab}\beta_{10}}\sin(\beta_{10}w/2) \cdot \frac{\frac{s\pi}{2L}}{\left(\frac{s\pi}{2L}\right)^2 - \left(\frac{\pi}{a}\right)^2}\begin{pmatrix} \sin\frac{\pi x_0}{a}\cos\frac{\pi L}{a} \\ -\cos\frac{\pi x_0}{a}\sin\frac{\pi L}{a} \end{pmatrix}_{s\ \text{even}}^{s\ \text{odd}} \tag{5.61}$$

The solution yields in particular the inner aperture field as expressed by the expansion (5.55), where the coefficients are now known. This is the key to the equivalent slot impedance. The reflection coefficient inside the waveguide in terms of the propagating TE_{10} mode becomes

$$\Gamma = \frac{H_x^{scat}}{H_x^{inc}} = \frac{\sin(\beta_{10}w/2)}{\omega\mu_0\sqrt{ab}} \sum_{p=1}^{N} \frac{E_p \frac{p\pi}{2L}}{\left(\frac{p\pi}{2L}\right)^2 - \left(\frac{\pi}{a}\right)^2} \cdot \begin{pmatrix} \sin\frac{\pi x_0}{a}\cos\frac{\pi L}{a} \\ -\cos\frac{\pi x_0}{a}\sin\frac{\pi L}{a} \end{pmatrix}_{p\ \mathrm{even}}^{p\ \mathrm{odd}} \tag{5.62}$$

from which we obtain the equivalent normalised slot impedance:

$$Z/Z_0 = \frac{2\Gamma}{1-\Gamma} \tag{5.63}$$

5.4.3 Results

Figure 5.24 shows the computed normalised resistance and reactance versus slot length compared with previous results by Khac [11], who used a pulse expansion with point matching. These results are for zero wall thickness.

Figure 5.25 shows computed results compared with measured results for a transverse slot in a reduced-height waveguide. The measured slot had rounded ends. The length was therefore reduced by $\Delta = 0.215w$ to represent an equivalent rectangular slot; cf. Figure 5.18.

The effect of the waveguide height is demonstrated in Figure 5.26. The resistance is approximately inversely proportional to the waveguide height. Figure 5.27

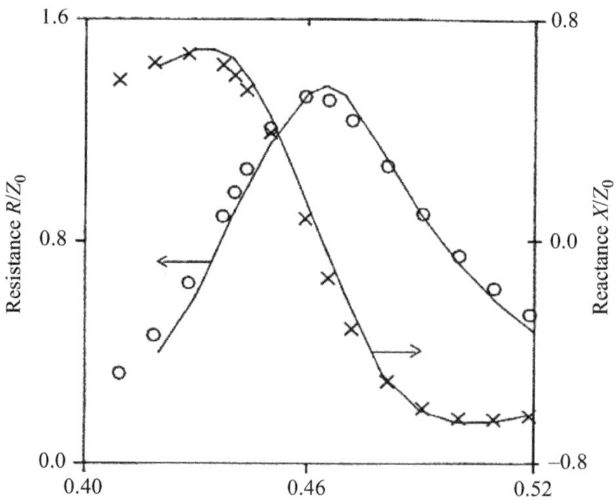

Figure 5.24 Transverse slot impedance in zero wall standard X-band waveguide versus normalised slot length 2L/λ. Rings and crosses denote computed results (present method). Lines are from [11]. © IEEE 1993. Reprinted from [43], with permission

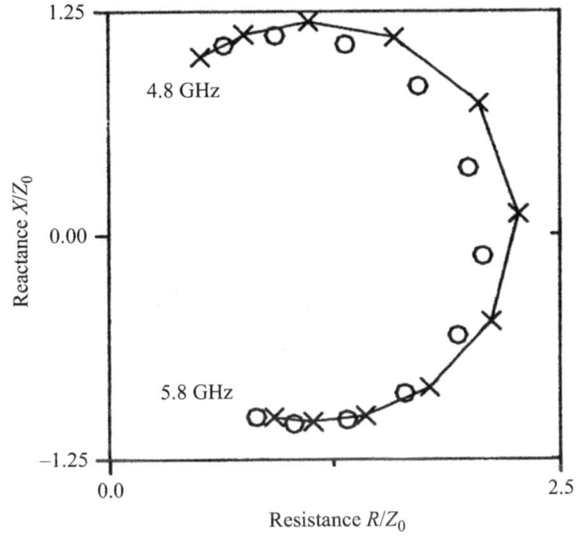

Figure 5.25 *Transverse slot impedance for a 28 mm long slot with zero offset from 4.8 to 5.8 GHz. $a = 39$ mm, $b = 10$ mm, $w = 3$ mm, $t = 1.9$ mm. Solid curve with crosses is computed, rings denote measured values. © IEEE 1993. Modified from [43], with permission*

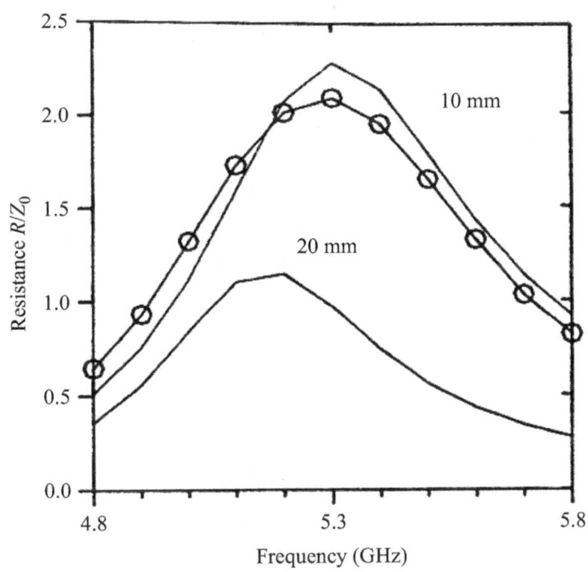

Figure 5.26 *Computed slot resistance for two waveguide heights $b = 10$ mm and $b = 20$ mm, other parameters as in Figure 5.25. Curve with rings is measured results for $b = 10$ mm. © IEEE 1993. Reprinted from [43], with permission*

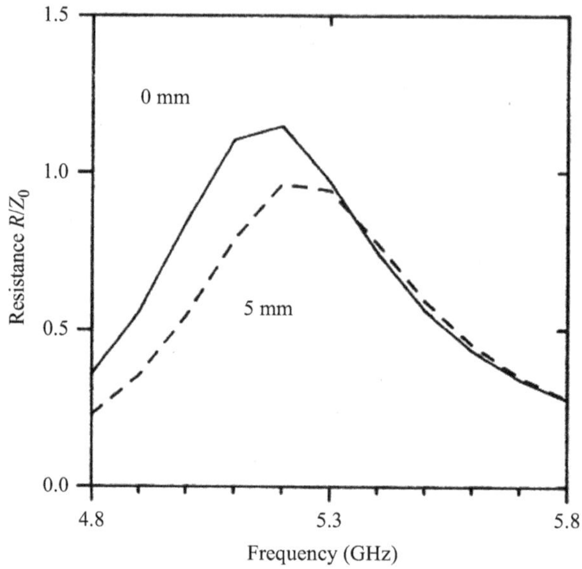

Figure 5.27 Computed slot resistance for two offset values, $x_0 = 0$ mm and 5 mm. $b = 20$ mm, other data as in Figure 5.26. © IEEE 1993. Reprinted from [43], with permission

shows the dependence on the offset. Only minor changes can be realised using this parameter.

There is no simple way of varying the excitation of a transverse slot. There is also no phase reversal technique available, such as alternating the offset direction or changing the sign of the inclination angle as in other types of slots. In arrays, this will lead to the generation of grating lobes if resonant slot spacing shall be used, that is, one guide wavelength for broadside radiation. Parallel plates in the form of baffles can be used to suppress the grating lobes in the E-plane of an array of transverse slots; more about this in Section 9.3.

References

[1] Watson W. H. 'Resonant slots'. *Journal of the Institution of Electrical Engineers*, Part IIA. 1946;**93**(4):747–777.

[2] Stevenson A. F. 'Theory of slots in rectangular waveguides'. *Journal of Applied Physics*. 1948;**19**(1):24–38.

[3] Wait J. R. 'On the conductance of slots'. *IRE Transactions on Antennas and Propagation*. 1956;**4**(2):124–127.

[4] Oliner A. A. 'The impedance properties of narrow radiating slots in the broad face of rectangular waveguide, Part I, II'. *IRE Transactions on Antennas and Propagation*. 1957;**5**(1):4–20.

[5] Yee H. Y. 'Impedance of a narrow longitudinal shunt slot in a slotted waveguide array'. *IEEE Transactions on Antennas and Propagation.* 1974;**22**(4):589–592.

[6] Stegen R. J. 'Slot radiators and arrays at X-band'. *IRE Transactions on Antennas and Propagation.* 1952;**1**(1):62–84.

[7] Elliott R. S. *Antenna Theory and Design.* Prentice-Hall, Englewood Cliffs, NJ, 1981.

[8] Lagerlöf R., Josefsson L. 'A phased array radar antenna for airborne early warning'. *Proceedings Military Microwaves.* London, July 11–13, 1990, pp. 257–261.

[9] Lewin L. *Advanced Theory of Waveguides.* Iliffe & Sons, Ltd., London, 1951.

[10] Falk K. 'Admittance of a longitudinal slot in a ridge waveguide'. *IEE Proceedings H – Microwaves, Antennas and Propagation.* 1988;**135**(4):263–268.

[11] Khac T. V. *A Study of Some Slot Discontinuities in Rectangular Waveguides.* Ph.D. Dissertation, Monash University, Australia, November 1974.

[12] Lyon R. W., Sangster A. J. 'Efficient moment method analysis of radiating slot in a thick-walled rectangular waveguide'. *IEE Proceedings H – Microwaves, Antennas and Propagation.* 1981;**128**(4):197–204.

[13] Yee H. Y., Richardson P. N. 'Slotted waveguide antenna arrays'. *IEEE Antennas and Propagation Society Newsletter.* 1982;**24**(6):4–8.

[14] Josefsson L. *Slots in Waveguide Arrays Analyzed by Expanding the Fields in Waveguide Modes and Floquet Modes.* UCLA Rpt. No. AP-202, EE Dept., USA, August 1983.

[15] Stern G. J., Elliott R. S. 'Resonant length of longitudinal slots and validity of circuit representation: theory and experiment'. *IEEE Transactions on Antennas and Propagation.* 1985;**33**(11):1264–1271.

[16] Josefsson L. 'Analysis of longitudinal slots in rectangular waveguides'. *IEEE Transactions on Antennas and Propagation.* 1987;**35**(12):1351–1357.

[17] Harrington R. F. *Field Computation by Moment Methods.* The MacMillan Co., 1968.

[18] George Green G. *An Essay on the Application of Mathematical Analysis to the Theories of Electricity and Magnetism.* T. Wheelhouse, Nottingham, 1828.

[19] Elliott R. S. *An Introduction to Guided Waves and Microwave Circuits.* (Appendix C: Stevenson's Green's functions for rectangular waveguide). Prentice-Hall, Englewood Cliffs, NJ, 1993.

[20] Collin R. E. 'On the incompleteness of *E* and *H* modes in wave guides'. *Canadian Journal of Physics.* 1973;**51**:1135–1140.

[21] Rahmat-Samii Y. 'On the question of computation of the dyadic Green's function at the source region in waveguides and cavities'. *IEEE Transactions on Microwave Theory and Techniques.* 1975;**23**(9):762–765.

[22] Van Bladel J. 'Contribution of the Ψ=constant mode to the modal expansion in a waveguide', *IEE Proceedings H – Microwaves, Optics and Antennas.* 1981;**128**(5):247–251.

[23] Eshrah I. A., Yakovlev A. B., Kishk A. A., Glisson A. W., Hanson G. W. 'The TE_{00} waveguide mode – The "complete" story'. *IEEE Antennas and Propagation Magazine.* 2004;**46**(5):33–41.

[24] Clemmow P. C. *The Plane Wave Spectrum Representation of Electromagnetic Fields.* Pergamon Press, Oxford, 1966.

[25] Stinson D. C. *Intermediate Mathematics of Electromagnetics*. Prentice-Hall, Englewood Cliffs, NJ, 1976.

[26] Meixner J. 'The behavior of electromagnetic fields at edges'. *IEEE Transactions on Antennas and Propagation.* 1972;**20**(4):442–446.

[27] Amitay N., Galindo V., Vu C. P. *Theory and Analysis of Phased Array Antennas.* Wiley Interscience, New York, 1972.

[28] Butler C. M., Wilton D. R. 'General analysis of narrow strips and slots'. *IEEE Transactions on Antennas and Propagation.* 1980;**28**(1):42–48.

[29] Richmond, J. H. 'On the edge mode in the theory of TM scattering by a strip or strip grating'. *IEEE Transactions on Antennas and Propagation.* 1980; **28**(6):883–887.

[30] Park K. P., Stern G. J., Elliott R. S. 'An improved technique for the evaluation of transverse slot discontinuities in rectangular waveguide'. *IEEE Transactions on Antennas and Propagation.* 1983;**31**(1):148–154.

[31] Rengarajan S. R., Josefsson L., Petersson R. 'Recent developments in broad wall slots in rectangular waveguides for array applications'. *Seventh International Conference on Antennas and Propagation*, ICAP 91 (IEE). 1991, pp. 729–732.

[32] Harrington R. F. *Time-Harmonic Elctromagnetic Fields.* IEEE Press, New York, 2001.

[33] Josefsson L. 'Analysis of longitudinal slots in rectangular waveguides including the effect of wall thickness'. *Proceedings of the URSI Electromagnetic Theory Symposium.* Budapest, Hungary. August 1986, pp. 367–369.

[34] Zhang M., Hirano T., Hirokawa J., Ando M. 'Method of Moments analysis of a waveguide round-ended wide slot by using numerical-eigenmode basis functions'. *IEEE Topical Conference on Wireless Communication Technology.* 2003, pp. 360–361.

[35] Stegen R. J. *Longitudinal Shunt Slot Characteristics*. Hughes Tech. Memo No. 261, November 1951.

[36] Derneryd, A. G., Lorentzon T. C. 'Design of a phase/frequency scanned array antenna with non-resonant slotted ridge waveguide elements'. *Proceedings of the IEEE Antennas and Propagation Symposium*, London, Ontario, June 1991, pp. 1728–1731.

[37] Karlsson I. 'Applications of waveguide arrays in commercial and military radars'. *Proceedings of the 1993 Antenna Applications Symposium,* Allerton Park, IL, USA, pp. 1–22, 1993.

[38] Marcuvitz N. (ed.) *Waveguide Handbook.* MIT Rad.Lab.Series, Vol. 10, 1951.

[39] Rengarajan S. R., Derneryd A. 'Improved scattering model for weak broadwall slots'. *Microwave and Optical Technology Letters.* 1993;**6**(8):504–507.

[40] Hashemi-Yeganeh S., Elliott R. S. 'Analysis of untilted edge slots excited by tilted wires'. *IEEE Transactions on Antennas and Propagation.* 1990; **38**(11):1737–1745.

[41] Dudley D. G. 'An iris-excited slot radiator in the narrow wall of rectangular waveguide'. *IRE Transactions on Antennas and Propagation.* 1961; **9**(4):361–364.

[42] Das B. N., Sinha M. 'Impedance characteristics of series slots'. *Proceedings of the Institution of Electrical Engineers.* 1974;**121**(11):1360–1362.

[43] Josefsson L. 'A waveguide transverse slot for array applications'. *IEEE Transactions on Antennas and Propagation.* 1993;**41**(7):845–850.

Chapter 6

The linear slotted waveguide array antenna

Based on the analysis of the single-slot antenna in the previous chapters we can now approach the problem of designing linear and planar slot arrays. We will first study longitudinal slot arrays and start with the simple case with one row of slots in a rectangular waveguide. For this case mutual coupling appears mainly in the H-plane and can in many cases be neglected. Some examples of computed and measured performance will be presented. Procedures for designing linear arrays of slots, including some examples, will be discussed.

6.1 Introduction

The longitudinal slot in a waveguide can be represented by an equivalent shunt admittance according to [1] as already mentioned in Chapter 1. In Chapters 4 and 5 mathematical expressions were derived for modelling this shunt admittance. It leads to the network representation of an array of slots according to Figure 6.1.

Figure 6.1 shows a resonant slot array where the waveguide is fed at one end and shorted at the other end. With a spacing from the last slot to the short of a quarter of a guide wavelength (corresponding to the phase delay φ_N) the short is transformed into an open circuit at the location of the slot. The other spacings shall be half a guide wavelength so that all slots radiate in phase. Note the alternating offsets (δ in Figure 6.1) that provide the missing 180° of phase shift. This means that all admittances appear in parallel as seen from the feeding end. A match is obtained when the sum of all admittances equals the waveguide admittance G_0:

$$G_0 = \sum_{i=1}^{N} y_i \tag{6.1}$$

At the design frequency the slots are resonant so that their admittances are real. With a uniform amplitude distribution (i.e. with equal slot conductances $= G_0/N$) a narrow broadside beam is obtained in the H-plane with first sidelobes about -13 dB below the main beam peak. With a tapered distribution the slot conductances shall be chosen according to the desired slot voltage distribution; the slot conductance is assumed to be proportional to the square of the slot voltage as an approximation. The relationship between the slot conductance and slot voltage will be further discussed in Section 6.3.

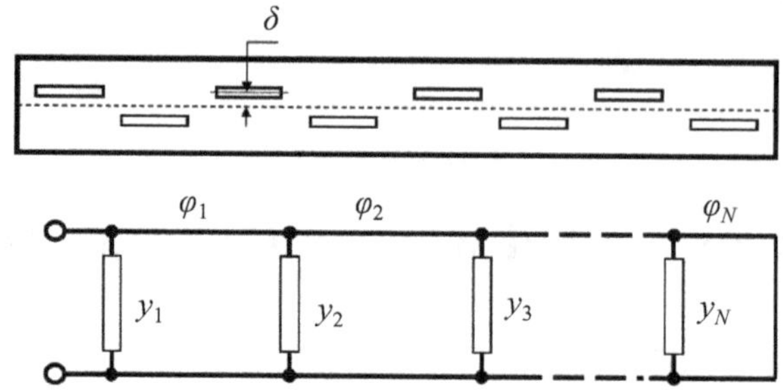

Figure 6.1 A linear slot array and its network representation with shunt admittances

A simple model for estimating the return loss bandwidth as a function of N, the number of radiating elements in an end-fed standing wave array consists of a transmission line periodically loaded by N shunt conductances as shown in Figure 6.1. The spacing between adjacent loads is a half wavelength at the centre frequency and the value of each conductance normalised to the characteristic admittance of the transmission line is kept as $1/N$ for simplicity. A short circuit is placed at the end of the transmission line at a distance of a quarter wavelength from the last load. Using the transmission line equations we obtain the normalised admittance across the ith conductance looking towards the short, where $i = 1$ is the load near the short circuit. Expressions for the total normalised admittance at section i given by Elliott [2] and reproduced in (6.2) and (6.3) are obtained from standard circuit theory and transmission line equations [3,4, p. 151].

$$\frac{Y_i}{G_0} = \frac{(Y_{i-1}/G_0)\cos(\beta d) + j\sin(\beta d)}{\cos(\beta d) + j(Y_{i-1}/G_0)\sin(\beta d)} + 1/N \tag{6.2}$$

$$\frac{Y_1}{G_0} = -j\cot(\beta d) \tag{6.3}$$

In (6.2), G_0 is the characteristic admittance of the transmission line and β is its phase constant. Using (6.2) recursively from $i = 1$ through N, one can determine the normalised admittance at the input port, and hence the reflection coefficient. This model does not include the waveguide dispersion, the frequency response of each shunt element or the mutual coupling between radiating elements. Therefore the results are to be used in a qualitative sense. Figure 6.2 shows the approximate value of the reflection coefficient in dB achievable for a specified percentage bandwidth for different values of N. If we use -10 dB reflection coefficient or approximately a VSWR of 2 as a reference, an array with eight elements in the end-fed configuration will produce about 8% bandwidth. In other words over 8% bandwidth, the reflection coefficient would be -10 dB or better. For greater values of N, the -10 dB reflection coefficient bandwidth is found to be lower, for example, a 16-element array exhibits only about 4% bandwidth.

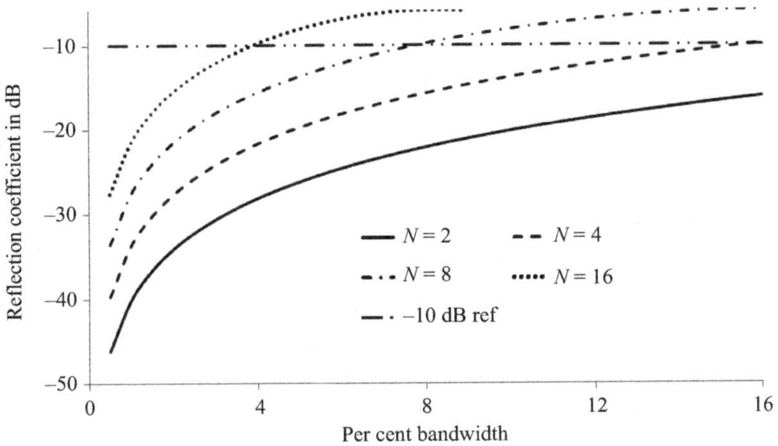

Figure 6.2 Reflection coefficient bandwidth of a standing wave array for different numbers N of slot elements (−10 dB ref is a baseline reference)

A centre-fed array is generally preferred over an end-fed array so as to achieve a greater bandwidth, since the distance between the feed and the short circuit is halved in a symmetric centre-fed array. Centre-fed arrays will be discussed in Chapter 7.

Selecting the sum of slot conductances higher than the waveguide conductance G_0 requires an impedance transformer at the input port. The scheme is called *overloading* and can be shown to increase the VSWR bandwidth of the array. This is discussed in more detail in Chapter 7, Section 7.5, and in Section 9.11.

6.1.1 Overview of design approaches

Linear slotted waveguide array designs have been reported since the 1950s [5–8] using approximate transmission line models with lumped shunt or series elements for the slots. These early designs assumed that the conductance of a shunt slot or the resistance of a series slot is proportional to the square of the slot voltage. Elliott has shown that the above-mentioned assumption can lead to errors [9]. A design technique for small arrays was presented by Elliott and Kurtz by incorporating the external mutual coupling between slots by using a dipole analogy [10]. Subsequently Elliott dispensed with the dipole analogy and derived the mutual coupling expressions using a half cosinusoidal distribution for the slot aperture electric field [11]. His derivation of the external mutual coupling expressions employed the reciprocity principle. This design procedure, also applicable to slots in dielectric filled waveguides, uses accurate models for isolated slots and for various coupling mechanisms. It has been widely used in the design of slot arrays. The Elliott design equations are derived in Section 6.3. In the first equation, radiating longitudinal slots are modelled in the form of equivalent shunt admittances in a transmission line. In the second equation, the active admittance of a longitudinal radiating slot, in the presence of all other slots in the array, is derived in terms of its self-admittance, that is, that of an isolated slot, and geometrical parameters such as offsets and lengths of all slots in the array. This expression includes external mutual coupling between all slots. Internal TE_{20} mode coupling between adjacent slots is derived later. The TE_{20} mode coupling between the slot closest to the short circuit and the short may be included using a similar approach. The design procedure uses an iterative technique as discussed in Sections 6.3, 6.5 and 6.6. Convergence is reached generally after a few iterations.

The design approach using the incremental conductance, discussed in Section 6.7, is especially useful in relatively large arrays in geometries where it is difficult to determine the self and mutual admittances. The incremental conductance is defined as the difference between the total input conductance of two arrays, one having N resonant slots and the other $N+1$. In both cases the slots are identical (same values of offset, length or tilt) with adjacent offsets or tilts alternating positive and negative values. Incremental conductance data is computed or measured for a range of values of slot offsets or tilts. It is then possible to design arrays that have slowly varying aperture distributions. It is possible to use this approximate technique as a starting design, which may be improved with more rigorous design techniques, if necessary.

The Elliott design technique models longitudinal radiating slots in the form of shunt admittances in a transmission line. For arrays having slots with large offsets in reduced height waveguides and compound slots characterised by offsets and tilts, a better slot model will be in the form of a Tee or Pi network. For such slots a design procedure using scattering parameters of the slot is discussed in Section 6.8.

The Elliott design procedure requires scattering characteristics of isolated longitudinal radiating slots as a function of slot offsets and lengths. Ideally a slot in a waveguide should be modelled as a four-port network, with two ports on the waveguide on either side of the slot and two ports for the two apertures of a slot when wall thickness is included in the model. For practical values of wall thickness, the voltages of both slot apertures are nearly the same. Therefore, the wall thickness is ignored in the Elliott design procedure except when determining the scattering data of isolated slots. It should be mentioned that the resonant length of a slot is sensitive to the wall thickness. A two-port model of a slot with both ports in the waveguide, with the knowledge of the slot voltage is adequate. The slot aperture distribution is assumed to be a half cosine distribution with a complex coefficient, and it is related to the type of excitation such as the incident TE_{10} mode, higher-order modes from adjacent slots or external radiation from all other slots. With the availability of many commercial computational electromagnetic (CEM) codes it is possible to compute slot data very accurately. The mutual coupling may be computed accurately as discussed in Section 6.4.

The design of slot arrays involves the determination of waveguide dimensions and slot parameters to achieve the required radiation characteristics, for example, gain, sidelobe level and cross-polarisation level, and the return loss as a function of frequency. On the other hand, in the analysis process one is interested in determining the radiation characteristics and the return loss, given the waveguide and slot parameters. Design techniques generally use transmission line models with lumped circuit loads representing the slots. While some analysis models also use similar loaded transmission line models, rigorous analyses employing finite element methods, integral equations/method of moments (MoM) and mode matching techniques provide very accurate solutions. The analysis procedure employing the MoM solution of the integral equations for the slot aperture fields is discussed in Chapter 8.

6.2 The accuracy of the slot array model

In early designs a few simplifications were made.

6.2.1 Slot model

Accurately measured slot data represents the real world and is often preferred if available. Computer simulations usually involve some simplifications, more or less justified. However, a very accurate slot field distribution is not always required. Simply stated, conductance or admittance is related to power and is thus stationary and insensitive to small variations in the assumed field shape [12, p. 428]. This was also illustrated in relation to Figure 5.14.

In array design and analysis the field distribution of individual slots is often assumed to be symmetric and sinusoidal. A non-symmetric component is sometimes observed, especially in reduced height waveguide; cf. Figures 5.9 and 5.10. This means that the simple shunt element model is not exactly valid and may have to be replaced with a Tee or Pi network to account for the asymmetrical scattering [13]. Alternatively a scattering model for the slot is useful in the design as explained in Section 6.8.

6.2.2 Network model

The network model (Figure 6.1) is assumed to carry only the fundamental TE_{10} mode. Higher-order modes are indeed excited near the slot but often assumed to decay rapidly. However, in some slot arrays, for example, when slot offsets are small, TE_{20} mode generated at a slot can couple a non-negligible amount into a neighbouring slot. This is discussed in Section 6.4. The problem with higher-order waveguide modes is also apparent in some planar arrays between adjacent coupling slots and at coupling junctions between a coupling slot and two radiating slots in the immediate vicinity. These scenarios are discussed in Chapter 8, Sections 8.4 and 8.5.

6.2.3 Ground plane

The slot models often assume an infinite, perfectly conducting ground plane. Edge diffraction effects were discussed in relation to Figures 4.12 and 4.24 and also in Section 8.9.

6.2.4 Mutual coupling

In Section 6.1 mutual coupling was only briefly mentioned. There is always some coupling in the *H*-plane but usually quite small. The examples in Figures 4.20 and

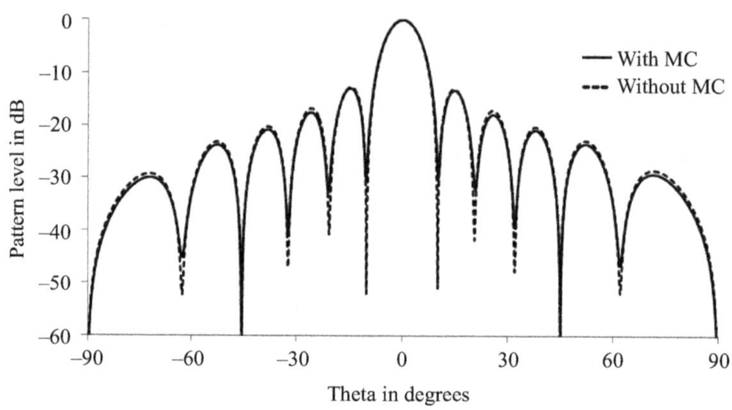

Figure 6.3 Computed H-plane pattern of two eight-element uniformly excited end-fed slotted waveguide arrays. Solid line represents the array designed including external mutual coupling; broken line corresponds to the array designed without mutual coupling

4.21 showed that the coupling level in the *H*-plane beyond the nearest slot was less than −40 dB. However, for stringent pattern shapes, for example, low sidelobes, the coupling in the *H*-plane may have to be included, cf. [14]. In the *E*-plane the coupling is much stronger and should be accounted for. Therefore the mutual coupling has typically to be included in the design of planar arrays, where both *E*-plane and *H*-plane coupling is present, see Chapter 7.

An illustration of the coupling effect on an eight-element equal amplitude *H*-plane linear slot array is shown in Figure 6.3. The array was designed first without including the mutual coupling effects, and in a subsequent design mutual coupling was included. A moment method analysis of both designs yielded the patterns shown in Figure 6.3. The MoM uses nine entire domain sinusoidal basis functions, similar to the ones discussed in Chapter 5, for each slot aperture. In addition, the MoM solution accounts for external mutual coupling and internal higher-order mode coupling. A further discussion on MoM for slot arrays is found in Chapter 8. As seen in Figure 6.3, the mutual coupling effect on the radiation pattern is very small and of no importance in this case. The example shown is for a half-height X-band (9.3 GHz) waveguide with slot widths 1.6 mm and a wall thickness of 0.5 mm and sum of conductances $G_0 = 1.0$. The principles for including the mutual coupling in the theoretical design are elaborated in Section 6.3.

Let us look at an example with a tapered excitation for lower sidelobes where a greater impact of the mutual coupling could be expected. A result for a −30 dB Dolph–Chebyshev pattern is shown in Figure 6.4. Here the mutual coupling effect is noticeable in the first sidelobe. Although the effect of the error is to reduce the first sidelobe level in this particular example, it may have an undesirable effect in other examples by increasing the sidelobe levels. Further details of the design of

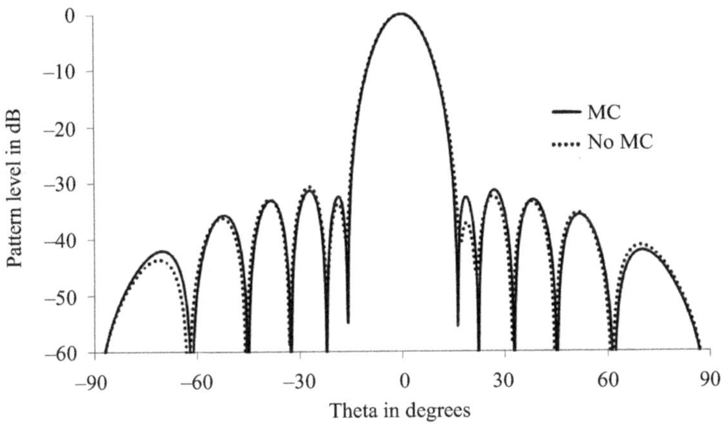

Figure 6.4 *Theoretically computed H-plane patterns of two eight-element slotted waveguide arrays with −30 dB Dolph–Chebyshev distribution. Solid line represents the array designed including external mutual coupling; broken line corresponds to the array designed without mutual coupling*

this array are found in Section 6.5. The sidelobe taper is due to the slot element pattern. Slight asymmetry in the pattern is found in the array designed without mutual coupling. This asymmetry is caused by the internal TE_{20} mode coupling generated at the slot closest to the short, which couples to the same slot after getting reflected by the short. The internal coupling will be discussed in Section 6.4. The asymmetry is not found for the array designed using the mutual coupling effects since the above-mentioned TE_{20} mode coupling was taken into account in the design. Therefore the slot voltages and the pattern are close to the ideal for this design.

6.3 The Elliott design equations

A method to iteratively design slotted waveguide array antennas including the external mutual coupling among the slots has been presented by [10]; see also [15]. The calculations were based on an equivalent dipole array, knowing that the current distribution on a resonant dipole is the same as the electric field distribution on a resonant slot. The method was later modified [11] to be valid directly for slot arrays. The modified design equations are easily derived from our knowledge about the slot modelling as a shunt element that we discussed in Chapter 4, Section 4.4.

6.3.1 The first design equation

Let us study the ith slot in the array. As previously stated the slot loading the waveguide can be represented by a shunt admittance, Figure 6.5(a) represents the waveguide, and Figure 6.5(b) an equivalent transmission line.

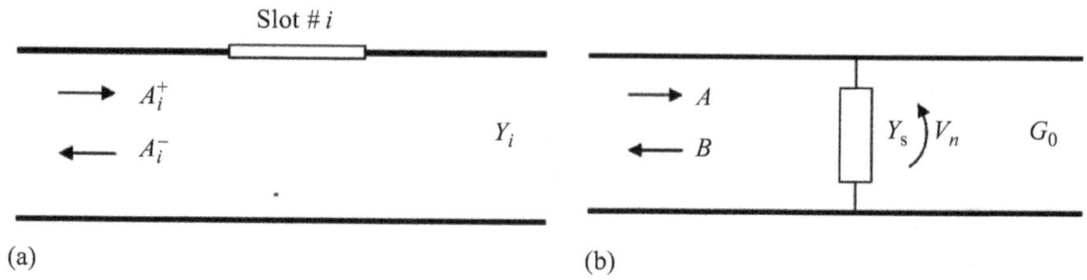

Figure 6.5 *(a) Slot in waveguide causes symmetric scattering as if loaded by a shunt admittance. (b) Equivalent transmission line, with characteristic admittance G_0, loaded by the equivalent slot admittance Y_s. V_n is the modal voltage at Y_s*

The reflection coefficient must be the same in the two representations, thus

$$\Gamma = A_i^-/A_i^+ = B/A \tag{6.4}$$

We can write for the second case

$$\Gamma = B/A = \frac{G_0 - (Y_s + G_0)}{G_0 + (Y_s + G_0)} = \frac{-Y_s}{Y_s + 2G_0} \tag{6.5}$$

Thus

$$\frac{Y_s}{G_0} = -\frac{2\Gamma}{1+\Gamma} = -\frac{2B/A}{1+B/A} = -\frac{2B}{A+B} = -\frac{2B}{V_n} \tag{6.6}$$

where V_n is the modal voltage $A+B$.

For equal reflected power in the two models we must have

$$\frac{1}{2}|A_i^-|^2 Y_i = \frac{1}{2}|B|^2 G_0 \tag{6.7}$$

That is,

$$B = \sqrt{Y_i/G_0}A_i^- \tag{6.8}$$

This gives

$$\frac{Y_s}{G_0} = -\frac{2}{V_n}\sqrt{\frac{Y_i}{G_0}}A_i^- \tag{6.9}$$

We have already an expression for the scattered mode amplitude A_i^- in (4.74). Thus we obtain

$$\boxed{\frac{Y_s}{G_0} = -K_1 f(x_0, L)\frac{V_s}{V_n}} \tag{6.10}$$

This expression relates the slot admittance Y_s, the slot voltage V_s and the modal voltage V_n in the first design equation of Elliott [11, Eq. (10)]. Equation (6.10) is valid for the isolated slot admittance as well as the active slot admittance. In the expression for the isolated slot admittance the slot voltage does not include mutual coupling whereas for the active admittance, mutual coupling is accounted for. An explicit expression for the mutual coupling is derived in the second design equation.

The constant

$$K_1 = -2\sqrt{\frac{Y_i}{G_0}} \frac{j\pi\sqrt{2}}{\beta_{10}k_0 a\sqrt{ab}}, \tag{6.11}$$

is the same as that in [11, Eq. (11)], while

$$f(x_0, L) = -\frac{\frac{\pi}{2kL}\cos\beta_{10}L}{\left(\frac{\pi}{2kL}\right)^2 - \left(\frac{\beta_{10}}{k}\right)^2}\cos\frac{\pi x_0}{a} \tag{6.12}$$

Equation (6.10) differs from Elliott's equation (10) in [11] by a minus sign because of sign differences in the normalisations for the TE_{10} mode in the two derivations. The sign difference does not cause any problem in the design of slot arrays since we always work with the ratio of active admittances to realise the aperture distribution, for example, (6.16). We also observe that the term (k/k_0) in the numerator inside the square root of Elliott's equation (11) in [11] for K_1 should not be present so that the corrected expression will be valid for dielectric filled waveguides also.

6.3.2 *The second design equation*

We will now derive the second design equation which includes the effect of the external mutual coupling. The situation is depicted in Figure 6.6.

We will use the superscript 'a' to indicate the active situation with several slots present; thus $\Gamma^a = \Gamma^{\text{active}} = A^-/A^+$. There are two contributions to the scattered wave A^-, namely the contribution from the isolated (single) slot and the contribution from the mutual coupling only. Thus

$$\Gamma^a = \left(A^-_{\text{isol}} + A^-_{\text{coupl}}\right)/A^+ \tag{6.13}$$

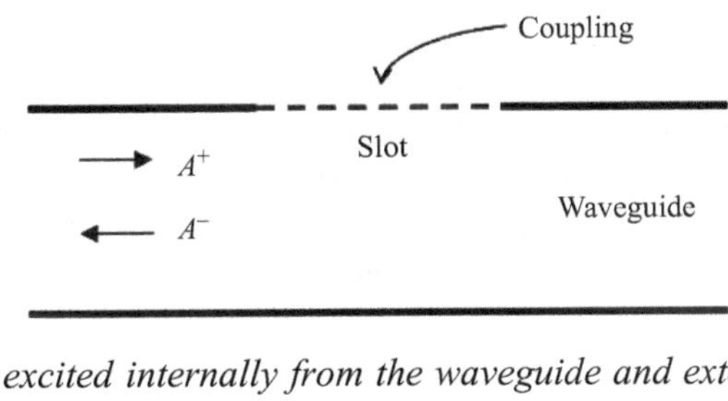

Figure 6.6 Slot excited internally from the waveguide and externally from the mutual coupling

The term A^- is proportional to the slot voltage V_s, which likewise can be divided into two components: $V_s = V_{sisol} + V_{scoupl}$.

We can write

$$\Gamma^a/\Gamma = \left(A^-_{isol} + A^-_{coupl}\right)/A^-_{isol} = V_s/V_{sisol} = V_s/\left(V_s - V_{scoupl}\right) = 1/\left(1 - V_{scoupl}/V_s\right) \tag{6.14}$$

In this expression Γ without the superscript a is the reflection coefficient for the isolated slot.

Now

$$\frac{Y^a}{G_0} = -\frac{2\Gamma^a}{1+\Gamma^a} = \frac{-2}{1+1/\Gamma^a}$$

Using (6.14) and (6.6) in the above expression we obtain

$$\boxed{\frac{Y^a}{G_0} = \frac{2}{\dfrac{2}{Y_s/G_0} + \dfrac{1}{\Gamma} \cdot \dfrac{V_{scoupl}}{V_s}}} \tag{6.15}$$

which essentially is the second design equation according to Elliott [11, Eq. (33)].

6.3.3 The iterative design algorithm for standing wave arrays

We will describe the design algorithm for standing wave arrays. In Section 6.6 the design algorithm for travelling wave arrays will be presented. The designer chooses the number of radiating slots, *N*, and excitation coefficients from system considerations such as the gain, and pattern shape. Excitation coefficients are chosen to be pure real for simplicity of implementation. The required input data for the design are the values of the waveguide inner dimensions, wall thickness, the slot width and the centre frequency for the design.

The design equations will be solved iteratively. Initially all the radiating slots are assumed to be resonant length. The values of slot offsets, alternating positive and negative, are assigned some initial values that are not critical, since they will be changed later. In the first iteration all the mutual coupling terms are ignored. In subsequent iterations the mutual coupling terms are fully accounted for, if mutual coupling is not negligible.

There are two unknowns for each slot, offset and length. Therefore in the linear array there are 2*N* unknowns. For all but the *N*th radiating slot two equations are generated, one for the real part and the other for the imaginary part of (6.10) in conjunction with (6.15). Equation (6.10) is expressed as a ratio of active admittances of the *n*th and *N*th slots as shown next.

$$\frac{Y_n^a}{Y_N^a} = \frac{f_n}{f_N}\frac{V_n^s}{V_N^s}\frac{V_N}{V_n} \tag{6.16}$$

The subscript n refers to the slot number n while N refers to the Nth slot, the one closest to the short used as a reference. Note that f_N and f_n are defined in (6.12). V_n^s and V_N^s are slot voltages while V_n and V_N are mode voltages. For standing wave arrays with a slot spacing of half guide wavelength, the mode voltage magnitudes are the same and their signs alternate and so do the signs of f_n s. For a travelling wave array, expressions for the mode voltages and the admittances are described in Section 6.6. For the Nth slot, that is, $n = N$, (6.16) is not relevant. Enforcing the active admittance of the Nth slot to be real yields one equation. One more equation is obtained by enforcing the total active admittance at the input port to the characteristic admittance of the TE_{10} mode, G_0 so as to achieve a match at the input port. Thus there are $2N$ non-linear equations in $2N$ unknowns for the array design. The simultaneous non-linear equations may be solved iteratively by a quasi-Newton technique [16]. Convergence is reached when the changes in the values of slot offsets and lengths between two successive iterations are less than the desired manufacturing tolerance, Δ, or if the desired performance is met. Figure 6.7 shows the flow chart for the design algorithm.

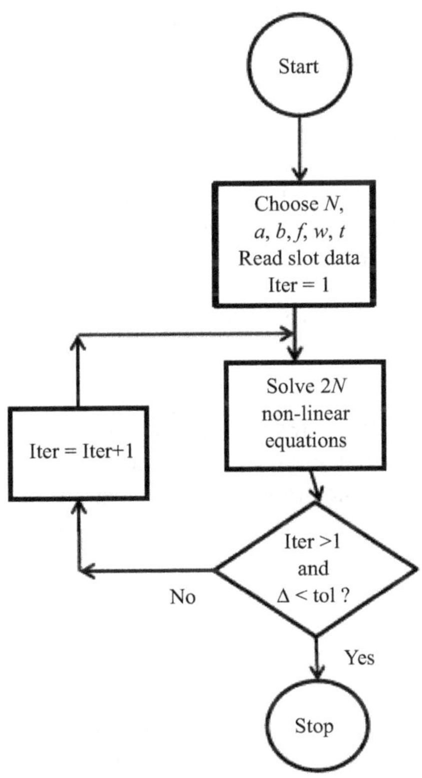

Figure 6.7 Flow chart of the design algorithm

6.4 Mutual coupling

6.4.1 Efficient computation of the external mutual coupling terms

The external mutual coupling expression has received much attention in the literature since it takes up most of the computational effort in the design of slot arrays, especially in planar arrays. Design of planar arrays will be discussed in Chapter 7. An approximation that works very well for slot lengths close to half wavelength in free space was presented by Mazzarella and Panariello [17]. They employ a Taylor series expansion around half wavelength long slots, with one or two terms for very good accuracy for the slot aperture distributions in the form of a half cosinusoid. For air-filled waveguides, the aperture distribution is similar to that of the current in a dipole antenna. For that case Mazzarella and Panariello provide fast and accurate expressions in terms of familiar sine and modified cosine integrals. The integral in the mutual coupling expression was carried out in the Fourier transform domain by Nauwelaers and van de Capelle [18]. By expressing the Fourier transform of the aperture fields in terms of a power series in the spectral variable, the integrals are carried out analytically in terms of spherical Hankel functions. Recursive relations are used to evaluate the spherical Hankel function. The results are interpreted in terms of multipole expansions. This method works very well when the spacing between slots is not too small.

A coordinate transformation in terms of difference coordinates has been employed to reduce the fourfold integral in the mutual coupling expression into a numerical double integral by Rengarajan and Gabrelian [19]. For narrow slots, a further approximation in terms of numerical single integrals is given later.

Elliott has shown that the voltage induced in slot n due to external mutual coupling from all other slots is given by the expression

$$V_{\text{scoupl}} = V_{\text{s ext}} = -j(\beta_{10}/k)(k_0 b)(a/\lambda)^3 \frac{1}{f_n^2} \frac{Y_s/G_0}{Y_s/G_0 + 2} \sum_{m=1}^{N}{}' V_m^s g_{mn}(L_m, L_n, X_{mn}, Z_{mn}) \tag{6.17}$$

where L_m and L_n are the half lengths of slots m and n, respectively, and X_{mn} and Z_{mn} are the spacings between slots m and n along the E-plane and H-plane directions, respectively [11, Eq. (29)]. The prime in the summation means that the $m = n$ term is excluded. The integral g_{mn} cast as a dimensionless quantity in terms of normalised lengths by Elliott [11] is related to the mutual admittance between apertures in a ground plane, Y_{mn}, in (4.115), repeated next.

$$Y_{mn} = \frac{-jb}{\pi a \eta} g_{mn} \tag{6.18}$$

For the convenience of calculation of Y_{mn} it has been assumed that $L_m = L_n$. The use of (6.17) in (6.15) will yield the second equation given by Elliott [11, Eq. (33)]. It may be seen that the mutual admittance derived in Chapter 4, (4.85) and (4.86) treat the case of slots on a ground plane. The presence of the waveguide under the slot n in (6.17) is responsible for the additional terms in (6.17) [11, Eqs. (29) and (33)].

We consider the mutual coupling integral g_{mn} specified in [11] and defined in (4.114) and (4.115) in Chapter 4. Equation (4.114) is used here to express g_{mn} in terms of slot length $2L$ and width w. Since slot lengths in arrays vary by a small amount, one may calculate g_{mn} for slots m and n by keeping their lengths equal by averaging the lengths of slot m and n. There are two terms in g_{mn} shown next, one containing a single integral and the other a double integral. We will simplify g_{mn2} to a single integral so that g_{mn} can be computed rapidly by numerically evaluating the single integrals.

$$g_{mn} = g_{mn1} + g_{mn2} \tag{6.19}$$

$$g_{mn1} = \frac{\pi}{2k_0 L}\int_{-L}^{L}\cos\left(\frac{\pi z}{2L}\right)\left\{\frac{\exp(-jkR_1)}{R_1} + \frac{\exp(-jkR_2)}{R_2}\right\}dz \tag{6.20a}$$

and

$$g_{mn2} = \left[k_0^2 - \left(\frac{\pi}{2L}\right)^2\right]\frac{1}{k_0}\int_{-L}^{L}\cos\left(\frac{\pi z}{2L}\right)\int_{-L}^{L}\cos\left(\frac{\pi z'}{2L}\right)\frac{\exp(-jkR)}{R}dz'dz \tag{6.20b}$$

where

$$R = \left[X_c^2 + (Z_c + z' - z)^2\right]^{1/2}, \tag{6.21a}$$

$$R_1 = \left[X_c^2 + (Z_c + L - z)^2\right]^{1/2} \tag{6.21b}$$

and

$$R_2 = \left[X_c^2 + (Z_c - L - z)^2\right]^{1/2}. \tag{6.21c}$$

Figure 6.8 shows the geometry of the coupling problem. Two sub-domains of the double integral in g_{mn2}, evaluated in terms of difference coordinate $\tau = z' - z$, are shown in Figure 6.9.

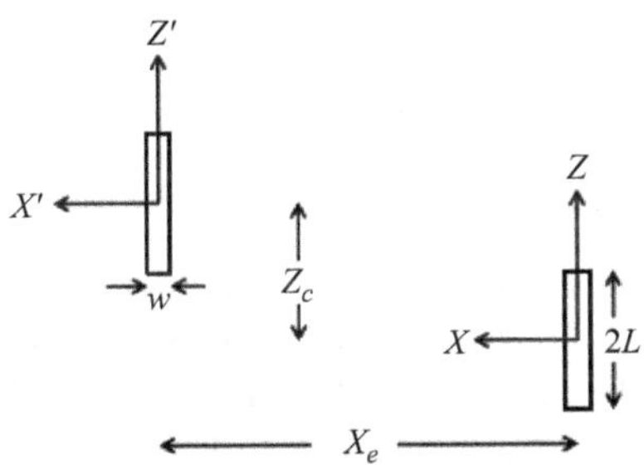

Figure 6.8 Geometry of the coupling problem

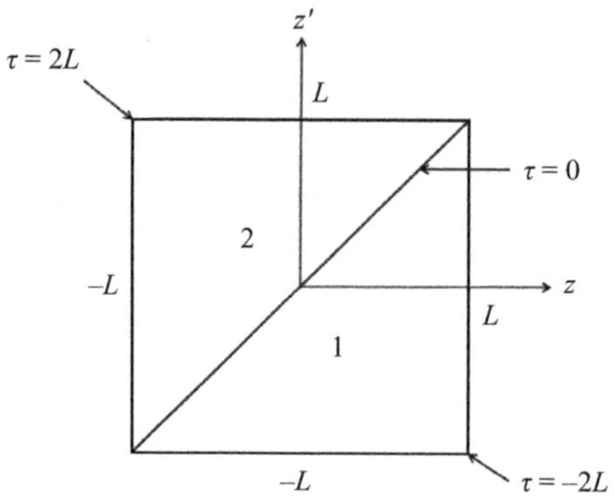

Figure 6.9 Two sub-domains of the coupling integral; τ < 0, and τ > 0

The double integral in g_{mn2} is denoted I_2 such that

$$I_2 = \int_{-L}^{L} \cos\left(\frac{\pi z}{2L}\right) \int_{-L}^{L} \cos\left(\frac{\pi z'}{2L}\right) \frac{\exp(-jkR)}{R} dz' dz \tag{6.22}$$

Using a coordinate transformation in terms of difference coordinates $\tau = z' - z$ we express I_2 as shown in (6.23). Since R is dependent on τ it is independent of z' for a fixed value of τ. Figure 6.9 shows the two domains of the integral labelled 1 and 2. I_2 is broken into two integrals, one for each domain as follows.

$$I_2 = \int_{-2L}^{0} \frac{\exp(-jkR)}{R} \int_{-L}^{L+\tau} \cos\left[\frac{\pi}{2L}(z' - \tau)\right] \cos\left(\frac{\pi z'}{2L}\right) dz' d\tau$$
$$+ \int_{0}^{2L} \frac{\exp(-jkR)}{R} \int_{-L+\tau}^{L} \cos\left[\frac{\pi}{2L}(z' - \tau)\right] \cos\left(\frac{\pi z'}{2L}\right) dz' d\tau \tag{6.23}$$

where $R = [X_c^2 + (Z_c + \tau)^2]^{1/2}$.

The z' integrals in (6.23) are evaluated analytically and the resulting expression for I_2 is shown next.

$$I_2 = \int_{-2L}^{2L} \frac{\exp(-jkR)}{R} \left[\cos\left(\frac{\pi\tau}{2L}\right)(L \mp \tau/2) \pm \frac{L}{\pi} \sin\left(\frac{\pi\tau}{2L}\right)\right] d\tau \tag{6.24}$$

In $\mp$ and $\pm$, the upper sign is used when $\tau > 0$ while the lower sign holds when $\tau < 0$.

6.4.2 The self-admittance

The mutual coupling analysis in Chapter 4 deals with rectangular apertures, each aperture fed by a waveguide with the same cross section as the aperture dimensions. Each aperture can be seen as a rectangular slot and the feeding waveguide is naturally present in the finite wall thickness of the waveguide (Figures 4.13 and 6.10). The analysis resulted in the mutual admittance between two apertures

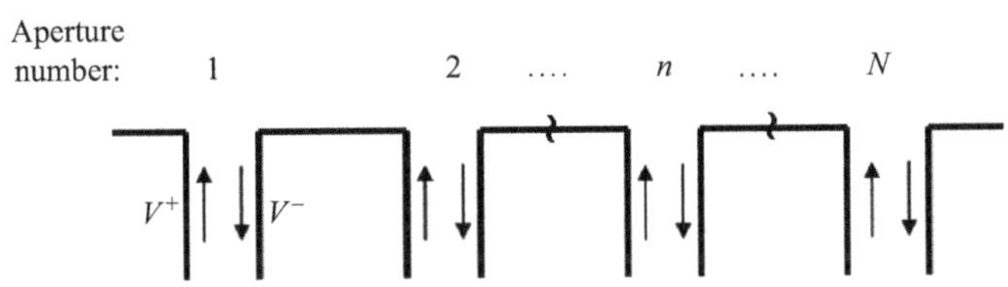

Figure 6.10 Waveguide array cross section with slot apertures and feeding waveguides

(alone, and in an array) and their dependence on their separation. An expression for the calculation of the mutual admittance Y_{mn} was given by (4.114).

The scattering matrix can be computed once we know the mutual admittance matrix $[Y^e]$ between the slot apertures. However, to fill $[Y^e]$ we also need the self-admittance Y^e_{nn} of a single slot-aperture before we can make the calculation:

$$[S] = ([Y^e] + [Y_0])^{-1}([Y_0] - [Y^e]) \tag{6.25}$$

where $[Y_0]$ is the diagonal matrix with the feeding waveguide admittance; cf. Chapter 4.

The self-admittance was in fact studied in the single slot analysis model of Chapter 5, (5.43). The difference is that here, in the coupling analysis model, we usually simplify and assume a sinusoidal electric field distribution in the slot aperture, thus putting the coefficients $p = s = 1$ in (5.43).

The self-admittance of rectangular apertures has been studied by Lewin [20] and Compton [21]. The Compton solution (his Equation (29)) is reproduced next:

$$Y^{\text{self}} = (2\pi)^2 \frac{2b}{a\omega\mu_0} \int_{-\infty}^{\infty} \int_{-\infty}^{\infty} \frac{k^2 - k_y^2}{\pi^2 k_x^2 k_z} \cdot \frac{\sin^2\left(\frac{k_x a}{2}\right)\cos^2\left(\frac{k_y b}{2}\right)}{\left(\pi^2 - k_y^2 b^2\right)^2} dk_x dk_y \tag{6.26}$$

Note that in Compton's expression b is the total width of the waveguide (= slot length) and a is the total height of the waveguide (= slot width), just the opposite of the definitions used in this book. It is easily shown that this solution is identical to our (5.43) with $p = s = 1$ inserted.

Compton transforms his solution into the following form, which can be evaluated numerically more easily.

$$Y^{\text{self}} = \frac{16\, jb}{a\omega\mu_0\pi} \int_{y=0}^{b} \int_{x=0}^{a} g(x)h(y) \frac{e^{-jk\sqrt{x^2+y^2}}}{\sqrt{x^2+y^2}} dxdy \tag{6.27}$$

The parameters are

$$\begin{cases} g(x) = \dfrac{\pi}{2}(a - x) \\ h(y) = D_1(b - y)\cos\dfrac{\pi y}{b} + D_2 \sin\dfrac{\pi y}{b} \end{cases} \tag{6.28}$$

where

$$\begin{cases} D_1 = \dfrac{1}{b^2}\left(\dfrac{k^2}{4\pi} - \dfrac{\pi}{4b^2}\right) \\[2ex] D_2 = \dfrac{1}{\pi b}\left(\dfrac{k^2}{4\pi} + \dfrac{\pi}{4b^2}\right) \end{cases} \tag{6.29}$$

In the graphs by Compton the aperture admittance values are normalised to the wave admittance of the medium outside the aperture. Lewin's graphs are normalised to the mode admittance of the feeding waveguide. In Figure 6.11 the

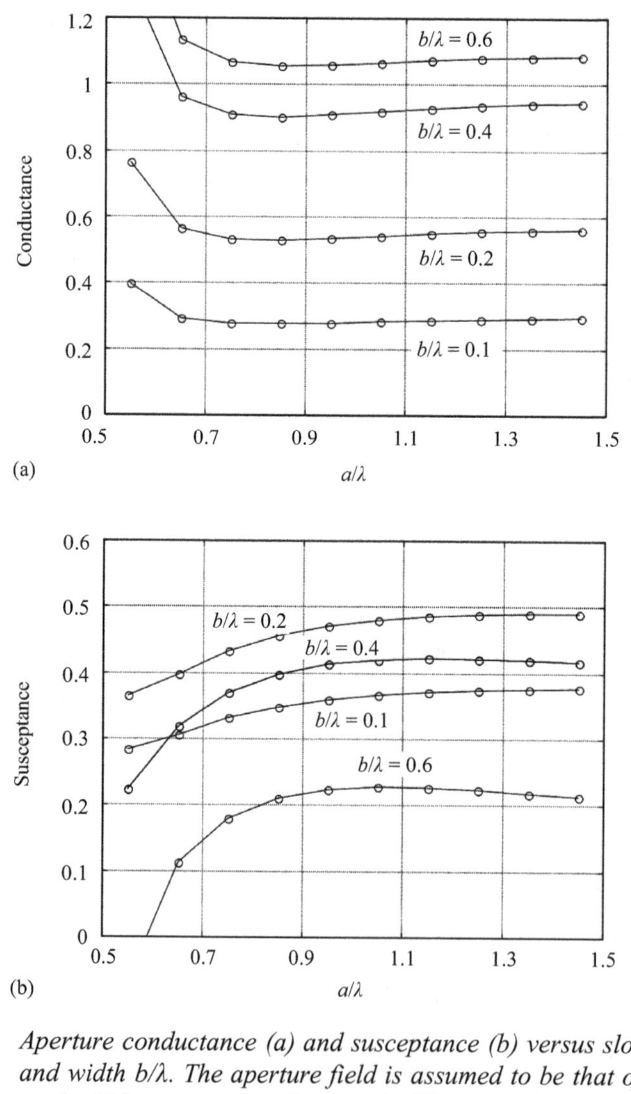

Figure 6.11 Aperture conductance (a) and susceptance (b) versus slot length a/λ and width b/λ. The aperture field is assumed to be that of a TE_{10} mode. Values are normalised to the TE_{10} modal admittance

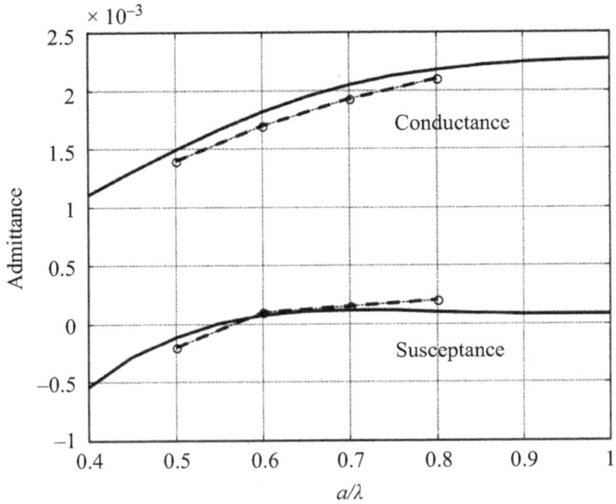

Figure 6.12 Calculated aperture admittance (Siemens) for a square aperture fed from a same size square waveguide, thus a = b. Dashed lines are measured. Reprinted from [24]

calculated admittance for some typical apertures sizes is plotted. (The definitions of aperture dimensions a/λ (length) and b/λ (width) are the ones chosen in this book.)

This relatively simple theory assuming just the fundamental TE_{10} mode gives quite good agreement with more complex models and measured results [22,23]. One further example is shown in Figure 6.12, where the admittance is given in Siemens (i.e. not normalised).

6.4.3 Internal TE_{20} mode coupling

The internal higher-order mode coupling between adjacent radiating slots, primarily from the evanescent TE_{20} mode, may be non-negligible. The TE_{20} mode scattered by a slot close to a short circuit, after reflection, may also couple a non-negligible voltage in that slot. Ignoring such coupling terms may contribute to errors in the slot aperture distribution, especially for small offset slots [25]. The contribution to slot voltage due to internal mutual coupling from the TE_{20} mode has been reported by Elliott and O'Loughlin [26]. We first evaluate the TE_{20} mode fields produced by the electric field in slot n given by

$$\vec{E}_n = \hat{\xi}_n \frac{V_n^s}{w} \cos\left(\frac{\pi \zeta_n'}{2L_n}\right) \tag{6.30}$$

using the waveguide Green's function (2.61). This can also be obtained from (4.72) for the TE_{20} mode of the rectangular waveguide.

The TE_{20} mode fields propagating in the $-z$ direction in the waveguide are given next.

$$\vec{E}_b = \hat{y}B_a \frac{\omega\mu_0}{2\pi/a}\sin(2\pi x'/a)\exp(\gamma_{20}z') \tag{6.31}$$

$$\vec{H}_b = B_a\left[\hat{z}\, j\cos(2\pi x'/a) - \hat{x}\frac{j\gamma_{20}}{2\pi/a}\sin(2\pi x'/a)\right]\exp(\gamma_{20}z') \tag{6.32}$$

The TE_{20} mode back scattering coefficient of slot n is

$$B_a = V_n^s h_n \frac{(2\pi/a)^2}{\gamma_{20}k_0\eta_0 abk} \tag{6.33}$$

where

$$h_n = 2\cos(2\pi\delta_n/a)\frac{\pi}{2kL_n}\frac{\cosh(\gamma_{20}L_n)}{(\gamma_{20}/k)^2 + \left(\dfrac{\pi}{2kL_n}\right)^2} \tag{6.34}$$

and $k = k_0$ if the waveguide is air-filled. The TE_{20} mode forward scattering coefficient is determined as $C_a = B_a$.

The TE_{20} mode induced voltage, $V_{n,20}^s$ in the nth slot is obtained by comparing the TE_{10} mode induced voltage V_{n10}^s as follows. Equation (7) in [11], combined with (6.5), yields (6.35)

$$V_{n10}^s = -\frac{\Gamma_n}{Kf_n}A_{10}^n \tag{6.35}$$

where f_n is defined as $f(x_0, L)$ in (6.12).

Similarly V_{n20}^s, the voltage induced by the TE_{20} mode wave of amplitude A_{20}^n is given by

$$V_{n20}^s = \frac{V_{n10}^s I_{20}}{I_{10}} \tag{6.36}$$

where

$$I_{10} = \iint_{\text{slot } n} H_{z10}\cos\left(\frac{\pi\xi_n}{2L_n}\right)ds' \tag{6.37}$$

and

$$I_{20} = \iint_{\text{slot } n} H_{z20}\cos\left(\frac{\pi\xi_n}{2L_n}\right)ds' \tag{6.38}$$

H_{z10} is the magnetic field of the TE_{10} mode wave and H_{z20} is that of the TE_{20} mode wave in the region of slot n. Combining (6.35) through (6.38) we obtain

$$V_{n\,\text{int}}^s = V_{n20}^s = -j\frac{\beta_{10}}{\gamma_{20}}\frac{e^{-\gamma_{20}d}}{f_n^2}\frac{Y_s/G_0}{Y_s/G_0+2}\left[h_n h_{n-1}V_{n-1}^s + h_n h_{n+1}V_{n+1}^s\right] \tag{6.39}$$

The internal coupled voltage is added to the external coupled voltage, (6.17), thus

$$V_{\text{scoupl}} = V_{s\,\text{ext}} + V^{s}_{n\,\text{int}} \tag{6.40}$$

6.4.4 Full wave analysis

The term 'full wave' is generally used to denote a rigorous electromagnetic analysis that solves Maxwell's equations subject to all boundary conditions such as the finite element method, the MoM solution to the integral equations and the finite difference method. However, the so-called full wave codes use some approximations to make the solution tractable. The procedure described earlier to include the TE_{20} mode coupling between adjacent slots can be extended to include as many evanescent modes as needed to obtain the desired accuracy in the design procedure. An MoM solution of the coupled integral equations for the electric fields of all slot apertures would provide a rigorous and accurate solution. Other computational techniques such as the finite element method would provide an accurate full wave solution as well. However, the full wave analysis techniques are generally not convenient for the *design* of slot arrays. Advanced design optimisation techniques using full wave moment method analysis are discussed in Chapter 8.

6.4.5 Mutual coupling compensation

When the scattering matrix S has been found from the aperture admittance matrix, see (6.25), we can study the impact of the coupling on the aperture excitations. Denote the aperture voltages travelling towards the apertures by the vector V^+ and those reflected back by the vector V^-, both referenced to the aperture plane (Figure 6.10). They are related as

$$V^- = S \cdot V^+ \tag{6.41}$$

and the total aperture excitation is

$$V^{ot} = (I + S) \cdot V^+ \tag{6.42}$$

where I is the identity matrix. Now, suppose we want to have a desired excitation V^{des}. We should then feed with

$$V^+ = (I + S)^{-1} \cdot V^{\text{des}} \tag{6.43}$$

The matrix $(I + S)^{-1}$ provides the desired correction [27,28]. It could in theory be realised as a network (Figure 6.13).

This corrective matrix would restore the excitation of the slot radiators to the desired values in the presence of mutual coupling. It would work even for different tapers and even phase steering of the array, etc. In practice it can hardly be realised in analogue form. However, for receive only as in digital beam forming it could be done [27].

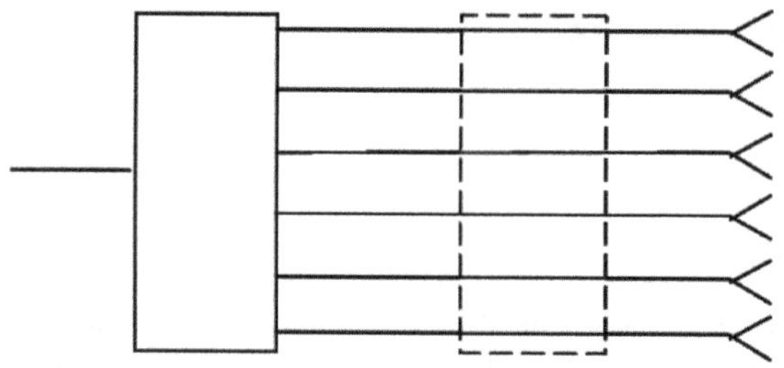

Figure 6.13 The insertion of a mutual coupling correction matrix

For one specific excitation a correction factor can be implemented as follows. The correction vector is $V^{\mathrm{corr}} = (I + S)^{-1} V^{\mathrm{des}} - V^{\mathrm{des}}$. In terms of a correction factor we get

$$corrfac = ((I + S)^{-1} V^{\mathrm{des}})./V^{\mathrm{des}} \tag{6.44}$$

where the symbol ./ stands for element by element division. The correction is dependent on the desired excitation.

This procedure could be applied to the design problem in the following way. Starting values for slot offsets and lengths provide the input to the mutual coupling calculation, yielding the scattering matrix as described earlier. Equation (6.43) then gives a new feeding vector which requires new offsets and lengths. Thus, the geometry is changed slightly, and an update of the mutual coupling is done. The procedure is iterated a few times.

6.5 Design example of a standing wave array

Standing wave arrays provide high efficiency but suffer from narrow return loss bandwidth. In standing wave arrays, the return loss bandwidth is in general more limiting than the bandwidth for pattern performance.

6.5.1 *Slot data needed for the design algorithm*

The design algorithm requires the scattering characteristics of isolated slots at the design frequency as the input data for the assumed waveguide internal dimensions a and b. The required slot data are the self-admittance of radiating slots as a function of slot offset and length. With the availability of CEM codes such as the MoM solution to the pertinent integral equations, the finite element method or the finite difference solution of Maxwell's equations subject to the boundary conditions, accurate slot data may be computed. It has been shown in Chapter 5 that the MoM using entire domain basis functions provides accurate results.

The self-admittance is determined from S_{11} of the two port scattering matrix of an isolated radiating slot as shown next.

$$\frac{Y}{G_0} = \frac{-2S_{11}}{1+S_{11}} \tag{6.45}$$

Thus the design data needed are (i) resonant lengths, $2L_{\text{res}}$, of longitudinal radiating slots computed at the design frequency for a range of values of slot offsets, $|\delta|$ up to a value typically less than $0.35a$. In addition, for each value of $|\delta|$ we need computed values of (ii) $|S_{11}|$ as a function of the normalised slot length, $\overline{L} = L/L_{\text{res}}$, in the range of 0.9–1.10 and (iii) the phase of S_{11} for the same range of offsets and normalised slot lengths.

In the iterative design, the required S_{11} for any offset and length is determined from a two-dimensional interpolation of $|S_{11}|$ and arg S_{11} from the computed (or measured) data as a function of offset and normalised length. The self-admittance of the slot may then be determined from (6.45). Alternatively the slot admittance may be plotted in terms of Stegen normalisation as shown in Figure 5.1(a). The conductance and susceptance curves may be fitted into polynomials for rapid calculations for any given length and offset.

6.5.2 Design of a −30 dB sidelobe-level Dolph–Chebyshev array

Table 6.1 shows the waveguide inner dimensions, a and b, slot width and slot thickness used for this design example. The relative slot voltages V_1^s through V_8^s for −30 dB Dolph–Chebyshev aperture distribution are also specified. In this theoretical example, the design frequency is 9.3 GHz.

Table 6.2 shows the resonant length as a function of the magnitude of the slot offset from the centreline, and Tables 6.3 and 6.4 show the magnitude and phase of S_{11} as a function of normalised slot length and offset for an isolated slot. The data were computed using the MoM solution to the integral equation of the slot aperture field.

The values of slot offsets and lengths obtained from two array designs, one without external mutual coupling and the other with mutual coupling are listed in Table 6.5. The slot offsets and lengths are found to be symmetric with respect to the centre but for alternating positive and negative offsets. The two sets of data have small differences near the centre, whereas significant differences are noted for weak slots at the ends. Amplitude and phase of slot excitations for the two designs obtained from full wave MoM analyses are also shown in Table 6.5. The design with mutual coupling produces excitations close to the required excitations for the

Table 6.1 Parameters of the eight-element Dolph–Chebyshev array

a = 22.86 mm		**b = 5.08 mm**		**Slot width = 1.5875 mm**		**Wall thickness = 0.508 mm**	
V_1^s	V_2^s	V_3^s	V_4^s	V_5^s	V_6^s	V_7^s	V_8^s
0.262	0.519	0.812	1.0	1.0	0.812	0.519	0.262

Table 6.2 Resonant length of radiating slots

Magnitude of slot offset, $\lvert\delta\rvert$, in mm	**Resonant length, $2l_{res}$, in mm**
0.444	14.99
0.889	15.01
1.333	15.05
1.778	15.09
2.222	15.15
2.667	15.23
3.111	15.31
3.556	15.40
4.000	15.50

Table 6.3 Magnitude of S_{11}

$\overline{L} = L/L_{res}$ $\lvert\delta\rvert$	**0.94**	**0.955**	**0.97**	**0.985**	**1.0**	**1.015**	**1.03**	**1.045**	**1.06**
0.444	0.0016	0.0019	0.0022	0.0024	0.0024	0.0023	0.0022	0.0020	0.0018
0.889	0.0065	0.0076	0.0086	0.0093	0.0096	0.0093	0.0086	0.0078	0.0071
1.333	0.0146	0.0169	0.0190	0.0206	0.0211	0.0204	0.0190	0.0173	0.0157
1.778	0.0255	0.0294	0.0331	0.0358	0.0366	0.0355	0.0332	0.0304	0.0276
2.222	0.0392	0.0449	0.0503	0.0542	0.0540	0.0538	0.0505	0.0464	0.0424
2.667	0.0557	0.0634	0.0705	0.0754	0.0768	0.0747	0.0702	0.0648	0.0594
3.111	0.0740	0.0837	0.0924	0.0984	0.1000	0.0975	0.0921	0.0854	0.0787
3.556	0.0941	0.1057	0.1159	0.1227	0.1245	0.1216	0.1153	0.1075	0.0995
4.000	0.1156	0.1288	0.1402	0.1476	0.1495	0.1461	0.1391	0.1304	0.1213

Table 6.4 Phase of S_{11} in degrees

$\overline{L} = L/L_{res}$ $\lvert\delta\rvert$	**0.94**	**0.955**	**0.97**	**0.985**	**1.0**	**1.015**	**1.03**	**1.045**	**1.06**
0.444	228.7	219.5	207.9	194.4	180.0	166.5	155.1	146.0	138.9
0.889	228.4	219.2	207.7	194.3	180.1	166.7	155.4	146.3	139.2
1.333	227.6	218.4	207.0	193.7	179.8	166.7	155.6	146.6	139.6
1.778	227.2	218.0	206.8	193.9	180.3	167.5	156.5	147.6	140.6
2.222	226.2	217.1	206.1	193.5	180.3	167.9	157.2	148.5	141.5
2.667	224.9	215.8	204.8	192.5	179.8	167.9	157.6	149.1	142.3
3.111	223.7	214.7	204.0	192.1	179.9	168.5	158.5	150.3	143.6
3.556	222.4	213.4	203.0	191.6	180.0	169.0	159.4	151.4	144.8
4.000	220.8	212.0	201.9	190.9	179.8	169.4	160.2	152.5	146.0

Dolph–Chebyshev array shown in Table 6.1 whereas the design without mutual coupling is found to have small amounts of error, especially near the edges. There is a slight asymmetry in the excitations obtained from the analyses because the structure is not strictly symmetric with a short at one end of the waveguide.

Table 6.5 Offsets and lengths in mm of slots of the −30 dB sidelobe Dolph–Chebyshev array, with and without mutual coupling (MC) (slot excitation phase is in degrees)

Slot #	Offset, δ withMC	Length, $2L$ withMC	Amplitude and phase of slot excitation (MC)	Offset, δ no MC	Length, $2L$ no MC	Amplitude and phase of slot excitation (no MC)
1	0.638	15.45	0.261, −1.1	0.58	15.44	0.246, 0.1
2	−1.26	15.59	0.515, −0.9	−1.17	15.52	0.493, −2.5
3	1.96	15.80	0.809, −0.5	1.86	15.69	0.796, −1.0
4	−2.42	15.98	1.0, −0.3	−2.34	15.85	1.0, 0.
5	2.42	15.98	1.0, 0.0	2.34	15.85	1.0, −0.1
6	−1.96	15.80	0.811, −0.6	−1.86	15.69	0.796, −0.6
7	1.26	15.59	0.517, −0.2	1.17	15.52	0.494, −1.9
8	−0.638	15.45	0.262, −0.1	−0.58	15.44	0.247, −4.31

Table 6.6 Directivity and beam squint of the end-fed Dolph–Chebyshev array (design with MC)

Frequency (GHz)	Directivity (dB)	Beam squint (°)
9.07	12.80	−1.3
9.3	13.03	0
9.50	13.025	0.9

The higher-order modes generated by the slot closest to the short get reflected by the short and couples to the slot. The asymmetry is slightly greater for the design without mutual coupling.

Figure 6.4 in Section 6.2 shows the computed *H*-plane pattern produced by the two designs at 9.3 GHz. The sidelobe levels decrease away from broadside because of the element pattern of the slot. There is a small difference between the two designs at some of the sidelobes. Table 6.6 shows the pattern characteristics at the centre frequency and at the band edges, where the return loss is 10 dB for the end-fed Dolph–Chebyshev array designed with mutual coupling. At the higher end of the band, the directivity is almost the same as that at 9.3 GHz. The increase in the electrical size of the aperture at 9.50 GHz is offset by the phase errors in the aperture distribution due to the design at a different frequency (9.3 GHz).

The centre-fed standing wave arrays do not exhibit beam squint with frequency. However, there is a small amount of beam squint with frequency in the end-fed array, with positive values corresponding to the squint towards the short at higher frequencies while negative values of squint are found at lower frequencies. At the design frequency we have a broadside beam. At other frequencies, the aperture phase distribution deviates from being equal, thereby creating a squint. The travelling wave arrays to be discussed in Section 6.6 exhibit significant beam squint as a function of frequency. The beam squint effect has been eliminated in a recent work by placing a meta-material, exhibiting negative values of permittivity and permeability in the frequency range of interest, between slots in the waveguide [29].

6.6 Design of a travelling wave array

The general characteristics of travelling wave arrays are discussed in Chapter 9, Section 9.2. A travelling wave array is designed for element spacing $d \neq \lambda_g/2$ and for small amounts of coupling between the propagating wave in the waveguide and each radiating element. The waveguide is terminated with a matched load. The reflection from each radiating element is small and the sum of all the scattered waves from the slots nearly cancel out at the input port, thereby providing a good match. Travelling wave arrays exhibit greater frequency bandwidth for return loss compared to the standing wave arrays but have less efficiency because of the power dissipated in the load. As the size of a travelling wave array increases, its return loss performance becomes better. However, since the main beam scans with frequency, for larger arrays, the gain bandwidth becomes small. In some applications, the beam scan with frequency may be used as a beam steering technique.

For large arrays, many elements near the feed have very small amounts of coupling, and these slots are weakly excited. The coupling characteristics of such slots are sensitive to their offsets and lengths and they require tight manufacturing tolerances. Thus the array size is a compromise between return loss bandwidth and realisable coupling values and gain bandwidth. It is generally recommended that the travelling wave array be designed to have the beam direction at least two beamwidths away from broadside.

Figure 6.14 shows the transmission line model of a linear travelling wave array with shunt admittances representing longitudinal broad wall slots and a match termination at the right end.

Travelling wave arrays may be designed with all radiating slots on the one side of the broad wall centreline or with alternating offsets [2,8,30]. In the former array the aperture phase distribution is approximately linearly progressing with a phase delay between adjacent elements in the propagation direction equal to $-\beta\, d$ where β is the phase constant of the TE_{10} mode and d is the element spacing. The beam pointing direction is essentially independent of element spacing d, given by

$$\theta_0 = \cos^{-1}(\beta/k_0) \tag{6.46}$$

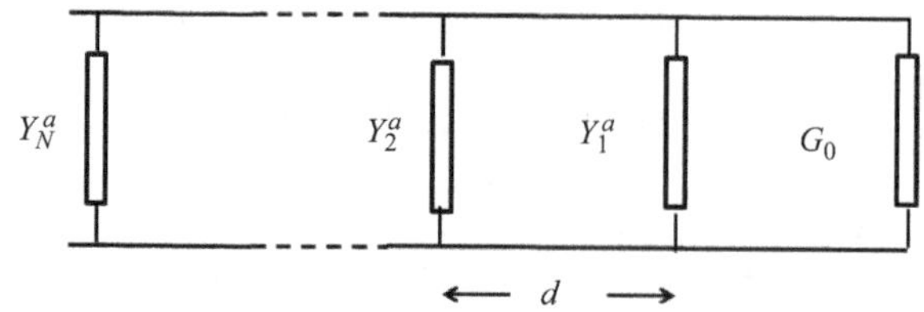

Figure 6.14 Transmission line model for a travelling wave array

where θ_0 is measured from the z-axis, pointed along the incident TE_{10} wave propagation direction. As an example, if $\beta/k_0 = 0.707$, $\theta_0 = 45°$. The element spacing should be greater than the typical slot length so that adjacent slots do not overlap in their z-coorodinates or introduce substantial amount of higher-order mode coupling but small enough to avoid grating lobes. For a travelling wave array the element spacing is chosen such that the reflected waves from all slots very nearly cancel which results in good input match.

For a travelling wave array consisting of slots with alternating offsets, the beam pointing angle is given by

$$\theta_0 = \cos^{-1}[\beta/k_0 - \lambda_0/(2d)] \tag{6.47}$$

Possible values of θ_0 are limited to a narrow angular range on either side of broadside as shown in Figure 6.15, where the required values of d in terms of

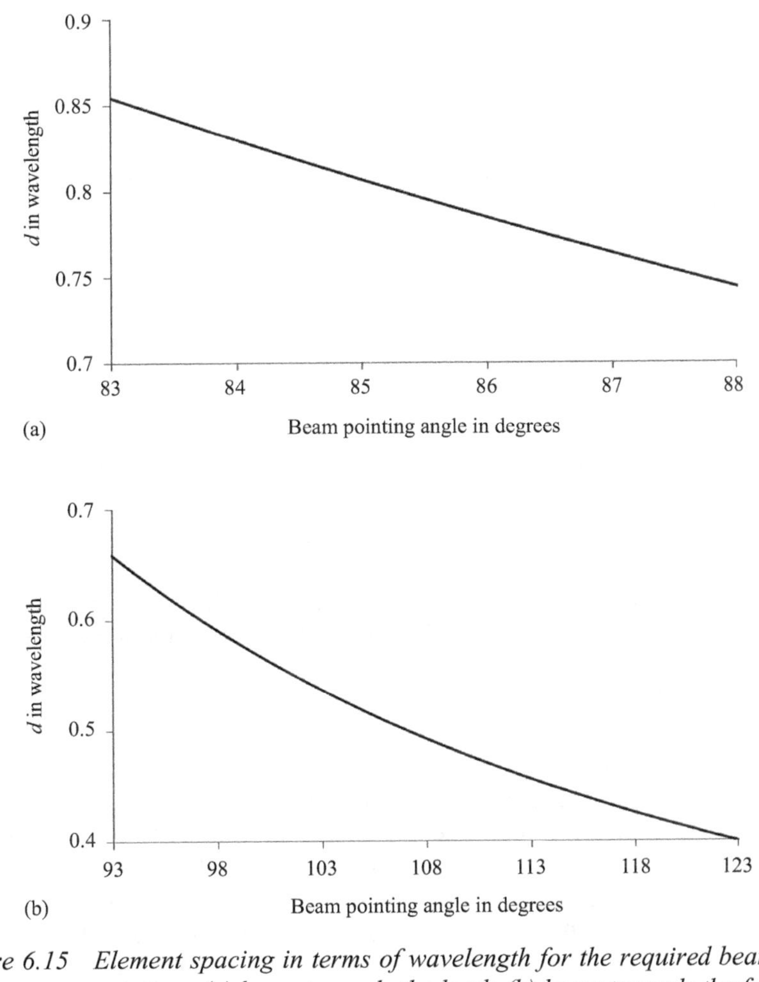

Figure 6.15 Element spacing in terms of wavelength for the required beam pointing: (a) beam towards the load; (b) beam towards the feed port

wavelength are shown for the assumed value of $\beta = 0.707\ k_0$. For $\theta_0 < 83°$, d becomes large enough to allow grating lobes whereas for large values of θ_0 the smaller spacing might create severe higher-order mode coupling effects.

6.6.1 Design equations for travelling wave array

The design equations (6.10) and (6.15) derived previously for the standing wave array are valid also for travelling wave arrays. However, the mode voltages are not equal in magnitude since the spacing is not equal to a half guide wavelength in this case. The mode voltages related to slot voltages and active admittances of slots as given by the transmission line equations below follow Elliott's derivations [2]. Thus, the mode voltages V_n and V_{n-1} at the location of slot n and slot $n-1$, respectively, are related as

$$V_n = V_{n-1}\left[\cos(\beta_{10}d) + j\frac{Y_{n-1}}{G_0}\sin(\beta_{10}d)\right] \tag{6.48}$$

where d is the spacing between elements. Y_{n-1} is the total admittance at the location of slot $n-1$, that is, the sum of its active admittance and the total admittance at the location of slot $n-2$ transformed through waveguide of length d. The total admittance at the location of slot n, Y_n $(n \geq 2)$ is given in (6.49a) while (6.49b) is valid for $n = 1$. Y_n is the active admittance of slot n, Y_n^a plus the input admittance of the loaded transmission line between slot n to the matched load.

$$\frac{Y_n}{G_0} = \frac{Y_n^a}{G_0} + \frac{(Y_{n-1}/G_0)\cos(\beta_{10}d) + j\sin(\beta_{10}d)}{\cos(\beta_{10}d) + j(Y_{n-1}/G_0)\sin(\beta_{10}d)} \tag{6.49a}$$

$$\frac{Y_1}{G_0} = \frac{Y_1^a}{G_0} + 1 \tag{6.49b}$$

Equations (6.16), modified to include mode voltages, and (6.48) may be combined to give the first design equation for a travelling wave array as given in (6.50) where $n \geq 2$.

$$\cos(\beta_{10}d) + j(Y_{n-1}/G_0)\sin(\beta_{10}d) = \frac{f_n}{f_{n-1}}\frac{V_n^s}{V_{n-1}^s}\frac{Y_{n-1}^a/G_0}{Y_n^a/G_0} \tag{6.50}$$

Equation (6.50) is the first design equation for the travelling wave array, similar to (6.16) for the standing wave array. The second design equation used for standing wave arrays, (6.15) with coupled voltages specified by (6.17) and (6.39) is valid for the travelling wave arrays also.

6.6.2 Design algorithm

The mode voltage at the location of the match termination may be assumed to be unity, without loss of generality. The power dissipated in the load is G_0 since the voltage is unity, and the mode voltage at the location of slot 1 is $V_1 = \exp(j\beta_{10}d)$. The offset of slot 1, δ_1 is chosen arbitrarily and its length is determined so as to make its active admittance determined by using (6.15), (6.17) and (6.39) pure real. The choice of δ_1 is not critical since it can be changed subsequently. We then determine the mode voltage of slot 2 using (6.48) and the offset and length of slot 2 are determined using (6.15), (6.17) and (6.39) so as to satisfy (6.50). The mutual coupling term is computed assuming that slots 3 to N have their desired excitations and are centred in the waveguide and that their lengths are equal to half wavelength in free space. These computations will be more accurate in subsequent iterations when the parameters get updated. This process continues until we reach slot N when we can determine V_N. The total normalised input admittance, Y_N^a/G_0, is determined using (6.49) recursively. The input reflection coefficient and incident wave power are now calculated. This process is iterated a few times starting from slot 1 again, until convergence is reached. Between iterations, δ_1 is adjusted to have a good compromise between the load power and the input match condition. A greater value of δ_1 may reduce the load power but increase the input reflection coefficient. As a rule of thumb, the number of slots and the spacing may be chosen such that the scattered wave voltage phasors from all the slots span the complex plane an integer number of times as suggested by Elliott [2]. However, since the scattering from different slots have different amounts; this is not a strict rule.

6.6.3 Example of a 21-element travelling wave array

A 21-element travelling wave array with a −30 dB Dolph–Chebyshev pattern was designed, built and tested by Elliott [2]. The longitudinal radiating slots of this array were offset on the same side of the broad wall centreline of a standard X-band waveguide. The amplitude distribution for the Chebyshev pattern is given in Table 6.7. The aperture phase distribution is assumed to be a linearly progressive phase delay corresponding to that of the TE_{10} waveguide mode. The array was designed using experimentally measured slot data and included the external mutual coupling. Elliott demonstrated the significance of the contribution of external mutual coupling to the slot admittance [2]. From (6.15) it is noted that the mutual coupling effect is significant for a weakly excited slot that has a small value of the reflection coefficient in the isolated case. In a travelling wave array with many weakly excited slots, the mutual coupling effects cannot be neglected. Internal higher-order mode coupling effects were not accounted for in the design.

Table 6.7 The amplitude distribution of a 21-element −30 dB Dolph–Chebyshev array

Element #	**1**	**2**	**3**	**4**	**5**	**6**	**7**
Amplitude	0.334	0.279	0.378	0.485	0.595	0.701	0.800
Element #	**8**	**9**	**10**	**11**	**12**	**13**	**14**
Amplitude	0.883	0.947	0.986	1.000	0.986	0.947	0.883
Element #	**15**	**16**	**17**	**18**	**19**	**20**	**21**
Amplitude	0.800	0.701	0.595	0.485	0.378	0.279	0.334

Table 6.8 Offsets and lengths of the radiating slots, designed by including TE_{20} mode coupling

Element #	**1**	**2**	**3**	**4**	**5**	**6**	**7**
Offset (mm)	3.000	1.134	1.714	2.225	2.736	2.690	2.413
Length (mm)	15.425	15.368	15.347	15.416	15.442	15.391	15.378
Element #	**8**	**9**	**10**	**11**	**12**	**13**	**14**
Offset (mm)	2.400	2.521	2.380	2.022	1.812	1.765	1.657
Length (mm)	15.420	15.434	15.383	15.349	15.367	15.378	15.337
Element #	**15**	**16**	**17**	**18**	**19**	**20**	**21**
Offset (mm)	1.381	1.128	0.989	0.873	0.682	0.366	0.829
Length (mm)	15.296	15.306	15.320	15.288	15.240	15.275	15.093

The measured reflection coefficient was low over 5% frequency band. However, measured patterns exhibited sidelobe levels in the range of −22 to −25 dB. Elliott attributed part of the discrepancy between theory and experiment to the internal coupling between slots via the TE_{20} mode that was not accounted for in the design [2]. In addition, manufacturing errors and inaccuracy in the slot data used in the design procedure and errors in measurement could also account for the discrepancy between theory and experiment.

In order to test the accuracy of the design procedure for the travelling wave array and to understand the higher-order mode coupling effects, we investigated the same 21-element array using accurate slot data generated by the MoM solution to the integral equation of the slot aperture electric field, similar to those shown in Tables 6.2–6.4 in Section 6.4 but calculated at 9.375 GHz. In addition, an accurate MoM analysis program was used to assess the accuracy of the design procedure and various phenomena that cause discrepancy between the design and analysis. This MoM analysis program had been verified against very carefully measured experimental data and also with the commercial code HFSS (www.ansys.com) which is used as an industry standard [31].

We designed two slot arrays. In the first design the internal higher-order mode coupling via the TE_{20} mode was included. The second one did not include the internal TE_{20} mode coupling to mimic the process that was used in [2]. The values of the slot spacing (0.389 λ_g or 17.4 mm) and the first slot offset were chosen to be the same as in [2].

Using (6.46) we obtain the beam direction measured from the z-axis to be 44.4°. Table 6.7 shows the relative values of slot voltages, that is, the aperture distribution, of the array to produce the −30 dB Dolph–Chebyshev pattern. The offsets and lengths of slots obtained by the accurate design that included the TE_{20} mode coupling are given in Table 6.8. The slot lengths in Table 6.8 assume square ended slots used in the MoM computations while those in [2] refer to round ended slots used in the measured slot data. In addition, our design included internal TE_{20} mode coupling while Elliott did not account for internal higher-order mode coupling.

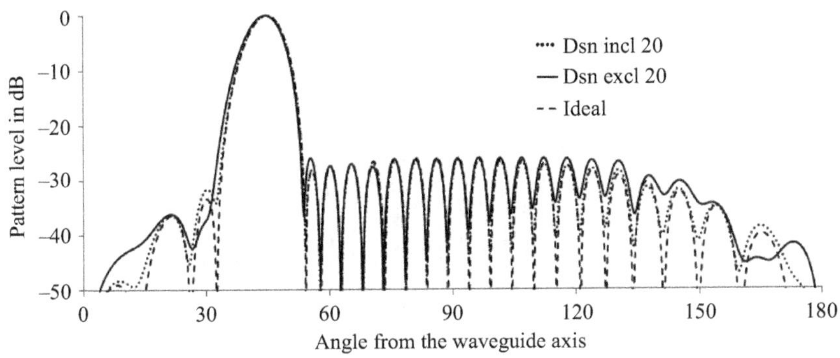

Figure 6.16 Computed far field pattern of the 21-element travelling wave array. Dotted line – design with TE_{20} mode; solid line – design without TE_{20} mode, both analysed using full wave MoM including all higher-order mode coupling; ideal – pattern of a 21 element collinear slot array with ideal aperture distribution

Figure 6.16 shows two radiation patterns produced by the full wave MoM analysis for the two array designs mentioned above. The MoM solution employs nine entire domain sinusoidal basis functions for each slot aperture (cf. Section 5.2) and is discussed in detail in Chapter 8 for the case of planar arrays. The pattern with the legend 'Dsn incl TE_{20}' refers to the array designed with the TE_{20} mode coupling while 'Dsn excl TE_{20}' refers to the array design that did not account for the internal TE_{20} mode coupling. Both MoM solutions account for external coupling between slots and accurately model the self-term of each slot as well as its excitation by the TE_{10} mode in the waveguide. Clearly the design that includes the TE_{20} mode coupling is closer to the ideal patterns and demonstrates the need for including the TE_{20} mode coupling between adjacent slots. The directivities of the three patterns had small differences and therefore in order to compare the patterns, all pattern peaks were normalised to 0 dB.

Figure 6.17 shows two radiation patterns produced by the MoM for the array that was designed including the TE_{20} mode coupling. The pattern with the legend 'TE_{20} MoM' was analysed with TE_{20} mode only for the internal coupling. The other, with the legend 'full wave MoM', was analysed with all higher-order modes for internal coupling. The two patterns had slightly different directivities. In order to compare the patterns, the peaks were normalised to 0 dB in both cases. Small deviation between the two patterns shows the contribution of higher-order modes other than TE_{20} that have not been accounted for in the design procedure. Among the contribution from all higher-order modes for this array, TE_{20} accounts for about a third while TE_{01}, TE_{11} and TE_{21} modes account for about half. Modes up to $m = 4$ and $n = 2$ captured most of all higher-order mode coupling effects. Most standing wave array designs work well with TE_{20} mode alone accounted for and occasionally one needs a few other modes such as TE_{01}, TE_{11}, and TE_{30}. Sidelobes in the angular region $\theta < 45°$ are substantially lower since the element pattern approaches

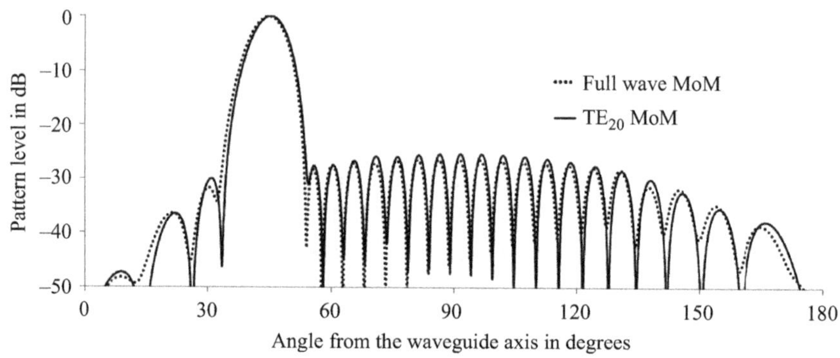

Figure 6.17 Computed far field patterns of the 21-element travelling wave array designed with TE_{20} mode coupling. Dotted line – full wave MoM analysis including coupling via all modes; solid line – TE_{20} mode MoM analysis including internal coupling between adjacent slots via TE_{20} mode only

a null as $\theta \to 0$, whereas sidelobes in the angular range $\theta > 45°$ keep increasing in levels up to 90° and then keep going down because of the element pattern.

Figure 6.18(a) shows the analysed aperture amplitude distributions using the full wave MoM, and TE_{20} MoM of the array designed by including the TE_{20} mode internal coupling. Both distributions are compared to the ideal −30 dB Dolph–Chebyshev distribution. The ideal aperture distribution is symmetric, whereas the other two exhibit slight asymmetry because of imperfections in the design. The maximum amplitude is found at different elements for the three cases. All aperture distributions have been normalised to a peak value of unity. We notice that all amplitude distributions are nearly the same with some deviations between the ideal and the other two distributions. The effect of higher-order modes other than the TE_{20} mode is found to be very small.

Figure 6.18(b) compares the computed values of the aperture phase distribution using MoM with the ideal linearly progressive phase of the TE_{10} mode wave subtracted out. The difference between the phase data shown in Figure 6.18(b) of the full wave MoM and TE_{20} MoM is found to be as high as 17° for some slots. The variation from the ideal linear phase is as high as −66° over a phase range of about 3,000°. It appears difficult to realise the ideal linear phase distribution because of the presence of a reflected TE_{10} mode wave. Figure 6.19 exhibits the input reflection coefficient, and the load power relative to the incident wave power, both in dB as a function of frequency. These values were also computed by the full wave moment method.

Approximate values of the beam pointing angle determined by (6.46) have been used to find the beam squint as a function of frequency plotted in Figure 6.20. The beamwidth is about 7.8° so this array can be used over 5% bandwidth without significant loss of gain. While an acceptable design of a travelling wave array can be accomplished using the procedure described here, further improvements can be

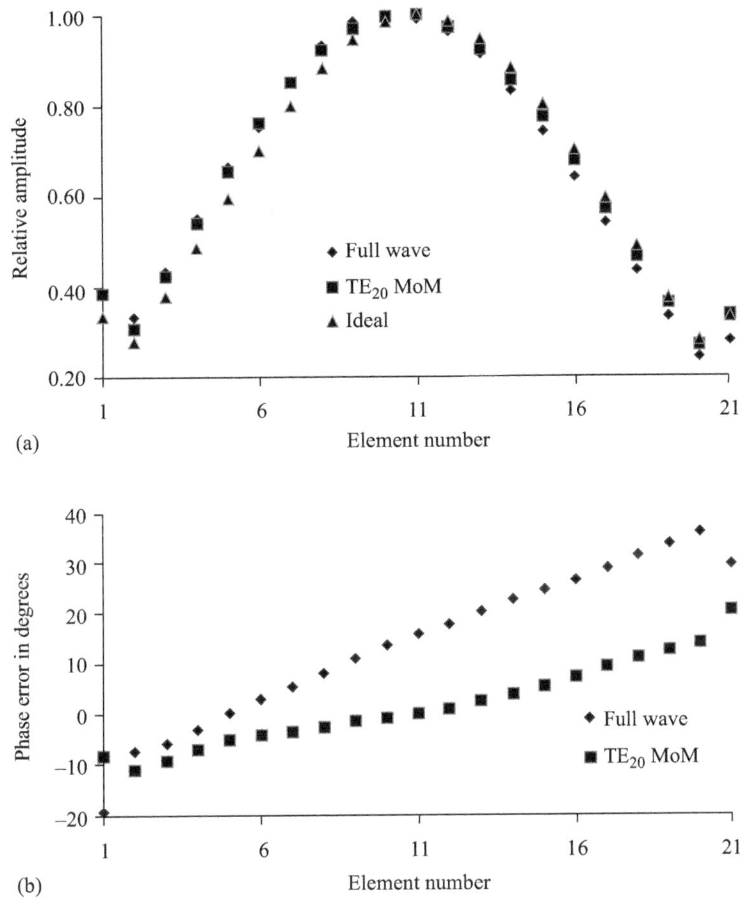

Figure 6.18 *Computed values of the aperture amplitude distribution and phase error relative to the ideal aperture phase: (a) amplitude distribution computed by two analysis methods; (b) phase error relative to the ideal aperture phase computed by two analysis methods. Squares correspond to the full wave MoM analysis, and diamonds correspond to the TE_{20} MoM analysis*

made using the global optimisation technique with a full wave MoM code, discussed in Chapter 8. Such an optimisation technique has been used to reduce the power dissipated in the load substantially in the traveling wave array [32]. Mode matching technique and generalised scattering matrix method discussed in Chapter 7 can also be used for accurate design and optimisations.

Elliott's equations were cast in a form for the analysis of a slot array by Hamadallah [33]. The external mutual coupling was incorporated in Hamadallah's analysis [33] but the internal higher-order mode coupling was not. Reasonably

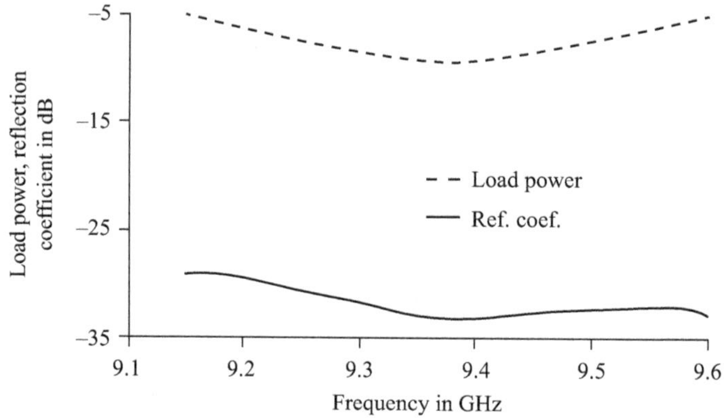

Figure 6.19 Reflection coefficient and load power relative to incident power in dB

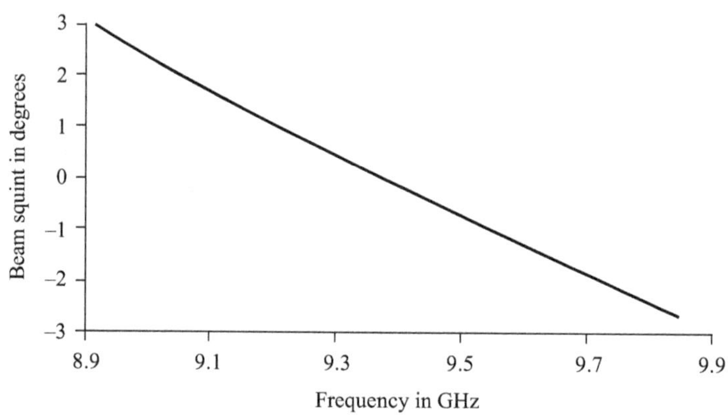

Figure 6.20 Beam squint in degrees as a function of frequency

good agreement was obtained between measured and computed patterns at the design frequency as shown in Figure 6.21.

Hamadallah's analysis equations are repeated next by including the TE_{20} mode coupling.

The slot voltages of an N-element linear array are calculated from the following matrix equation:

$$[G][V^s] = \frac{k_2}{k_1} V_1 [E] \tag{6.51}$$

where

$$G_{mn} = g_{mn} + \frac{k \exp(-\gamma_{20} d)}{\gamma_{20}(k_0 b)(a/\lambda)^3} h_n\{\delta_{m\,n-1} h_{n-1} + \delta_{m\,n+1} h_{n+1}\} \text{ if } m \leq n \tag{6.52}$$

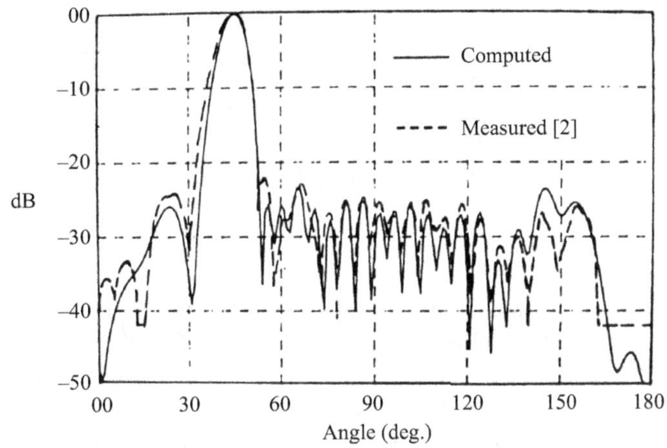

Figure 6.21 Computed and measured patterns of the 21-element travelling wave array. © IEEE 1989, reprinted from [33]

and, if $m > n$,

$$G_{mn} = g_{mn} + \frac{k\exp(-\gamma_{20}d)}{\gamma_{20}(k_0b)(a/\lambda)^3}h_n\{\delta_{m\,n-1}h_{n-1} + \delta_{m\,n+1}h_{n+1}\}$$

$$- jk_2f_mf_n\sin\{(m-n)\beta_{10}d\} \tag{6.53}$$

$\delta_{m\,n-1}$ and $\delta_{m\,n+1}$ are Kronecker delta functions (=1 if the two subscripts are equal, otherwise 0).

$[V^s]$ is the column matrix containing the N slot voltages.

The nth element in the column matrix $[E]$ is given by

$$E_n = \frac{\sin[\beta_{10}\{d_0 + (n-1)\}d\}]}{\sin(\beta_{10}d_0)}f_n \tag{6.54}$$

for a standing wave array. The distance from the short to the centre of the closest slot is d_0.

For a travelling wave array with a match termination

$$E_n = \exp[j\beta_{10}(n-1)d]f_n \tag{6.55}$$

$$k_2 = \frac{2kG_0}{j\beta_{10}(k_0b)(a/\lambda)^3} \tag{6.55a}$$

and

$$k_1 = \frac{1}{j(a/\lambda)}\sqrt{\frac{2G_0}{\eta(\beta_{10}/k)(ka)(kb)}} \tag{6.55b}$$

All other parameters are defined in Sections 6.3 and 6.4.

The mode voltage V_1 may be set to unity as a reference.

A solution of (6.51) yields all slot voltages. In order to determine the load power of a travelling wave array and the input reflection coefficient, we need to determine the active admittance of each slot and the total admittance at each slot section looking towards the load. Equations (6.48)–(6.50) may be used to determine the slot voltages. For a standing wave array (6.49b) becomes

$$\frac{Y_1}{G_0} = \frac{Y_1^a}{G_0} - j\cot(\beta_{10}d_0) \tag{6.56}$$

While this analysis technique is useful for determining the performance of a linear standing wave or travelling wave array, it suffers from the assumptions made in Elliott's design technique. These assumptions are listed here once again. The slot aperture distribution is a half cosinusoid equiphase. The slot is modelled as a shunt admittance in an equivalent transmission line. Wall thickness is ignored except for the use of scattering data of isolated slots such the resonant length, and self-admittance. Full wave analysis techniques such as HFSS, MoM, mode matching, and generalised scattering matrix method can provide more accurate analyses and optimisations.

Modern antenna pattern measurements, for example, spherical near-field technique and CEM codes such as MoM and HFSS show extremely close agreement, often in the order of 0.1–0.2 dB even at low sidelobe levels [31]. This is a testament to the maturity and accuracy of both the analysis and measurement techniques. The full wave MoM results shown here would correspond to very accurately measured experimental results such as what one obtains from a well-maintained and well-calibrated modern spherical wave range. By comparing the full wave MoM results and reduced MoM model in which TE_{20} mode coupling alone is accounted for among the higher-order mode coupling, we have a good understanding of the reasons for the discrepancy between the design and analysis (or measurement).

A technique to reduce the movement of beam pointing direction with frequency of a travelling wave array was patented by [34] and is described in Section 9.2.4 in Chapter 9.

6.7 Array design using the concept of incremental conductance

This concept, introduced originally by Watson, is described by Silver [35] for the design of edge slot arrays, shown in Figure 6.22, based on experimental measurements of incremental conductance since it was difficult to determine the slot self-admittance and mutual coupling effects [36]. This is a simple approximate method and provides adequate accuracy for linear arrays that are not too short with nearly uniform aperture distribution or slowly varying distribution. For other arrays, this method may provide a good starting approximation which can be improved using better computational models. Although many CEM codes can be used to obtain self-admittance data, mutual coupling computation is more complicated.

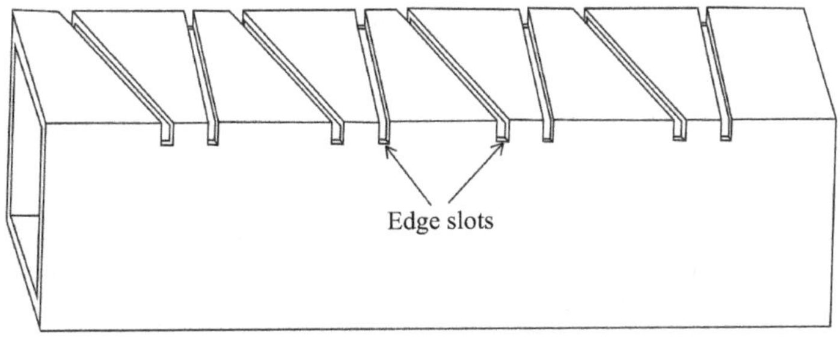

Figure 6.22 A linear array of side wall slots

The term 'incremental conductance' originates from the measurement of the input admittance of a large array of uniform spacing. The difference between the input admittance of an array of N identical slots and a second array of $N+1$ slots is the incremental admittance. All slots have equal lengths and the same magnitude of tilt with respect to the vertical direction but alternate in sign. For a particular value of slot length, the incremental admittance becomes pure real, yielding incremental conductance. This is a difficult experimental process but it is easier to compute with a CEM code, if available.

Let us consider a standing wave array of N edge slots. All slots have the same tilt angles with respect to the vertical but alternating signs and same lengths. The measured difference between the input admittance of the array with N and $N+1$ elements gives the incremental admittance if N is reasonably large, about 25 or greater. Using a good CEM code for an infinite periodic linear array, it is possible to solve the aperture distribution of two slots in a unit cell which, when repeated in the exterior region, would produce an infinite array coupling effect [35]. However inside the waveguide, there is only one unit cell with two elements. A two-element unit cell is needed when adjacent slots have alternating offsets or alternating tilts as in the case of edge wall slot arrays. For each tilt angle, one obtains the reflection coefficient inside the waveguide for a range of values of slot lengths. From the reflection coefficient, one can determine the active admittance of a single slot. Thus we generate the active admittance data for a range of values of slot tilts and lengths.

The relationship between the slot voltage, the mode voltage and the active admittance of a side wall slot, similar to (6.10), may be cast in the following form similar to Elliott's design equation (1).

$$\frac{Y_n^a}{G_0} = K f_n(\theta_n, L_n) V_n^s / V_n \tag{6.57}$$

where θ_n is the tilt angle of slot n with respect to the vertical and $2L_n$ is the slot length. The incremental conductance is approximately the same as the active conductance for slots near the centre of an array. Incremental conductance can be obtained from a CEM code as a function of slot tilt. The design equation for a standing wave array is expressed in the form of the ratio (6.58) since the mode voltage magnitudes are equal with alternating signs for half wavelength spacing; cf. (6.16).

$$\frac{Y_n^a}{G_0} \Big/ \frac{Y_1^a}{G_0} = \frac{|f_n|}{|f_1|} \frac{V_n^s}{V_1^s} \tag{6.58}$$

Equation (6.58) serves as the design equation for a linear array. One needs to obtain the values of active admittances of slots so as to realise the required aperture distribution and an input match. The aperture distribution may be complex but the amount of phase variation achievable with rotated slots is limited. This method may be used for an initial design of an array and subsequently one can improve the design using a full wave analysis code and an optimisation procedure. It has been shown that for linear uniformly excited arrays this method provides good accuracy whereas for nonuniform excitations better methods are needed; see e.g. [37].

6.8 Array design in terms of scattering parameters

The representation of a longitudinal slot in the form of a shunt element becomes poor for large offsets and for small height waveguides. A Tee or Pi-type circuit would be a better representation for such slots. Compound slots with an offset from the waveguide centreline and tilt with respect to the waveguide axis may also have to be modelled in the form of Tee or Pi circuits. However, it is easier to work with scattering wave representations for such slots. In the following we derive a design procedure for a linear array of slots in the standing wave configuration. A design procedure employing scattering wave representation of slots was used to find the coupling levels of a linear array of compound coupling slots in the feed waveguide of a planar array [38]. Such a design procedure is also applicable to a travelling wave array. A generalised design procedure using the scattering matrix formulation of a finite array has been presented in the literature [39]. The following scattering wave representation considers TE_{10} mode in the waveguide and a half cosinusoidal aperture distribution for the slots. It can be generalised to include higher-order modes in the waveguide and additional modes for the slot aperture distribution in a manner similar to the generalised scattering matrix method discussed in Section 7.9 in Chapter 7.

Figure 6.23 shows a linear array of longitudinal radiating slots, but they could also be compound slots. It is assumed that the isolated slot data in terms of the scattering matrix representation as a function of physical parameters, for example, offset, length and tilt, have been computed and made available during the design process. We will assume that there is a short at the end but the design procedure is valid also for a match termination also. For simplicity we ignore the internal higher-order mode coupling. It is easy to incorporate it in the mutual coupling term. We explain the procedure assuming TE_{10} mode waves incident on each slot from both directions and scattered TE_{10} mode waves.

1. We assume the parameters of the Nth slot, length, offset, and tilt in the case of a compound slot. These estimates are not critical since they will be changed

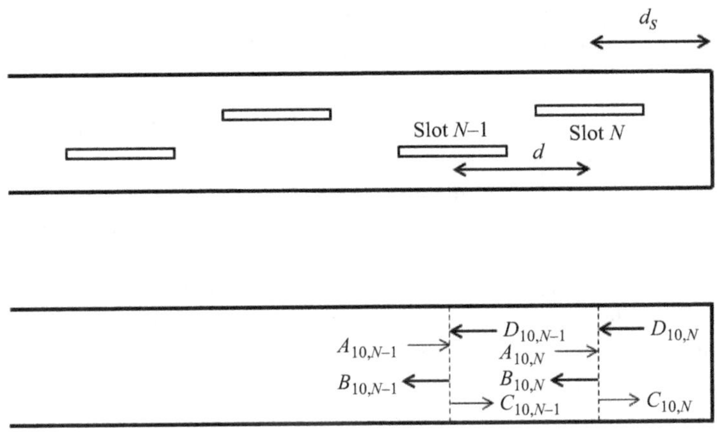

Figure 6.23 Incident, transmitted and scattered TE_{10} mode waves of slots

later. The scattering parameters of the isolated slot N are determined from computed slot data for the assumed values of offset and length.

2. We then obtain the contribution $V_{s\,N\,coup}$ due to external mutual coupling from all other slots assuming that all other slots are centred in the waveguide with their voltages specified by the aperture distribution and lengths equal to half wavelength in free space. These values will be updated later in the iterative process. The isolated and active reflection coefficients of slot N are related as shown in (6.14) which is repeated below.

$$\Gamma_{aN}/\Gamma_N = V_{sN}/(V_{sN} - V_{sNcoupl}) \tag{6.59}$$

The slot voltage due to mutual coupling, $V_{sNcoupl}$, may be obtained for compound slots in a manner similar to the derivation for longitudinal slots.

3. Steps 1 and 2 are repeated until slot N is resonant in an active sense, i.e. in the presence of mutual coupling. We then determine the active two-port scattering matrix $[S]_{Na}$. We set $A_{10,N} + C_{10,N}$ to 1 as a reference.
4. If there is a match termination at the end, $D_{10,N} = 0$ whereas if there is a short,

$$D_{10,N} = -[A_{10,N} + C_{10,N}]\exp(-2\beta_{10}d_s) \tag{6.60}$$

5. By converting $[S]_{Na}$ to a transmission matrix, we can determine the TE_{10} wave amplitudes and phases at the input side of slot N, $A_{10,N}$ and $B_{10,N} + D_{10,N}$. The slot voltage V_{sN} is also determined.
6. We now determine the parameters of slot $N - 1$ such that its voltage V_{sN-1} relative to V_{sN} satisfies the required aperture distribution after including the mutual coupling. Its scattering parameters in the absence of mutual coupling are obtained from computed slot data.
7. Use the procedure stated in step 2 for slot $N - 1$ to calculate, V_{sN-1}, and $[S]_{N-1a}$.
8. The TE_{10} mode waves incident and leaving the output side of slot $N - 1$ are obtained from the corresponding wave amplitudes and phases at the input side

of slot N using approriate phase advance or phase delay. Then by converting $[S]_{N-1a}$ to a transmission matrix, we calculate the wave amplitudes and phases at the input side of slot $N-1$.

We continue this process until we reach the first slot at the feed end. The input reflection coefficient is given by $\Gamma_{in} = [D_{10,1} + B_{10,1}]/A_{10,1}$. If Γ_{in} is high, in the next iteration we adjust the parameters of slot N so that it couples a greater amount of power to external radiation. After a few iterations the process converges and produces good return loss and correct slot excitations.

The scattering wave model is a convenient design approach for an array of compound slots, discussed in Chapter 8 instead of a Tee or Pi-type circuit representation used by [40]. In Section 6.9 a variation of Elliott's design procedure employing a Tee equivalent circuit for the longitudinal radiating slot is discussed [41].

6.9 Improvement to Elliott's design procedure

Elliott's design procedure uses a shunt admittance model for the longitudinal radiating slots. For slots in reduced height waveguides and for slots with large offsets, the shunt model has been found to introduce errors. A Tee network for each radiating slot may be derived from the computed scattering parameters of an isolated slot. Such a Tee network consists of a dominant term which is the shunt admittance and two series reactances as shown in Figure 6.24.

The series and shunt elements of the Tee network are readily obtained from the two-port scattering matrix parameters of the slot. The ABCD matrix of the Tee network is obtained by multiplying the ABCD matrices of the series and shunt elements in Figure 6.24 as given in (6.61) [42]. By comparing the elements of the ABCD matrix of the Tee network to the scattering parameters of the slot, we obtain the values of the network elements given in (6.62) [42].

$$\begin{bmatrix} A & B \\ C & D \end{bmatrix} = \begin{bmatrix} 1 & Z \\ 0 & 1 \end{bmatrix} \begin{bmatrix} 1 & 0 \\ Y & 1 \end{bmatrix} \begin{bmatrix} 1 & Z \\ 0 & 1 \end{bmatrix} = \begin{bmatrix} 1+ZY & 2Z+Z^2Y \\ Y & 1+ZY \end{bmatrix} \qquad (6.61)$$

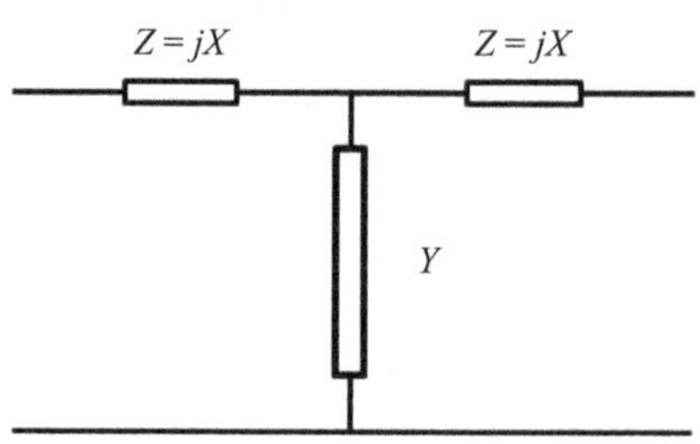

Figure 6.24 Tee network model of a longitudinal radiating slot

$$Y = \frac{1}{Z_0}\frac{(1 - S_{11})(1 - S_{22}) - S_{12}S_{21}}{2S_{21}} \tag{6.62}$$

$$1 + ZY = \frac{(1 + S_{11})(1 - S_{22}) + S_{12}S_{21}}{2S_{21}} \tag{6.63}$$

In (6.62) the characteristic impedance of the transmission line, Z_0, is the same as that of the TE_{10} mode in the waveguide. If the aperture distribution has a small amount of odd symmetric variation in the longitudinal direction, the series element is small and it is reactive as shown by Costanzo *et al.* [41]. In that case a two-step design is carried out [41]. Initially the slot array is designed by ignoring the series elements and by assuming a shunt model for the slots. Subsequently, the slot array is designed by including different spacings between adjacent slots, instead of the conventional half guide wavelength spacing. The mutual coupling calculation is carried out using Elliott's model. The circuit model used by Constanzo *et al.* is shown in Figure 6.25.

The line lengths obtained from the above-mentioned procedure for an eight-element slot array in a dielectric filled waveguide deviated about 3% from half guide wavelength. The authors have not compared the performance of their design with the original Elliott's design. Because of the presence of the series elements and the variation of slot lengths relative to half guide wavelength, the new design is expected to be better. However, if the series impedances are not very small this procedure may not be accurate. Generalised scattering matrix models for array analysis and design will be discussed in Chapter 7.

The Elliott's design procedure or variations discussed in this chapter serve as a framework for designing any type of array antenna. One needs to include all the physics such as the characterisation of each radiating element, the mutual coupling effects and excitations. Examples in the literature include slot arrays fed by boxed striplines and microstrip dipole array fed by microstrip transmission line [43–45].

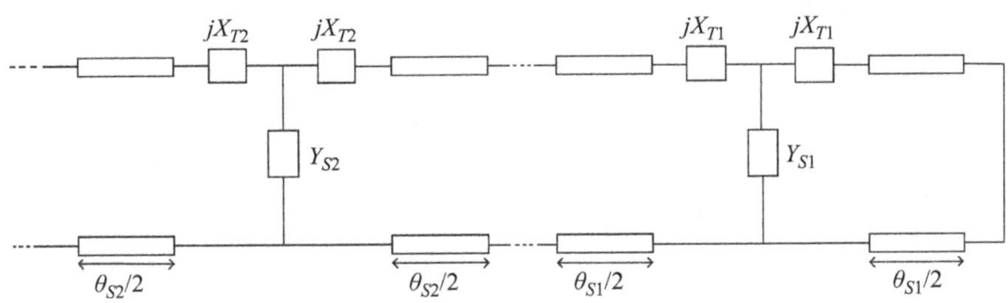

Figure 6.25 Circuit model of a linear array of slots modelled as Tee networks with different spacings. © IEEE 2010, reprinted from [41]

References

[1] Stevenson A. F. *Series of Slots in Rectangular Waveguides.* National Research Council of Canada Radio Reports 12 and 13, 1944.

[2] Elliott R. S. 'On the design of traveling-wave-fed longitudinal shunt slot arrays'. *IEEE Transactions on Antennas and Propagation.* 1979;**27**(5): 717–720.

[3] Montgomery C. G., Dicke R. H., Purcell E. M. *Principles of Microwave Circuits*, MIT Radiation Laboratory Series, vol. 8. McGraw-Hill, New York, 1948.

[4] Collin R. E. *Foundations for Microwave Engineering*. McGraw-Hill, New York, 1966.

[5] Stegen R. J. 'Slot radiators and arrays at X-band'. *IRE Transactions on Antennas and Propagation*. 1952;**1**(1):62–84.

[6] Stegen R. J., Reed R. H. 'Arrays of closely-spaced nonresonant slots'. *IRE Transactions on Antennas and Propagation.* 1954;**2**(3):109–113.

[7] Kaminow I. P., Stegen R. J. *Waveguide Slot Array Design.* Hughes Aircraft Company, Culver City, CA, Tech. Memo No. 348, 1954.

[8] Dion A. 'Nonresonant slotted array'. *IRE Transactions on Antennas and Propagation.* 1958;**6**(4):360–365.

[9] Elliott R. S. 'A note on the relation between slot conductance and slot voltage'. *IEEE Transactions on Antennas and Propagation.* 1984;**32**(10): 1133–1134.

[10] Elliott R., Kurtz L. 'The design of small slot arrays'. *IEEE Transactions on Antennas and Propagation.* 1978;**26**(2):214–229.

[11] Elliott R. S. 'An improved design procedure for small arrays of shunt slots'. *IEEE Transactions on Antennas and Propagation.* 1983;**31**(1):48–53.

[12] Harrington R. F. *Time-Harmonic Electromagnetic Fields.* IEEE Press, New York, 2001.

[13] Gulick J., Stern G., Elliott R. S. 'The design of linear and planar arrays of waveguide-fed longitudinal slots'. *IEEE International Antennas and Propagation Symposium Digest.* 1986;**2**:685–688.

[14] Kay A. F., Simmons A. J. 'Mutual coupling of shunt slots'. *IRE Transactions on Antennas and Propagation.* 1960;**8**(4):389–400.

[15] Elliott R. S. *Antenna Theory and Design.* IEEE Press, Hoboken, NJ, 2003.

[16] Press W. H., Teukolsky S. A., Vetterling W. A., Flannery B. P., Metcalf M. *Numerical Recipes: The Art of Scientific Computing, 3rd Edition.* Cambridge University Press, Cambridge, 2007.

[17] Mazzarella G., Panariello G. 'On the evaluation of mutual coupling between slots'. *IEEE Transactions on Antennas Propagation.* 1987;**35**(11):2189–2192.

[18] Nauwelaers B., van de Capelle A. 'Series expressions for mutual coupling between magnetic currents'. *Proceedings of the International Conference on Aerospace Application,* Torino, Italy, 1989, pp. 199–202.

[19] Rengarajan S. R., Gabrelian E. 'Efficient and accurate evaluation of external mutual coupling between compound broad wall slots'. *IEEE Transactions on Antennas and Propagation.* 1992;**40**(6):733–737.

[20] Lewin L. *Advanced Theory of Waveguides.* Iliffe and Sons, Ltd., London, 1951.

[21] Compton R. T. *The Aperture Admittance of a Rectangular Waveguide Radiating into a Lossy Half-Space.* The Ohio State University Antenna Laboratory Report. No. 1691-1, 1963.

[22] Mautz J. R., Harrington R. F. 'Transmission from a rectangular waveguide into half space through a rectangular aperture'. *IEEE Transactions on Microwave Theory and Techniques.* 1978;**26**(1):44–45.

[23] MacPhie R. H., Zaghloul A. I. 'Radiation from a rectangular waveguide with infinite flange – exact solution by the correlation method'. *IEEE Transactions on Antennas and Propagation.* 1980;**28**(4):497–503.

[24] Cohen M., Crowley T., Levis K. *The Aperture Admittance of a Rectangular Waveguide Radiating into Half Space.* The Ohio State University Antenna Laboratory Report. ac 21114 S. R. No. 22, 1953.

[25] Rengarajan S. R., Nardi D. 'On internal higher order mode coupling in slot arrays'. *IEEE Transactions on Antennas and Propagation.* 1991;**39**(5): 694–698.

[26] Elliott R. S., O'Loughlin W. R. 'The design of slot arrays including internal mutual coupling'. *IEEE Transactions on Antennas and Propagation.* 1986; **34**(9):1149–1154.

[27] Steyskal H., Herd J. 'Mutual coupling compensation in small array antennas'. *IEEE Transactions on Antennas and Propagation.* 1990;**38**(12):1971–1975.

[28] Derneryd A. G. 'Compensation of mutual coupling effects in array antennas'. *IEEE International Antennas and Propagation Symposium Digest.* vol. 2, 1996, pp. 1122–1125.

[29] Oh, S.-S, Shafai L. 'Beam squint minimization in series-slotted waveguide-array antennas using double negative materials'. *Proc. International Symposium on Antennas and Propagation Digest.* 2005. pp. 507–510.

[30] Yee H. Y. 'Slot antenna arrays', in Johnson R. C., Jasik H. (eds.). *Antenna Engineering Handbook.* McGraw-Hill, New York, 1984.

[31] Rengarajan S. R., Zawadzki M. S., Hodges R. E. 'Waveguide-slot array antenna designs for low-average-sidelobe specifications'. *IEEE Antennas and Propagation Magazine.* 2010;**52**(6):89–98.

[32] Rengarajan S. R., 'GA-MoM Optimization of Slot Arrays'. *Proceedings of the International Applied Computational Electromagnetics Society Symposium.* 2018 (to be published).

[33] Hamadallah M. 'Frequency limitations on broad-band performance of shunt slot arrays'. *IEEE Transactions on Antennas and Propagation.* 1989;**37**(7): 817–823.

[34] Karlsson E. R. Waveguide element for an electrically controlled radar antenna. US patent no. 4788552, 1986.

[35] Silver S. *Microwave Antenna Theory and Design.* MIT Radiation Laboratory Series. vol. 12, McGraw-Hill, New York, 1949.

[36] Yee H. Y. 'The design of a large waveguide array of shunt slots'. *IEEE Transactions on Antennas and Propagation.* 1992;**40**(7):775–781.

[37] Bucci O. M., Mazzarella G., Panariello G. 'A note on edge-shunt slot array design'. *IEEE International Antennas and Propagation Symposium Digest.* vol. 3, 1997. pp. 1946–1949.

[38] Rengarajan S. R., Derneryd A. G. 'Application of compound coupling slots in the design of shaped beam antenna patterns'. *IEEE Transactions on Antennas and Propagation.* 1993;**41**(1):59–65.

[39] Rubio J., Corcoles J., de Aza M. A. G. 'Inclusion of feeding effects in the generalized scattering matrix formulation of a finite array'. *IEEE Antennas and Wireless Propagation Letters.* 2009;**8**:819–822.

[40] Montesinos-Ortego I., Zhang M., Sierra-Perez M., Hirokawa J., Ando M. 'Systematic design methodology for one-dimensional compound slot-arrays combining Method of Moments, equivalent circuit model and forward matching procedure'. *IEEE Transactions on Antennas Propagation.* 2013; **61**(1):453–458.

[41] Costanzo S., Casula G. A., Borgia A., *et al.* 'Synthesis of slot arrays on integrated waveguides'. *IEEE Antennas and Wireless Propagation Letters.* 2010;**9**:962–965.

[42] Pozar D. M. *Microwave Engineering.* John Wiley and Sons, Hoboken, NJ, 2012.

[43] Shavit R., Elliott R. 'Design of transverse slot arrays fed by a boxed stripline'. *IEEE Transactions on Antennas and Propagation.* 1983;**31**(4): 545–552.

[44] Robertson R., Elliott R. 'The design of transverse slot arrays fed by the meandering strip of boxed stripline'. *IEEE Transactions on Antennas and Propagation.* 1987;**35**(3):252–257.

[45] Elliott R. S., Stern G. J. 'The design of microstrip dipole arrays including mutual coupling'. *IEEE Transactions on Antennas and Propagation.* 1987; **35**(3):252–357.

Chapter 7

Design of planar slotted waveguide array antennas

In this chapter, we will present Elliott's design procedure for planar standing wave slot arrays including a detailed design example with computed and measured results. For large planar arrays a modification to Elliott's design procedure using an infinite array model is sometimes preferred. Large standing wave arrays can be broken up into sub-arrays with a parallel feed network to improve bandwidth. Important parameters are the total normalised slot conductance of radiating waveguides and the total normalised resistance of feed waveguides. Additional examples of slot array designs will be presented, including a procedure for designing a travelling wave feed to excite radiating waveguides with either standing wave or travelling wave slot arrays. Other design and analysis methods in the literature will also be reviewed.

7.1 Introduction

Planar slot arrays are employed in applications requiring pencil beams by stacking several linear arrays or waveguide sticks containing radiating slots next to each other. They are preferable to conventional parabolic reflector antennas since they can be designed to provide greater control of the pattern, especially of the sidelobe levels. They are used in numerous radar and communication systems in ground and space applications. The use of all metal waveguides in slot arrays is attractive in deep space applications where severe radiation environment is encountered. Since slot arrays occupy small volumes, they are easy to deploy.

Figure 7.1 shows a planar array consisting of six radiating waveguides stacked next to each other with six longitudinal radiating slots cut in the broad wall of each. Each radiating waveguide is shorted at a distance of a quarter guide wavelength from the centre of the radiating slot at each end. The radiating slots are cut on a flat plate which will form the top broad walls of all the radiating waveguides stacked next to each other. Therefore a planar array like this is also called a flat plate array. The radiating waveguides are fed by coupling slots cut on top broad wall of a feed waveguide placed underneath the radiating waveguides and orthogonal to them. Resonant centred-inclined coupling slots are commonly used in the design. Other types of coupling slots will be discussed in Chapter 8.

Figure 7.1 A planar array of longitudinal radiating slots cut in rectangular waveguides

In planar arrays, mutual coupling between radiating slots in the E-plane is significant and needs to be included in the design and analysis. Inclusion of mutual coupling is generally a challenging process. Planar slot arrays have been reported in the literature since 1958 [1–4]. The design of a planar array was presented by Elliott [5] that included the model of the coupling slots between the feed waveguide and the radiating waveguides. The method of accounting for the internal higher-order mode coupling, discussed in Chapter 6 may be applied to the design of planar arrays as well. We will discuss Elliott's procedure for planar slot arrays in detail since it provides an accurate procedure for the designs of slot arrays. For large arrays the computational burden of Elliott's method becomes impractical, because of the large amount of external mutual coupling computations. The method based on the infinite array mutual coupling model is a viable alternative. Large arrays are often broken into sub-arrays. The sub-array architecture is discussed with some examples, followed by travelling wave feeds used for planar arrays. The chapter concludes with a discussion on various design and analysis methods found in the current literature.

7.2 Elliott's design procedure for planar standing wave arrays

The design procedure for planar standing wave arrays described by Elliott in 1988 [5] is presented here with the addition of internal higher-order mode coupling effects [6]. This is the first rigorous procedure introduced for the design of planar arrays and it has been used successfully in numerous designs. The design procedure allows one to determine the offsets and lengths of the longitudinal radiating slots and lengths and tilt angles of the centred-inclined coupling slots, which are commonly used to excite the radiating waveguides. It is assumed that the excitation coefficients of radiating slots are pure real but the design procedure is applicable

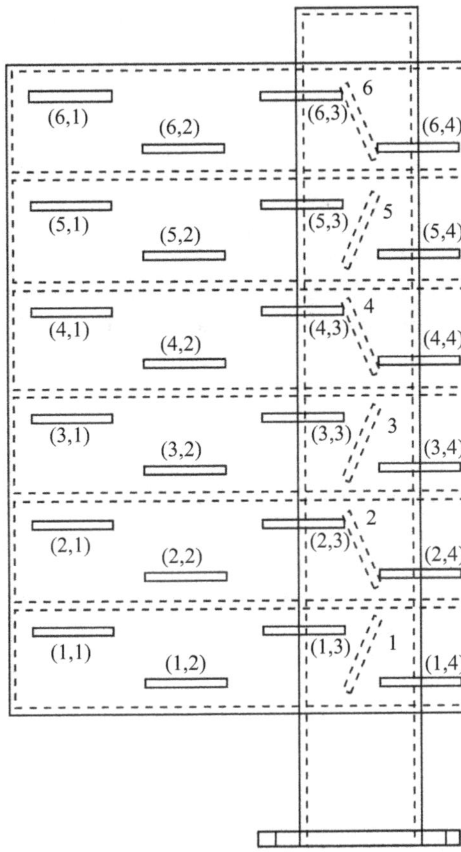

Figure 7.2 Ordering of radiating and coupling slots in a planar array

for complex excitations as well. However, it should be remembered that the amount of phase control achievable for resonant elements is limited, generally within ±60° from the resonant values.

Let the radiating waveguides be numbered from $i = 1$ through M with the first one near the input end of the feed waveguide. There is a short circuit at a distance equal to a half guide wavelength from the centre of the last coupling slot. The centred-inclined coupling slot behaves like a series impedance because of its odd-symmetric scattering, as explained in Chapter 8. Therefore a short is placed a half guide wavelength away from the last coupling slot in the standing wave array. Radiating slots in each radiating waveguide are numbered with two indices, the first one, i being the number of the radiating waveguide while the second, $j = 1$ through N, denotes the radiating slot number in waveguide i. The first radiating slot is on the left of the feed waveguide when the feed waveguide propagation direction is upward as shown in Figure 7.2. For simplicity the value of N is assumed to be the same for each radiating waveguide but the design procedure is valid even if it is not the case.

The first design equation for linear arrays given in (6.16) becomes (7.1) for planar arrays.

$$\frac{Y_{ij}^a/G_0}{Y_{11}^a/G_0} = \frac{|f_{ij}|V_{ij}^s}{|f_{11}|V_{11}^s}\frac{\chi_1}{\chi_i} \tag{7.1}$$

The subscripts i and j denote the radiating waveguide number and the slot number in that waveguide respectively. The reference slot 11 is the first slot in the first radiating waveguide. The magnitude of the mode voltage of all slots in a radiating waveguide is the same because of the half guide wavelength spacing between adjacent slots at the design frequency. Therefore mode voltages do not appear in (7.1). A resonant coupling slot acts as a transformer with the coupling coefficient χ_i given by

$$\chi_i^2 = S_{11}/(1 - S_{11}) \tag{7.2}$$

where S_{11} is the reflection coefficient of the four-port centred-inclined coupling slot with all the ports match-terminated. In Chapter 8 coupling slot models are discussed in detail. Clearly the active admittance of a radiating slot is proportional to the slot voltage or the excitation coefficient whereas it is inversely proportional to the coupling coefficient of the coupling slot in that radiating waveguide. Obviously if the coupling coefficient is greater, a smaller value of active admittance is needed to realise a given slot excitation.

The terms f_{ij} and Y_{ij}^a/G_0 in (7.1) have been previously defined in (6.12) and (6.15), respectively, with a single subscript for each. They are reproduced in (7.3) and (7.4) in terms of double subscripts corresponding to those of each radiating slot. Equation (7.4) is a combination of (6.15), (6.17) and (6.39).

$$f_{ij} = f_{ij}(x_{0ij}, L_{ij}) = -\cos(\pi x_{0ij}/a)\frac{\dfrac{\pi}{2kL_{ij}}\cos(\beta_{10}L_{ij})}{\left(\dfrac{\pi}{2kL_{ij}}\right)^2 - \left(\dfrac{\beta_{10}}{k}\right)^2} \tag{7.3}$$

where $x_{0ij} = a/2 + \delta_{ij}$.

$$\frac{Y_{ij}^a}{G_0} = \frac{1}{G_0/Y_{ij} + \dfrac{1}{2f_{ij}^2}MC_{ij}} \tag{7.4}$$

The mutual coupling term in (7.4) is obtained by combining (6.17) and (6.39), and it is given by

$$MC_{ij} = j(\beta_{10}/k)(k_0b)\frac{(a/\lambda)^3}{V_{ij}^s}\sum_{m=1}^{M}\sum_{n=1}^{N}{}' V_{mn}^s g_{mn,ij}(L_{mn}, \delta_{mn}, L_{ij}, \delta_{ij})$$
$$+ j\frac{\beta_{10}}{\gamma_{20}}e^{-(\gamma_{20}d)}\left[h_{ij}h_{i,j-1}V_{i,j-1}^s + h_{ij}h_{i,j+1}V_{i,j+1}^s\right] \tag{7.4a}$$

where

$$h_{ij} = 2\cos(2\pi\delta_{ij}/a)\frac{\pi}{2kL_{ij}}\frac{\cosh(\gamma_{20}L_{ij})}{(\gamma_{20}/k)^2 + \left(\frac{\pi}{2kL_{ij}}\right)^2} \tag{7.4b}$$

and $g_{mn,ij}$ is the expression for external mutual coupling between radiating slots mn and ij, defined previously in (6.18) and (4.115). The prime in the summation indicates that $m \neq i$ and $n \neq j$ simultaneously. In a planar array, the E-plane coupling between slots in different waveguides generally becomes substantial.

The normalised self-admittance of the longitudinal radiating slot ij, Y_{ij}/G_0, is obtained from the two-port scattering parameter of the slot with the reference plane passing through the centre of the slot. S_{11ij} is given by (7.5) which is the same as (6.45) with double subscripts.

$$Y_{ij}/G_0 = -2\frac{S_{11ij}}{1 + S_{11ij}} \tag{7.5}$$

7.2.1 Design algorithm

Figure 7.3 shows the flow chart for the design algorithm. The procedure starts with an initial determination of the array geometry, the number of radiating waveguides, M, the number of radiating slots, N, waveguide interior dimensions, a, and b, slot spacing, d, slot width, w, wall thickness, t, the aperture distribution and the design frequency, f. Waveguide dimensions are chosen such that the half guide wavelength of the feed waveguide at the centre frequency is equal to the 'a' dimension of the radiating waveguide plus its sidewall thickness. This allows the radiating waveguides to be stacked next to each other and excited in phase by keeping the coupling slots half guide wavelength apart, with alternating positive and negative tilt angles providing the required 180° phase reversal. Scattering parameters of isolated radiating slots are computed for a range of values of slot offsets and lengths normalised to the resonant length for each offset, just as it was done for linear arrays in Section 6.5. For coupling slots, the resonant length and the value of $|S_{11}|$ at resonance are generated for a range of tilt angles. For simplicity coupling slots are all assumed to be resonant.

For radiating waveguides 2 through M, there are $2N + 1$ unknowns, the offsets and lengths of N slots and the coupling coefficient of the coupling slot exciting that radiating waveguide normalised to the coupling coefficient of the first radiating waveguide. By separating (7.1) into real and imaginary parts, two equations are obtained for each slot, thereby producing $2N$ equations for N slots in each of these radiating waveguides. An additional equation is obtained when one sets the total normalised active conductance, $\bar{G}_{tot}$ of each radiating waveguide to a specific value in the design, such as 2, thus producing $2N + 1$ equations in as many unknowns for each of the radiating waveguides 2 through M. Since the normalised value of the coupling coefficient is unity for the first waveguide, there is one less unknown or a

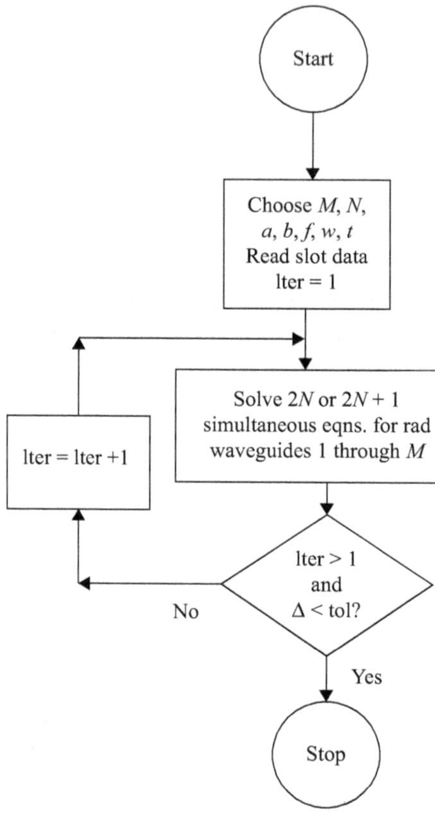

Figure 7.3 Flow chart for the design of a planar array

total of $2N$ unknowns for this waveguide. For the slot 11 ($i = 1$ and $j = 1$), (7.1) is not relevant. However, we obtain one equation for slot 11 by choosing the imaginary part of the active admittance of Y_{11}^a to be zero. Thus for the first radiating waveguide we have $2N$ equations in $2N$ unknowns. Thus there are as many equations as the number of unknowns for each radiating waveguide. However, the simultaneous equations are nonlinear. The total normalised slot conductance $\bar{G}_{\text{tot}}$ mentioned above is discussed in [7] and in Section 7.4.

At the start of the first iteration, the initial values of alternating slot offsets for all radiating slots are chosen arbitrarily and their lengths have been set to equal their resonant values. The choice of these values is not critical since they will be updated to more accurate values in subsequent iterations. Recall that in Section 6.3 there was a discussion on the iterative design for linear standing wave arrays where it was stated that $2N$ equations in $2N$ unknowns were solved using a quasi-Newton method. The same technique is used in the planar array by solving the non-linear equations of one radiating waveguide at a time [8]. After solving for the unknowns in all radiating waveguides, we iterate the process. The iteration is stopped when changes in linear dimensions of slots between successive iterations are less than achievable values of manufacturing tolerance. In order to determine the manufacturing tolerance required to achieve the desired antenna performance, designers evaluate how changes in slot parameters such as lengths, offsets and tilts on the antenna characteristics affect the antenna performance, for example, gain, sidelobe level and cross-polarisation. Based on such studies it may be possible to set the tolerance needed to achieve the required performance. At low frequencies, tolerance may not be an issue. The stopping criterion for iterations may be when change in linear dimensions is below a predetermined level, obtained by a process similar to that used to determine the tolerance. Experience has shown that good design solutions are obtained typically after five iterations. The iterative design process yields the offsets and lengths of all slots and all the coupling coefficients relative to that of the first coupling slot. By enforcing a match at the input port or by equating the total normalised resistance of the feed waveguide to unity, we determine the coupling coefficient χ_1 from (7.6), and hence all χ_i as follows.

$$\bar{R}_{in} = \chi_1^2 \sum_{i=1}^{M} \bar{G}_{\text{tot},i} \bar{\chi}_i^2 = 1 \tag{7.6}$$

where

$$\bar{\chi}_i = \chi_i / \chi_1 \tag{7.7}$$

is the normalised coupling coefficient of the coupling slot i relative to that of the coupling slot 1 and

$$\bar{G}_{tot,i} = \sum_{j=1}^{N} Y_{ij} / G_0 \tag{7.8}$$

Using (7.2), the scattering parameter S_{11i} of the ith coupling slot in a four-port coupler is determined. Then from the original computed data of coupling slots one can interpolate the resonant length and the tilt angle corresponding to each S_{11i}.

7.2.2 *Example of a 6 × 6 planar array*

We demonstrate the design procedure for the 6 × 6 planar array shown in Figure 7.1. A separable aperture distribution with excitation coefficients in each principal direction is given by 0.513:0.814:1:1:0.814:0.513. The array factor of a linear array multiplied by the element pattern of a typical slot with this aperture distribution yields sidelobe levels of −19.5 and −22.5 dB in the *E*-plane and *H*-plane, respectively. The design is carried out at 9.3 GHz. The interior dimensions of the feed waveguide are 22.86 mm × 10.16 mm while those of the radiating waveguides are 21.713 × 5.08 mm. The choice of half-height radiating waveguides helps in minimising the effect of internal higher-order mode coupling problem in the junction regions involving each coupling slot and a pair of straddling radiating slots in the immediate vicinity [9]. The guide wavelength of the feed waveguide at 9.3 GHz is

45.46 mm. The wall thickness of each radiating waveguide is 0.508 mm while the coupling slot thickness is 1.27 mm. When radiating waveguides are stacked next to each other spacing between centres of adjacent guides is equal to $21.71 + 2 \times 0.508 = 22.73$ mm. This is exactly equal to half guide wavelength in the feed waveguide at 9.3 GHz and is also the spacing between adjacent coupling slots. There is a short at a distance of half guide wavelength from the coupling slot farthest from the feed end. The radiating slots have alternating offsets with half guide wavelength spacing, which is 24.055 mm. At each end of every radiating waveguide, there is a short at a distance of a quarter guide wavelength from the nearest slot centre.

We obtain computed slot data using the moment method solution to the pertinent coupled integral equations of isolated radiating or coupling slot. Such data may be obtained from any computational electromagnetics code or from experimental measurements. For a range of tilt angles of coupling slots, resonant lengths and $|S_{11}|$ are determined with all the four ports match-terminated. The reference planes of all the four ports pass through the centre of the coupling slot. Table 7.1 shows computed data for coupling slots. For radiating slots, we determine the resonant length for a range of values of offsets. Then for each offset we determine $|S_{11}|$ and arg S_{11} for a range of values of slot lengths normalised to the resonant length. Radiating slot data in Tables 7.2–7.4 are similar to those in Tables 6.2–6.4 in Chapter 6. In the design, the total normalised slot conductance in each radiating waveguide was set at 2.8 so that magnitudes of slot offsets would not be too small. The total normalised

Table 7.1 Resonant length in mm and $|S_{11}|$ of coupling slots

Tilt (°)	$\|S_{11}\|$	**Resonant length = $2L_{res}$**	**Tilt (°)**	$\|S_{11}\|$	**Resonant length = $2L_{res}$**
8.0	0.0015	15.82	18.0	0.0071	15.86
10.0	0.0024	15.83	20.0	0.0086	15.87
12.0	0.0033	15.83	22.0	0.102	15.88
14.0	0.0045	15.84	24.0	0.119	15.90
16.0	0.0058	15.85	26.0	0.137	15.91

Table 7.2 Resonant length of radiating slots

Offset (mm)	**Resonant length = $2L_{res}$ in mm**
0.333	15.57
0.667	15.60
1.0	15.65
1.333	15.73
1.667	15.82
2.0	15.93
2.333	16.07
2.667	16.22
3.0	16.40

Table 7.3 Magnitude of S_{11} as a function of offset (mm) and normalised length (mm)

$\bar{L} = L/L_{res}$ $\lvert\delta\rvert$	**0.94**	**0.955**	**0.97**	**0.985**	**1.0**	**1.015**	**1.03**	**1.045**	**1.06**
0.333	0.0027	0.0031	0.0034	0.0037	0.0038	0.0037	0.0034	0.0032	0.0029
0.667	0.0107	0.0122	0.0136	0.0146	0.0148	0.0144	0.0136	0.0126	0.0115
1.000	0.0239	0.0271	0.0300	0.0320	0.0326	0.0318	0.0300	0.0278	0.0256
1.333	0.0421	0.0474	0.0521	0.0553	0.0562	0.0548	0.0519	0.0483	0.0446
1.667	0.0646	0.0722	0.0789	0.0833	0.0845	0.0826	0.0786	0.0735	0.0683
2.000	0.0908	0.101	0.109	0.115	0.116	0.114	0.109	0.102	0.0957
2.333	0.121	0.133	0.142	0.149	0.150	0.147	0.141	0.134	0.126
2.667	0.153	0.166	0.177	0.184	0.185	0.182	0.176	0.168	0.159
3.000	0.187	0.201	0.212	0.219	0.221	0.217	0.211	0.202	0.192

Table 7.4 Phase of S_{11} in degrees as a function of offset and normalised length

$\bar{L} = L/L_{res}$ $\lvert\delta\rvert$	**0.94**	**0.955**	**0.97**	**0.985**	**1.0**	**1.015**	**1.03**	**1.045**	**1.06**
0.333	225.5	216.4	205.4	193.0	180.1	167.9	157.5	148.9	142.0
0.667	225.0	216.0	205.1	192.8	180.1	168.2	157.8	149.3	142.5
1.000	224.2	215.2	204.5	192.5	180.2	168.6	158.5	150.1	143.3
1.333	222.8	213.9	203.3	191.7	179.8	168.6	158.9	150.8	144.2
1.667	221.5	212.7	202.4	191.3	179.9	169.3	160.0	152.1	145.6
2.000	220.0	211.3	201.4	190.8	180.1	170.1	161.1	153.6	147.2
2.333	217.9	209.4	199.9	189.9	179.9	170.5	162.1	154.9	148.8
2.667	216.1	207.9	198.9	189.4	180.1	171.4	163.5	156.7	150.8
3.000	213.8	205.9	197.4	188.6	180.0	171.9	164.6	158.2	152.6

Table 7.5 Tilt angles and lengths of coupling slots obtained from the design

Coupling slot number, *i*	**Tilt angle (°)**	**Slot length (mm)**
1	−10.9157	15.8300
2	16.1483	15.8523
3	−19.2417	15.8683
4	19.2417	15.8683
5	−16.1483	15.8523
6	10.9157	15.8300

resistance in the feed waveguide was set to unity to achieve a match. Table 7.5 shows the tilt angles and lengths of the six coupling slots. The values of offsets and lengths of radiating slots are shown in Table 7.6. All data in Tables 7.5 and 7.6 were obtained using the design procedure mentioned in Section 7.2.1.

Table 7.6(a) Offsets and lengths of radiating slots obtained from the design

Slot number	Offset (mm)	Length (mm)	Slot number	Offset (mm)	Length (mm)	Slot number	Offset (mm)	Length (mm)
11	1.271	15.980	21	1.283	15.856	31	1.161	15.877
12	−2.326	16.141	22	−1.846	15.872	32	−1.820	16.143
13	2.240	16.362	23	2.584	16.304	33	2.466	16.367
14	−2.988	16.457	24	−2.375	16.108	34	−2.247	16.343
15	1.860	16.195	25	2.009	16.024	35	1.903	16.084
16	−1.428	15.895	26	−1.167	15.736	36	−1.219	15.981

Table 7.6(b) Offsets and lengths of radiating slots obtained from the design

Slot number	Offset (mm)	Length (mm)	Slot number	Offset (mm)	Length (mm)	Slot number	Offset (mm)	Length (mm)
41	1.219	15.980	51	1.167	15.736	61	1.428	15.895
42	−1.903	16.084	52	−2.009	16.024	62	−1.860	16.195
43	2.248	16.343	53	2.375	16.108	63	2.988	16.457
44	−2.466	16.367	54	−2.584	16.304	64	−2.240	16.362
45	1.820	16.143	55	1.846	15.872	65	2.326	16.141
46	−1.161	15.877	56	−1.283	15.856	66	−1.271	15.979

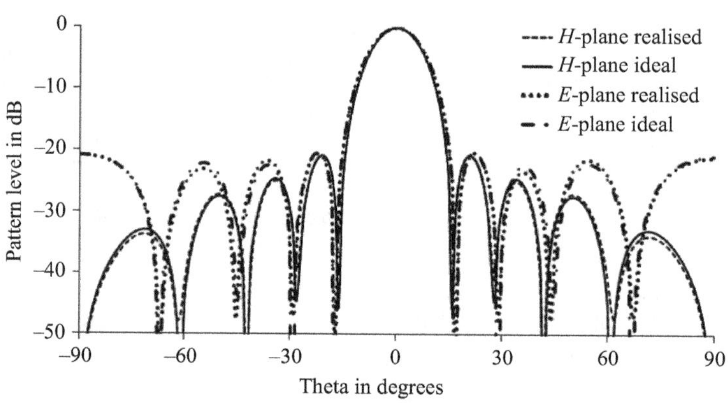

Figure 7.4 E-plane and H-plane patterns, ideal and designed

Slot ends become rounded or semi-circular when fabricated by milling. All square ended slot lengths are increased by an amount equal to 0.215 slot width, to account for square end to round end equivalence based on equal area criterion. The equivalence between round ended and square ended slot has been discussed in Section 5.3.3. See also (5.54).

Figure 7.4 shows the E-plane and H-plane patterns of the 6×6 array computed using the method of moments (MoM) analysis used in this work described in

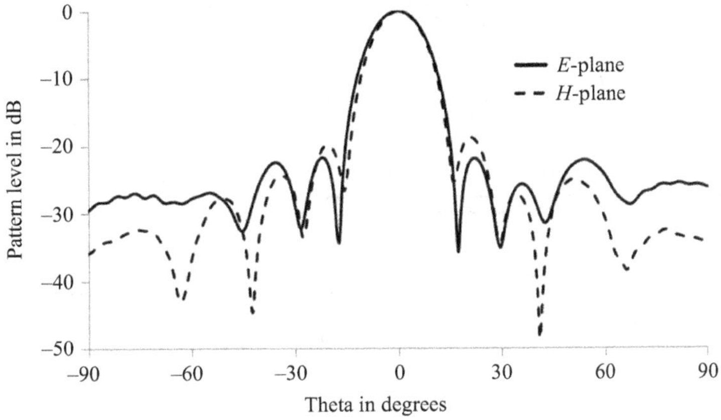

Figure 7.5 Measured patterns of the 6 × 6 array

Chapter 8. It is assumed that the slots are embedded in an infinite ground plane and therefore radiation is confined to the front half space only. The ideal patterns in Figure 7.4 are obtained with the specified aperture distribution with the actual slot locations. The electric field of each slot is assumed to have a half-cosine amplitude and uniform phase distribution in the ideal pattern whereas in the realised pattern the computed aperture distribution for each slot is used. Small discrepancies between the ideal and realised patterns are attributed to imperfections in the design procedure and implementation. Figure 7.5 shows the measured pattern of an array fabricated with the use of design data shown in Tables 7.5 and 7.6. The discrepancy between computed and measured patterns, especially in the far out sidelobe region, is attributed primarily due to diffraction from the edges of the finite ground plane. Manufacturing tolerances may also account for some discrepancy between theory and measurement.

7.3 Design of large arrays using the infinite array mutual coupling model

In large arrays with slowly varying aperture distributions, all slots except the ones near the edges are surrounded by approximately identical neighbouring elements but for alternating signs for the offsets. The external mutual coupling expression for a representative slot n is then obtained using a simplified infinite array model. Infinite array mutual coupling model has been used in a number of antenna arrays. In microstrip reflectarrays the infinite array mutual coupling model has been used successfully for many designs and analyses [10]. It has been shown that the infinite array model for slot arrays yields errors for only a couple of rows of slots near the *E*-plane edges. The accuracy of the model has been established for moderate to large arrays [11,12]. In this model, slot n is embedded in two doubly periodic infinite arrays, one with positive offsets and

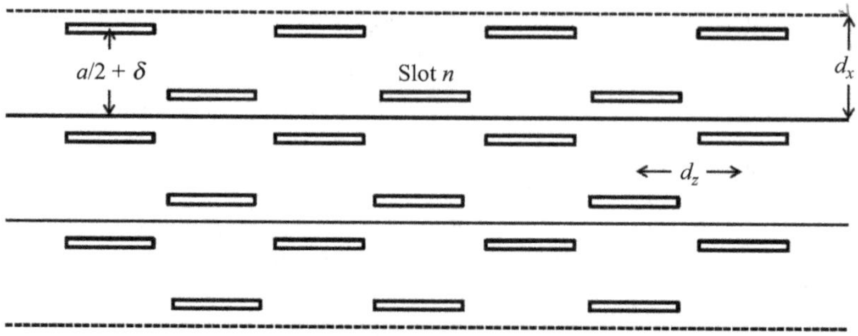

Figure 7.6 Infinite array mutual coupling environment for slot n

the other with negative offsets, both in rectangular lattice with period $2d_z$ along the radiating waveguides and d_x across. Figure 7.6 shows a few of the slots in the infinite array model. The magnitude of the offset of all slots, δ_n, and their half lengths, L_n, are those of slot n.

7.3.1 Infinite array mutual coupling expression [11,13]

The term associated with the external mutual coupling experienced by slot ij expressed in (7.4) is given next where the prime in the summation means that $m \neq i$ and $n \neq j$.

$$MC_{ij}^{(1)} = j(\beta_{10}/k)(k_0 b)\frac{(a/\lambda)^3}{V_{ij}^s}\sum_{m=1}^{M}\sum_{n=1}^{N}{}' V_{mn}^s g_{mn,ij}(L_{mn}, \delta_{mn}, L_{ij}, \delta_{ij}) \tag{7.9}$$

In the infinite array environment, it is assumed that all slot voltages are equal or slowly varying. In the latter case all slot voltages are approximated to be equal. Therefore all slot lengths are the same and the magnitudes of slot offsets are equal. Then (7.9) becomes

$$MC_{ij}^{(1)} = j(\beta_{10}/k)(k_0 b)(a/\lambda)^3 \sum_{m=-\infty}^{\infty}\sum_{n=-\infty}^{\infty}{}' g_{mn,ij}(L_{ij}, \pm\delta_{ij}, L_{ij}, \delta_{ij}) \tag{7.10}$$

The sign for the offset of slot mn is chosen based on its location because of alternating slot offsets. We express (7.10) as follows.

$$MC_{ij}^{(1)} = -\frac{4\pi\eta_0}{w^2}(\beta_{10}/k)(k_0 b)(a/\lambda)^3 T \tag{7.11}$$

where

$$T = \sum_{m=-\infty}^{\infty}\sum_{n=-\infty}^{\infty}{}' I_{mn,ij}(L_{ij}, \pm\delta_{ij}, L_{ij}, \delta_{ij}) \tag{7.12}$$

and $I_{mn,ij}(L_{ij}, \pm\delta_{ij}, L_{ij}, \delta_{ij}) = Y_{mn}L_{ij}w/2$. Y_{mn} is specified in (4.113) for a pair of slots with single subscripts whereas we have double subscripts for the planar array. It is depicted in Figure 4.15 with slot lengths a and widths b whereas in (7.12) the half length and width are L_{ij} and w, respectively.

Equation (7.12) may be expressed as

$$T = \iint_{\text{slot } ij} H_\varsigma \cos\left(\frac{\pi\zeta_{ij}}{2L_{ij}}\right) ds' \tag{7.13}$$

where the integral is carried out over slot ij, ζ_{ij} is its local coordinate along the axis of the slot, and cosine weighting corresponds to its aperture distribution. H_ς is the magnetic field along the location of slot ij, produced by half cosinusoidal magnetic currents of unit peak strength at the location of all other slots, radiating in free space in the absence of the ground plane. Clearly each term inside the summation in (7.12) corresponds to the integral in (7.13) when the contribution to H_ς due to unit magnetic current in slot mn only is considered.

$$H_\varsigma = \frac{1}{j4\pi\omega\mu_0}\left(\frac{\partial^2}{\partial\varsigma^2} + k_0^2\right) \iint_{\text{slot } mn} S \cos\left(\frac{\pi\zeta_{mn}}{2L_{ij}}\right) ds' \tag{7.14}$$

where

$$S = \sum_m \sum_n \left[\frac{\exp(-jk_0R_1)}{R_1} + \frac{\exp(-jk_0R_2)}{R_2}\right] \tag{7.15}$$

$$R_1 = \left[\{\varsigma - (\varsigma' + 2md_z)\}^2 + \{\xi - (\xi' + nd_x)\}^2\right]^{1/2} \tag{7.16}$$

$$R_2 = \left[\{\varsigma - (\varsigma' + (2m+1)d_z)\}^2 + \{\xi - (\xi' - 2\delta + nd_x)\}^2\right]^{1/2} \tag{7.17}$$

and the integral is carried out over all the slot apertures except that of ij. In the double summation in (7.15) all positive and negative integer values and zero are used for m and n except $m = 0$ and $n = 0$ in (7.16). Using Poisson's sum formula the double summations S may be expressed in terms of double infinite Floquet series [13] as shown in (7.18), where all positive and negative integer values and zero are used for p and q.

$$S = \frac{-j2\pi}{2d_zd_x} \sum_p \sum_q \frac{\exp\left[-jk_0\left\{\frac{p\lambda_0}{2d_z}(\varsigma - \varsigma') + \frac{q\lambda_0}{d_x}(\xi - \xi')\right\}\right]}{k_0\left[1 - \left(\frac{p\lambda_0}{2d_z}\right)^2 - \left(\frac{q\lambda_0}{d_x}\right)^2\right]^{1/2}}$$
$$\times\left[1 + (-1)^p \exp\left(-jq\frac{4\pi\delta}{d_x}\right)\right] \tag{7.18}$$

Using (7.18) in (7.14) we obtain

$$H_\varsigma = \frac{-\lambda_0}{8\pi\omega\mu_0 d_z d_x}\left(\frac{\partial^2}{\partial\varsigma^2}+k_0^2\right)\iint_{\text{slot}\,mm}\sum_p\sum_q \frac{\exp\left[-jk_0\left\{\frac{p\lambda_0}{2d_z}(\varsigma-\varsigma')+\frac{q\lambda_0}{d_x}(\xi-\xi')\right\}\right]}{\left[1-\left(\frac{p\lambda_0}{2d_z}\right)^2-\left(\frac{q\lambda_0}{d_x}\right)^2\right]^{1/2}}$$

$$\bullet\left[1+(-1)^p\exp\left(-jq\frac{4\pi\delta}{d_x}\right)\right]\cos\left(\frac{\pi\zeta_{mn}}{2L_{ij}}\right)ds' \tag{7.19}$$

The integral over slot *mn* yields

$$w\frac{\frac{\pi}{L_{ij}}}{\left(\frac{\pi}{2L_{ij}}\right)^2-\left(\frac{p\pi}{d_z}\right)^2}\cos\left(\frac{p\pi L_{ij}}{d_z}\right) \tag{7.20}$$

After using (7.20) in (7.13), the integral over slot *ij* and the operation $\left(\frac{\partial^2}{\partial\varsigma^2}+k_0^2\right)$ yields

$$k_0^2\left[1-\left(\frac{p\lambda_0}{2d_z}\right)^2\right]w\frac{\frac{\pi}{L_{ij}}}{\left(\frac{\pi}{2L_{ij}}\right)^2-\left(\frac{p\pi}{d_z}\right)^2}\cos\left(\frac{p\pi L_{ij}}{d_z}\right) \tag{7.21}$$

From (7.19) through (7.21) and (7.13) we obtain (7.22).

$$T = \frac{-\lambda_0 k_0^2 w^2}{8\pi\omega\mu_0 d_z d_x}\sum_p\sum_q\left[1-\left(\frac{p\lambda_0}{2d_z}\right)^2\right]\frac{\left[1+(-1)^p\exp\left(-jq\frac{4\pi\delta}{d_x}\right)\right]}{\left[1-\left(\frac{p\lambda_0}{2d_z}\right)^2-\left(\frac{q\lambda_0}{d_x}\right)^2\right]^{1/2}}$$

$$\times\frac{\left(\frac{\pi}{L_{ij}}\right)^2}{\left[\left(\frac{\pi}{2L_{ij}}\right)^2-\left(\frac{p\pi}{d_z}\right)^2\right]^2}\cos^2\left(\frac{p\pi L_{ij}}{d_z}\right) \tag{7.22}$$

Using (7.22) in (7.11) we obtain the expression for the external mutual coupling for an infinite array model. It is possible to obtain scattering data for a slot embedded in an infinite array as a function of offset and length and use such data to design a slot array. For a large array such a design works very well and for a moderate size array it may still yield acceptable results.

7.4 Choice of total slot conductance in radiating waveguides and resistance in feed waveguides

In the design of a planar standing wave array, the total slot conductance of each radiating waveguide, G, is one of the design parameters. A linear array can be designed with a specified value of G and with an impedance transformer to match the input port. The frequency behaviour of linear arrays has been analysed by considering equally spaced constant series or shunt loaded transmission lines [7]. The frequency response of the shunt or series load is ignored since the bandwidth of the array is usually small compared to that of a load. The analysis employs the transmission matrix of each unit cell composed of a length of transmission line equal to the spacing between successive slots and the load impedance or admittance [14]. A larger value of G yields a greater bandwidth for return loss whereas the bandwidth for pattern performance such as the sidelobe level is lower. A smaller value of G will produce the opposite effect. In the case of a planar array, the total normalised shunt conductance G and the total normalised series resistance R of the feed waveguide are design parameters. If R is not equal to 1, there is a need to use a transformer. Derneryd and Petersson [7] have calculated the effects of overloading on the return loss and sidelobe performance over 5% bandwidth for a slot array of 10 wavelength diameter circular aperture designed to provide a −30 dB sidelobe level. We reproduce their results in Figures 7.7–7.9.

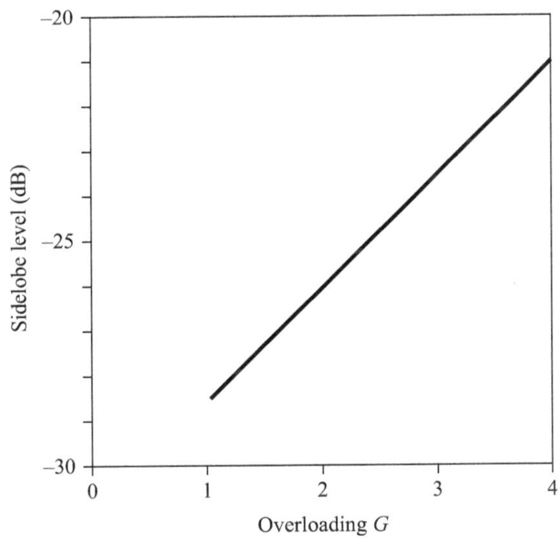

Figure 7.7 First sidelobe level in the H-plane at the upper frequency limit as a function of overloading in radiating waveguides. Reprinted from [7], with permission

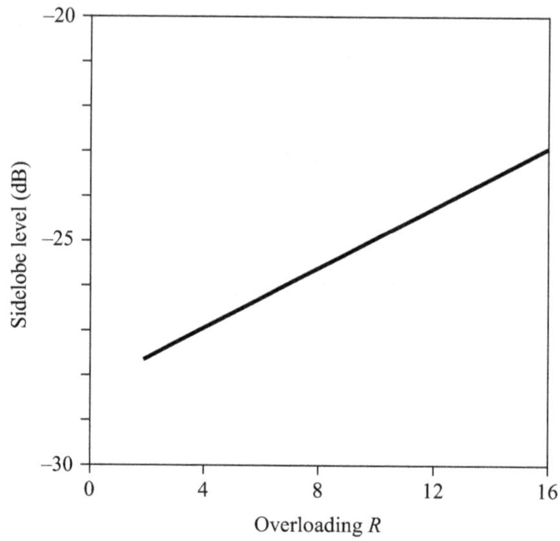

Figure 7.8 First sidelobe level in the E-plane at the upper frequency limit as a function of overloading in the feed waveguide. Reprinted from [7], with permission

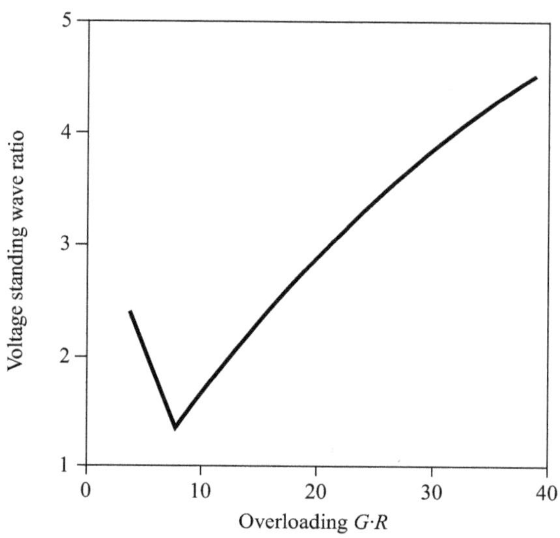

Figure 7.9 Voltage standing wave ratio at the upper frequency limit as a function of overloading of the waveguides. Reprinted from [7], with permission

Figures 7.7 and 7.8 show that in order to produce good pattern performance, especially low sidelobes over the desired bandwidth G and R have to be small. However, a better VSWR over the desired bandwidth is obtained for a value of the product $GR = 8$ as shown in Figure 7.9. A choice of $G = 2$ yields the best compromise for the performance of VSWR and sidelobe level over frequencies. Therefore it is recommended that the values of G and R be chosen as 2 and 4, respectively. These values are guidelines for designs since Derneryd and Petersson's approximate analysis did not consider the frequency response of radiating and coupling slots and that of the impedance transformer.

7.5 Sub-array architectures

The standing wave arrays are generally designed at the centre of the required frequency band. At frequencies away from the design frequency the slot excitations deviate from their ideal design values. Therefore the return loss and pattern characteristics such as gain and sidelobe levels degrade. For large arrays the return loss bandwidth is more limiting than the bandwidth for pattern performance. It was shown in Figure 6.2 that the return loss bandwidth deteriorates as the number of slots in a linear standing wave array increases. A similar behaviour is found for planar arrays also. It is possible to improve the performance of large arrays by breaking them into several sub-arrays, with each sub-array fed by a corporate power divider. The sub-array architecture is based on a compromise between the superior performance achievable by small sub-arrays and the cost and complexity of the required power divider network when an array is made up of a large number of sub-arrays. In this section we show some examples of sub-array architectures. Section 9.11.5 in Chapter 9 also has some discussions on the sub-array architecture.

7.5.1 A planar array of 160 × 160 elements consisting of 10 × 10 element sub-arrays

Glacier and land ice surface topography interferometer (GLISTIN), consisting of two large Ka band 4 m × 1 m slotted waveguide array antennas was proposed for a synthetic aperture radar for mapping ice sheets and glaciers [15]. For demonstrating the technology a 1 m × 1 m array was developed and built. It consisted of 160 × 160 elements and was made up of 16 so-called sticks each with a separate waveguide feed port [15]. Received signals from the sixteen sticks was combined to produce digital beams in the receive mode in the H-plane or elevation. A 10 × 160 element array was used as a transmit stick [15]. Figure 7.10 shows the back of the 17 sticks. Each stick is made up of four 40 × 10 element modules, one of which is shown in Figure 7.11. The back of the 40 × 10 module has four ports for adjusting each 10 × 10 sub-array for possible phase differences between sub-arrays introduced by the manufacturing imperfections. The return loss and pattern performance of this array was found to be satisfactory for the radar application even though the strong mutual coupling between the radiating slots in the E-plane reduces the bandwidth substantially. Section 7.6.1 presents the design and measured results of this system.

7.5.2 A slot array with 8 × 8 radiating elements consisting of two sub-arrays

Figure 7.12 shows a radiometer array consisting of 8 × 8 radiating slots divided into two sub-arrays [16]. Each 8 × 4 sub-array is centre-fed in each direction and

Figure 7.10 Back of the 1 m × 1 m demo array antenna consisting of 16 receive sticks and one transmit stick (at the bottom). © IEEE 2009, reprinted from [15], with permission

hence the maximum length of line in two orthogonal directions corresponds to one and two guide wavelengths respectively. This architecture with small sub-arrays was able to meet low sidelobe levels and low levels of reflection coefficients over approximately 5% bandwidth. This array designed with an even symmetry for the slot offsets in the *H*-plane helped in reducing the grating lobes called butterfly lobes produced by the conventional slot arrays with alternating positive and negative offsets. Butterfly lobes are discussed in Section 9.9. Section 7.6.2 presents results for sidelobe levels of the slot arrays exhibiting symmetry in the two halves of the *H*-plane and for the array with alternating slot offsets along the entire *H*-plane.

7.5.3 A slot array with four-element sub-arrays

Figure 7.13 shows a two-layer 16 × 16 slot array consisting of 64 sub-arrays designed to operate at 60 GHz that was built using the diffusion bonding process [17]. Each sub-array has four longitudinal offset slots in two radiating waveguides. Figure 7.14 shows a schematic view of a quadrant of the array.

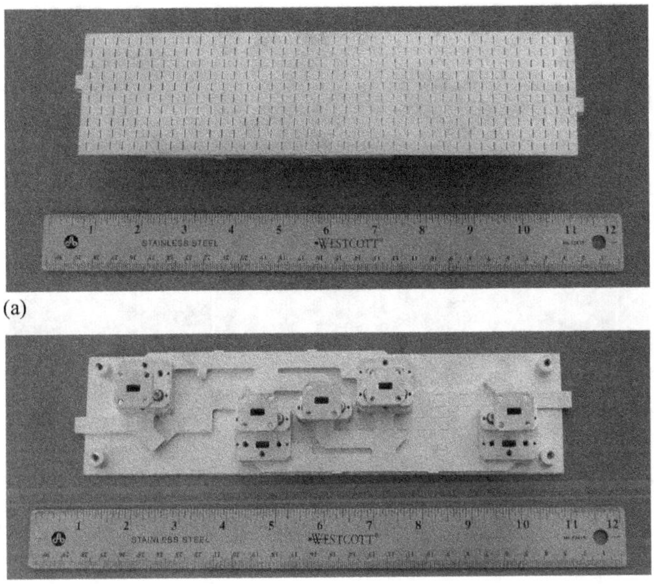

(a)

(b)

Figure 7.11 One of four modules of a stick, each consisting of four 10 × 10 sub-arrays: (a) Radiating side of the module. Four modules are combined in one stick. (b) Rear side of one module showing power dividers. © IEEE 2009, reprinted from [15], with permission

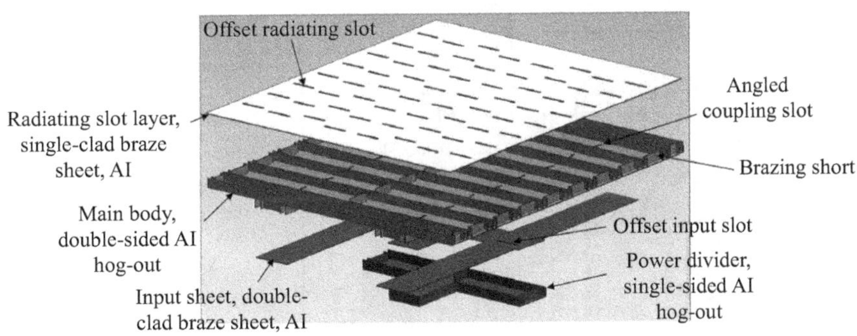

Figure 7.12 An 8 × 8 array divided into two sub-arrays. © IEEE 2010, reprinted from [16], with permission

The lower layer contains the corporate power divider made up of *H*-plane T junctions. A single *H*-plane T junction shown in Figure 7.15(a) divides the incident power into the desired power ratio. In this array all T junctions are designed to provide equal power division. The centred septum length l_f and the iris width w_f are chosen to provide a match at the input port as long as the branch ports

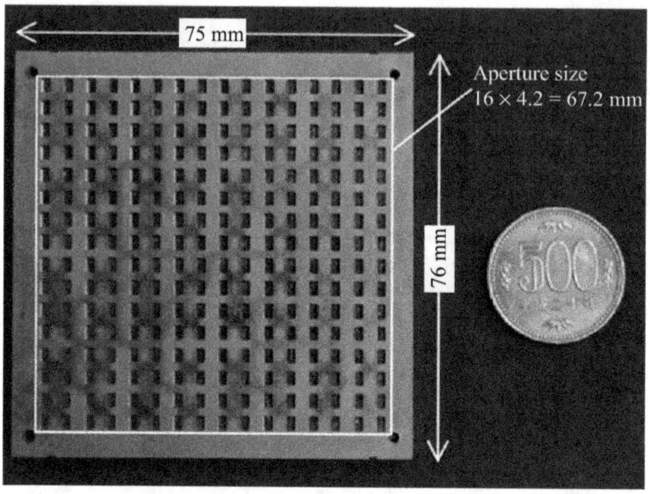

Figure 7.13 A 16 × 16 array divided into two sub-arrays. © IEEE 2011, reprinted from [17], with permission

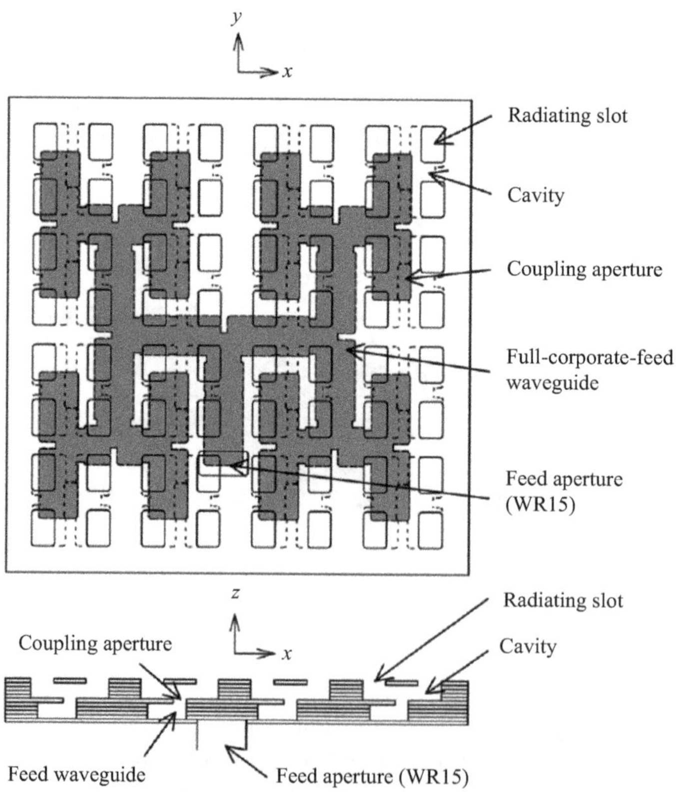

Figure 7.14 A schematic view of a quadrant of the array with H-plane T junctions in the lower layer in black and the upper layer consisting of radiating slots. © IEEE 2011, reprinted from [17], with permission

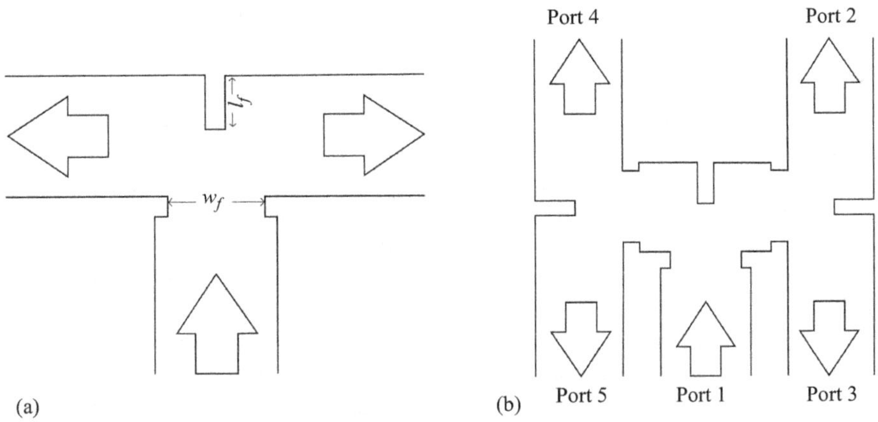

Figure 7.15 H-plane power dividers: (a) a single H-plane T-junction divider; (b) three H-plane T's for four-way power. © IEEE 2011, reprinted from [17], with permission

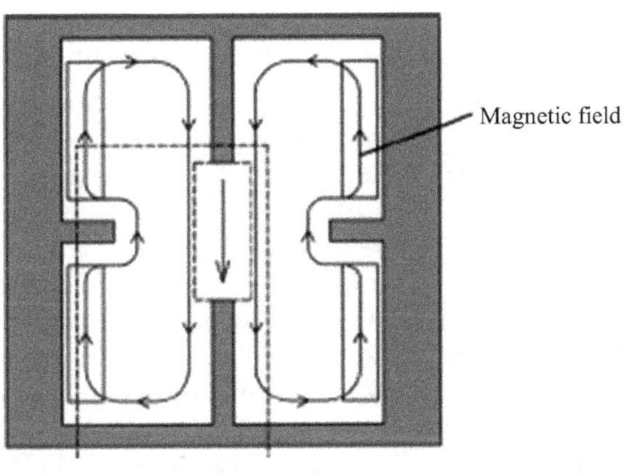

Figure 7.16 Excitation of two radiating waveguides, each containing two radiating slots. © IEEE 2011, reprinted from [17], with permission

are matched. It is generally possible to design such a T junction with greater than 20 dB return loss over a broad bandwidth. Figure 7.15(b) shows that a combination of three *H*-plane T's can provide one to four way power division. With the use of additional *H*-plane T junctions one can achieve the required power division for the entire array.

Figure 7.16 shows the method of exciting each of the two radiating waveguides in each sub-array. The two radiating waveguides are excited by a coupling window which is a longitudinal offset slot in the waveguide in the lower layer. The coupling window with its magnetic field in the longitudinal direction will excite each of the two radiating waveguides equally in amplitude but with 180° out of phase. Therefore, in order to excite the four slots with equal phase, the offsets of slots in the same waveguide are on the same side of the centreline whereas the offsets of slots in the two waveguides are on opposite sides. The slot spacing is between 0.812 and 0.925 wavelengths in the frequency range of operation. In the array design, external mutual coupling was modelled based on an infinite array of four-element sub-arrays. Uniform aperture distribution was desired but because of limitations in the infinite array mutual coupling model the sub-arrays near the edges had significant error. This array achieved a return loss better than 14 dB over 8.3% bandwidth and better than 80% efficiency. Efficiency is the ratio of the antenna gain to the ideal directivity of the antenna assuming uniform aperture distribution. An efficiency of 80% corresponds to a value of gain equal to 1 dB below the ideal directivity. It is possible to achieve such a broadband performance in a slot array if many small sub-arrays are used. However, for large arrays containing thousands of elements, the power divider may become very complicated and one may have to use somewhat larger sub-arrays.

7.6 Examples of planar slot array designs

7.6.1 Large array design using the infinite array model [11,15]

In Section 7.5.1 we discussed a large Ka band slot array consisting of 16 receive sticks, each with 160 radiating slots in the E-plane and 10 radiating slots in the H-plane. Each stick has four modules of 40×10 radiating slots each module consisting of four sub-arrays of 10×10 radiating slots. Figures 7.11 and 7.12 show the back of the full array and the 40-element module. This array was designed for uniform aperture distribution. The infinite array mutual coupling model was employed in the design. Internal higher-order coupling effects were found to be not significant for this array and therefore all radiating slots were identical but for alternating signs of offsets. Similarly all coupling slots were also identical but for alternating tilts. In each sub-array containing 10×10 radiating slots $\bar{G}_{\text{tot}}$ was set to 2.8 since a lower value of $\bar{G}_{\text{tot}} = 2$ would provide rather small offsets. In addition, the bandwidth for the pattern performance is not critical here since the return loss bandwidth is more sensitive than the pattern bandwidth (see Section 7.4). The inner dimensions of the radiating waveguides are 5.707 mm $\times$ 1.43 mm. Slot width and wall thickness are both equal to 0.508 mm. At the design frequency of 35.66 GHz a slot length of 4.163 mm and offset of 0.4 mm yields a normalised active conductance of 0.28 so that $\bar{G}_{\text{tot}}$ in each radiating waveguide would be 2.8. The total normalised resistance of each feed waveguide has been chosen to be 4.0 and a shunt series coupling slot is used to match it to an input waveguide in another layer. Therefore the coupling coefficient of each coupling slot is given by $\bar{G}_{\text{tot}}\chi^2 = 0.4$ or $\chi^2 = 1/7$. The coupling slot models are discussed in Chapter 8.

A value of $\chi^2 = 1/7 = \frac{S_{11}}{1-S_{11}}$ corresponds to the S_{11} of each resonant coupling slot $= 0.125$. The feed and input waveguide inner dimensions are 5.707 mm $\times$ 2.85 mm. The coupling slot is found to have a length of 4.140 mm and a tilt of 24°

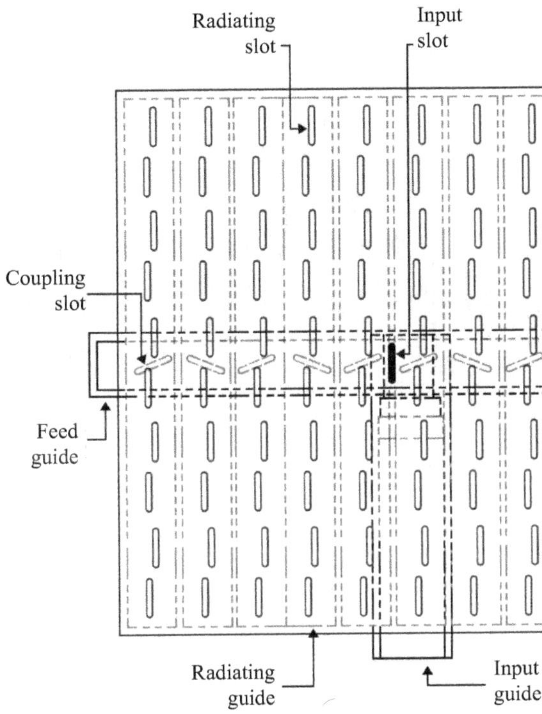

Figure 7.17 A sub-array containing 100 radiating slots, 10 coupling slots and an input shunt-series coupling slot

with respect to the axis of the feed waveguide. At the centre of the feed waveguide of each 10×10 sub-array, there is a shunt-series coupling slot that matches a normalised resistance of 4.0 to the input port on one side. The other side of the input waveguide is shorted at a distance of a quarter guide wavelength from the centre of the slot. The shunt series coupling slot has an offset of 2.081 mm and length equal to 4.242 mm. It is centred transverse to the feed waveguide. Figure 7.17 shows the slots in a sub-array.

Figures 7.18 and 7.19 show the *E*-plane and *H*-plane patterns of a 10×40 module, respectively. The computed patterns are obtained from the moment method analysis of an infinite array of 10×10-element sub-arrays with uniform excitations of all sub-arrays. From the moment method solution we determine the pattern of each sub-array and then determine the pattern of four sub-arrays in the 10×40 module. Measured patterns are in very good agreement with computed plots except in the far-out sidelobe regions. The discrepancies may be attributed to the fact that the computation assumes an infinite ground plane and therefore it does not account for the edge diffraction from the finite ground plane. In addition, imperfections in the fabrication process, especially the 1 mil (25 μ) mechanical tolerance in the manufacturing process may account for some error. Reflection

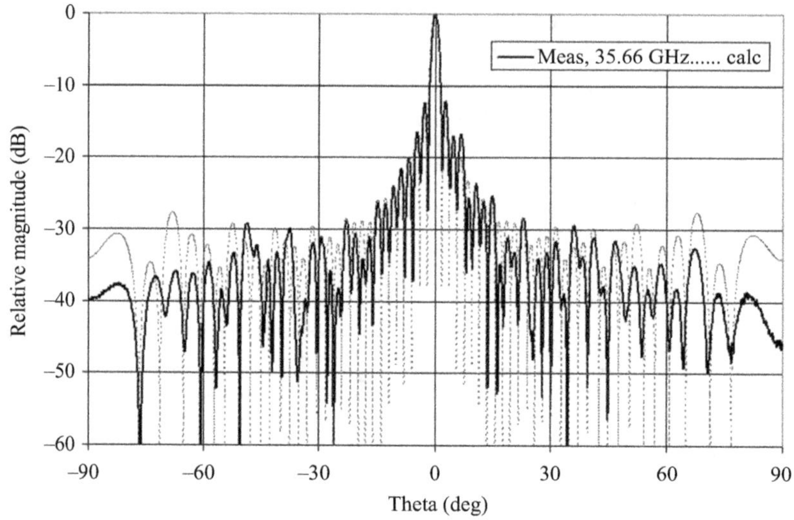

Figure 7.18 Azimuth (E-plane) patterns of a 10 × 40 module at 35.66 GHz. © IEEE 2009, reprinted from [15], with permission

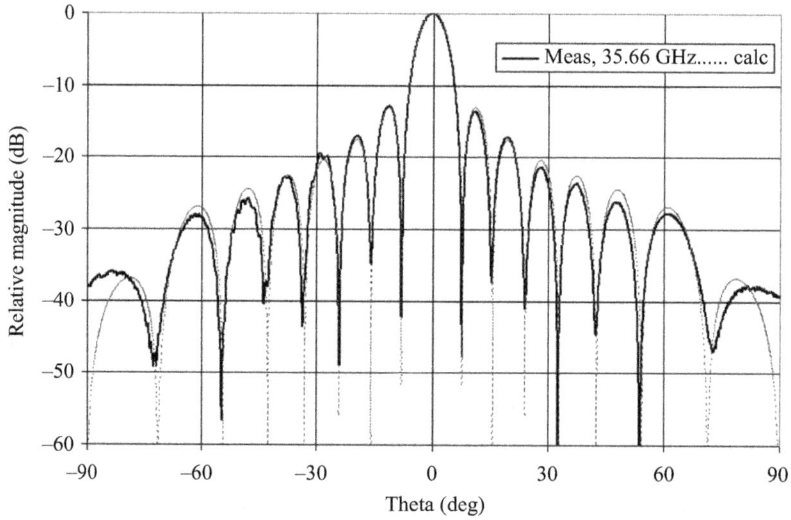

Figure 7.19 Elevation (H-plane) patterns of a 10 × 40 module at 35.66 GHz. © IEEE 2009, reprinted from [15], with permission

coefficient plots shown in Figure 7.20 indicate some discrepancy, especially at the frequency where the 10 × 40 module is tuned, partly because of errors in the fabrication process. Figure 7.21 shows the *H*-plane patterns of 16 sticks and the pattern computed by the moment method. The light traces corresponding to the patterns of

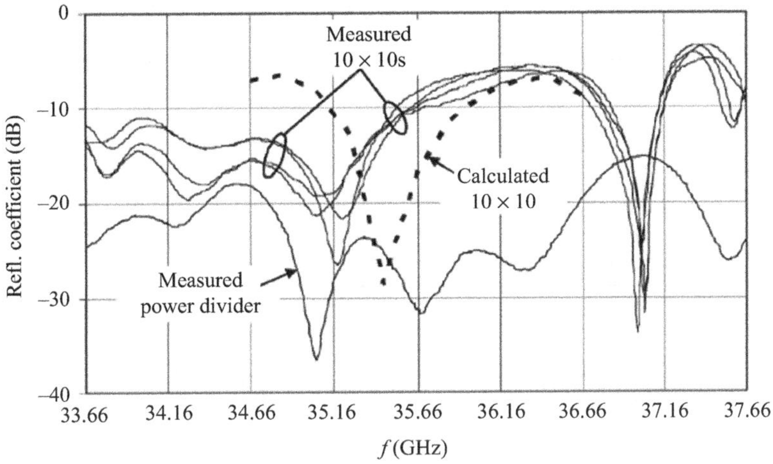

Figure 7.20 Reflection coefficient of each sub-array and the power divider of a 10 × 40 module. © IEEE 2009, reprinted from [15], with permission

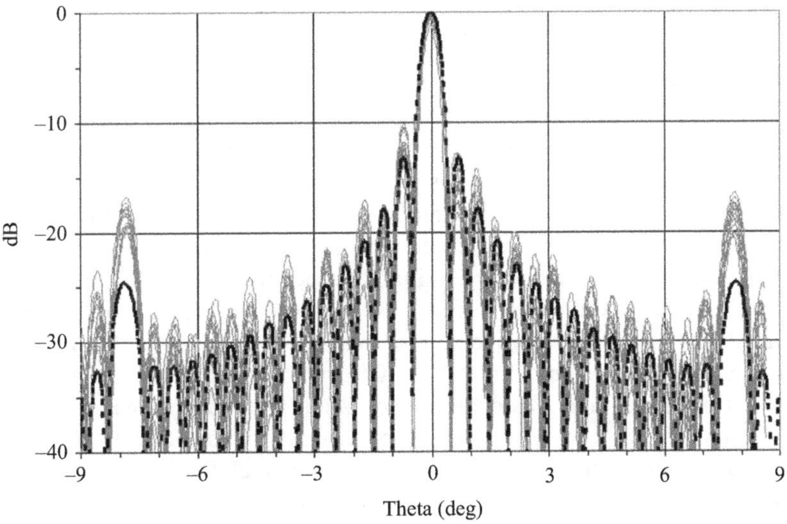

Figure 7.21 E-plane patterns of 16 sticks (light) and computed (dark) at 35.66 GHz. © IEEE 2009, reprinted from [15], with permission

different sticks are not identical because of the repeatability problem associated with the fabrication process. Grating lobes near $\pm 8^\circ$ are due to the wall thickness between sub-arrays. A key requirement of the array design and fabrication was to align the 16 beams in the *E*-plane radiated by each of the 16 sticks, within 1/10th of the beamwidth or 0.042°. This objective was met.

Table 7.7 Average sidelobe specification

Elevation angles (°)	Average SLL (dB)
20–30	–34
30–40	–36
40–70	–38
70–90	–39
90–150	–40
150–180	–30

7.6.2 An 8 × 8 slot array for radiometer application

Figure 7.13 shows an 8 × 8 array employed in a radiometer. Three such antennas, all 8 × 8 arrays operating at S, C, and X bands were developed, operating at 2.6, 5.2 and 10 GHz, respectively. The three antennas had the same specifications and similar designs. Waveguide wall thickness for the three frequencies were 0.762, 1.016 and 1.27 mm and machining tolerances were 0.127, 0.0508 and 0.0254 mm, respectively. The 3-dB beamwidth of the antenna had to be <13°. The sidelobe levels averaged over all azimuthal values is specified in different elevation angular regions in Table 7.7. The pattern performance had to be satisfied over 4% bandwidth. The return loss had to be no less than 10 dB with the average value better than 15 dB in the 4% frequency band.

A separable 35 dB Taylor $\bar{n} = 4$ aperture distribution sampled from the continuous distribution was employed in both principal planes of the 8 × 8 array. The amplitude taper in each direction was −13.86 dB with −27.7 dB dynamic range. The 64-element slot array is divided into two sub-arrays of 8 × 4 elements with two feed waveguides. Each feed waveguide consists of eight coupling slots and eight radiating waveguides with four radiating slots per radiating waveguide. Elliott's design procedure described in Section 7.2 was used to design both sub-arrays, sequentially in the iterative process. The two sub-arrays couple through external mutual coupling between radiating slots. The total normalised slot conductance in each radiating waveguide was chosen to be 2 and the total normalised slot resistance in each feed waveguide was set to 1.5. A smaller value of the total normalised slot resistance has been chosen to achieve a good sidelobe performance over the bandwidth. A shunt-series coupling slot is used in the input waveguide to excite each feed waveguide with the specified return loss performance.

The conventional array architecture shown in Figure 7.22 introduced grating lobes called butterfly lobes at about $\theta = 80°$ in the H-plane. Such grating lobes have been treated widely in the literature [18,19] and also discussed in Section 9.9 in this book. The butterfly lobes may be eliminated in slot arrays with centred longitudinal slots excited by compound irises [20] or by alternating asymmetric ridges in the waveguide [21]. Iris excited slots will be discussed in Chapter 8. In the array configuration shown in Figure 7.23 the butterfly lobes are reduced by designing a symmetric arrangement of slots with respect to a vertical centre line.

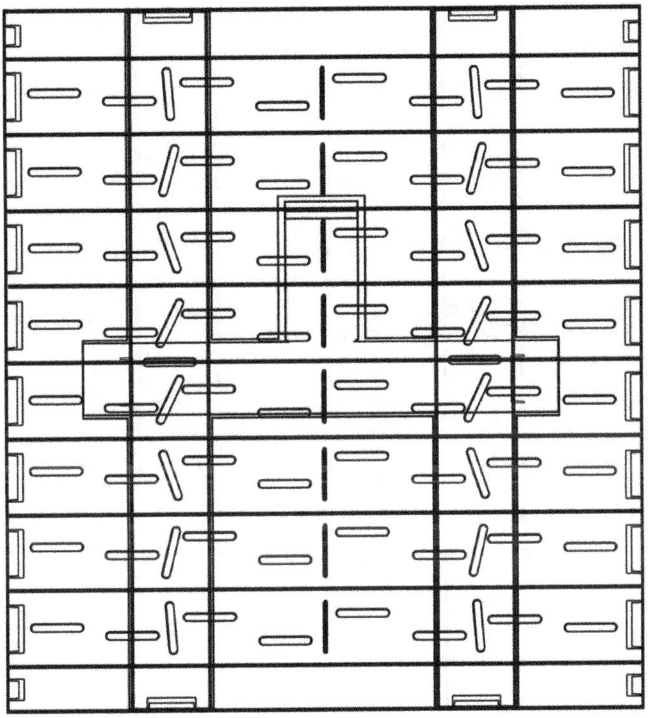

Figure 7.22 Conventional array architecture with alternating slot offsets. © IEEE 2010, reprinted from [16], with permission

Uniform phase distribution of slot apertures was achieved with 180° phase difference between the excitations of the two feed waveguides as shown in Figure 7.23. Tilt angles of coupling slots in the two sub-arrays are in the same direction in Figure 7.22 whereas they are in opposite directions in Figure 7.23. Table 7.8 shows the sidelobe levels achieved for the two arrays shown in Figures 7.22 and 7.23. The symmetric design yields lower sidelobes over the angular ranges beyond 30° whereas it increases the sidelobe level slightly for angular region below 30°.

Far-field patterns of the S-band antenna, both computed by the finite element technique HFSS and measured using the spherical near-field technique are found to have very good agreement over nearly 60 dB dynamic range [16]. They are reproduced in Figures 7.24 and 7.25. The spherical far field pattern is mapped onto a circle, with θ varying from 0° to 180° radially and ϕ varying from 0° to 360° azimuthally. $\theta = 180°$ point in the far field sphere is stretched into the circumference. Average sidelobe levels in different angular ranges obtained from computed and measured data are compared in Table 7.9. Excellent agreement between two sets of data is noted and the design margin is found to be more than adequate.

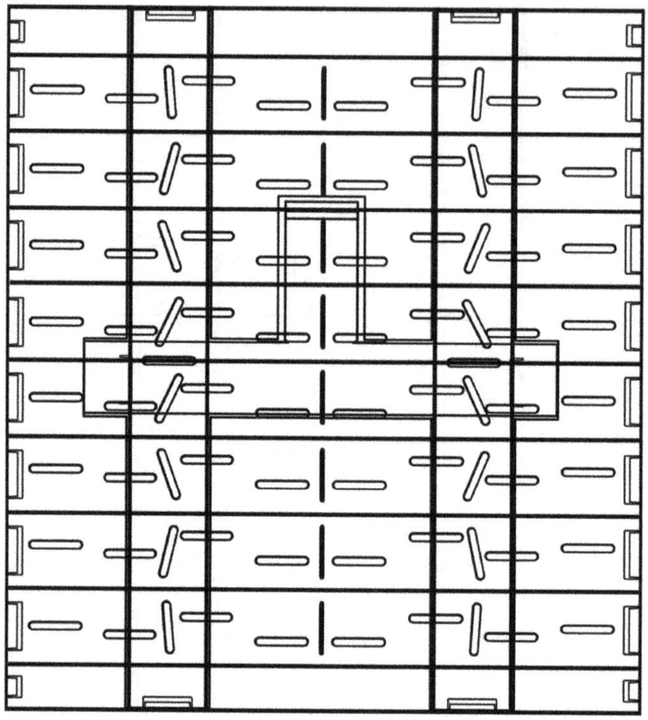

Figure 7.23 *Symmetric array architecture. © IEEE 2010, reprinted from [16], with permission*

Table 7.8 *Average sidelobe for the X-band antennas (© IEEE 2010, reprinted from [16], with permission)*

Elevation angles (°)	Average SLL (dB) (Figure 7.22)	Average SLL (dB) (Figure 7.23)
20–30	−37.98	−36.06
30–40	−37.60	−39.16
40–70	−40.56	−44.07
70–90	−41.36	−45.70

7.7 Design of a travelling wave feed for a planar array

A planar array of slots can be designed to have a travelling wave feed [22] in one or both principal directions. Figure 7.26 shows a travelling wave feed in which centred-inclined coupling slots excite 80 radiating waveguides, each containing 8 radiating slots. The feed waveguide is match terminated at the far end. The radiating waveguides consist of longitudinal radiating slots in the standing wave

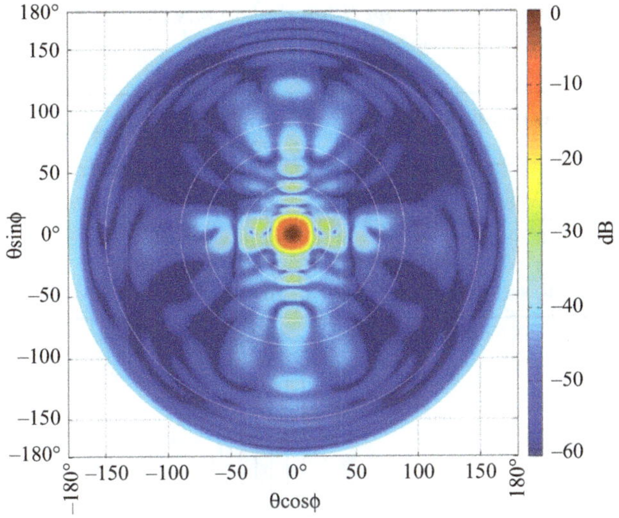

Figure 7.24 Far-field pattern of the S-band antenna computed by HFSS at 2.6 GHz. © IEEE 2010, reprinted from [16], with permission

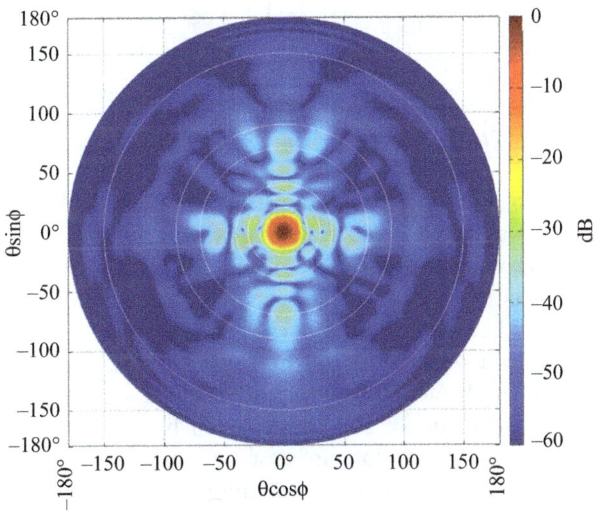

Figure 7.25 Far-field pattern of the S-band antenna measured at 2.6 GHz. © IEEE 2010, reprinted from [16], with permission

configuration. Coupling slots in the feed waveguide may be designed independent of the radiating slots if we assume a certain amount of total active conductance, G_{tot}, in each radiating waveguide at the design frequency. For the specified value of G_{tot}, for example, $G_{tot} = 2$, each coupling slot may be considered as a four-port

Table 7.9 Average sidelobe for the S-band antennas (© IEEE 2010, reprinted from [16] with permission)

Elevation angles (°)	Specification (dB)	Average SLL in dB computed by HFSS	Average SLL in dB measured
20–30	−34	−36.7	−36.4
30–40	−36	−38.4	−38.3
40–70	−38	−43.5	−43.0
70–90	−39	−46.0	−45.8
90–150	−40	−50.8	−52.1

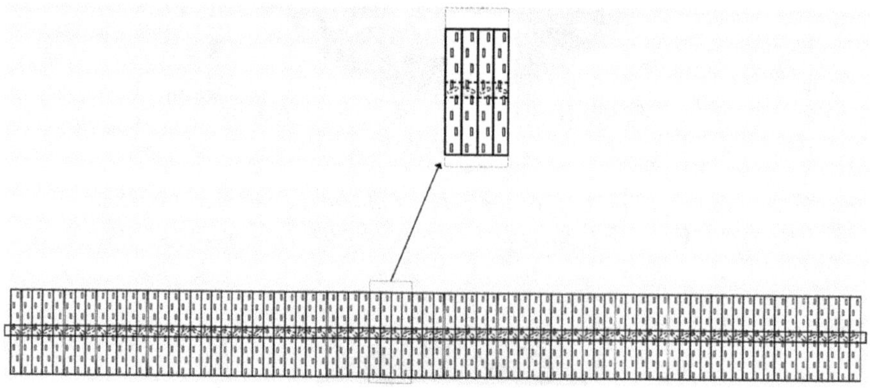

Figure 7.26 A planar array of longitudinal slots with a travelling wave feed. © IEEE 2011, reprinted from [22], with permission

network with the two ports in the radiating waveguides match-terminated. If we have a centred feed waveguide with travelling wave type of radiating slots in the forward-fire backward-fire arrangement (discussed in Chapter 9) and if each radiating waveguide is properly matched, each four-port coupling slot sees match terminations in the radiating waveguide. Therefore one needs to design two-port coupling slots in the feed waveguide for a match at the feed port and to achieve the desired excitations at each radiating waveguide. It is assumed that the excitation coefficients are pure real but for the phase added by the propagating travelling wave and therefore all coupling slots are assumed to be at resonance. The procedure given here for the centred-inclined coupling slots can be modified for other types of coupling slots such as shunt-series coupling slots that are offset longitudinal in the feed waveguide and transverse to the radiating waveguide as well as compound coupling slots that are offset and inclined in the feed waveguides.

7.7.1 Design procedure for a travelling wave feed in terms of scattering wave representation of a coupling slot

A centred-inclined slot coupler at resonance can be represented by a four-port coupler where all ports are planes passing through the centre of the slot, two in the

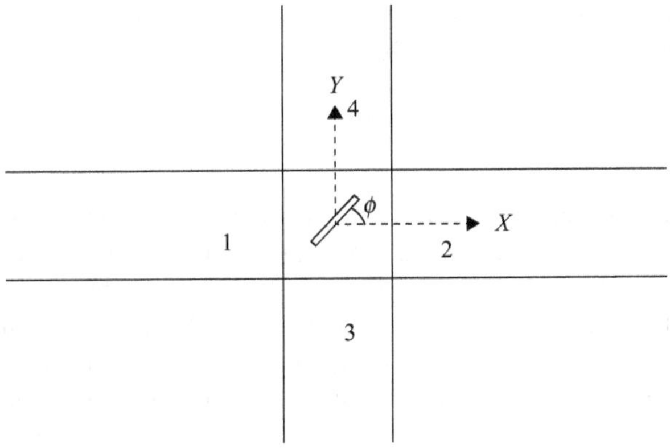

Figure 7.27 Slot coupler as a four-port device

feed waveguide and two in the radiating waveguide. The scattering matrix of such a coupler at resonance (see Figure 7.27) is given by [23]

$$[S] = \begin{bmatrix} S_{11} & S_{12} & S_{13} & S_{14} \\ S_{21} & S_{22} & S_{23} & S_{24} \\ S_{31} & S_{32} & S_{33} & S_{34} \\ S_{41} & S_{42} & S_{43} & S_{44} \end{bmatrix} = \begin{bmatrix} r & p & q & -q \\ p & r & -q & q \\ q & -q & p & r \\ -q & q & r & p \end{bmatrix} \tag{7.23}$$

where $p = (1 - r)$ and $q = [r(1 - r)]^{1/2}$. If the S-matrix for a positive value of tilt angle ϕ is $S(r,p,q)$, then for a negative value of ϕ the S-matrix becomes $S(r,p,-q)$. The tilt angle ϕ is shown in Figure 7.27 with ports 1 and 2 in the feed waveguide and 3 and 4 in the radiating waveguide. If ports 3 and 4 are terminated in reflection coefficients $\Gamma_3 = \Gamma_4 = \Gamma$ (=0 if ports 3 and 4 are matched), the coupler may be represented as a two-port network.

$$[S_2] = \begin{bmatrix} r' & 1 - r' \\ 1 - r' & r' \end{bmatrix} \tag{7.24}$$

where

$$r' = r + 2q^2[1/\Gamma - 1 + 2r]^{-1} \tag{7.25}$$

The scattering matrix (7.24) may be expressed in terms of a transmission matrix, relating the incident and reflected waves, a_{1n}, b_{1n} at the input port to the corresponding waves, a_{2n}, b_{2n} at the output port of the nth slot.

$$\begin{bmatrix} a_{1n} \\ b_{1n} \end{bmatrix} = [T_n] \begin{bmatrix} a_{2n} \\ b_{2n} \end{bmatrix} \tag{7.26}$$

where

$$[T_n] = \frac{1}{1 - r'} \begin{bmatrix} -r' & 1 \\ 1 - 2r' & r' \end{bmatrix} \tag{7.27}$$

The transmission matrix of a waveguide of length d equal to the spacing between adjacent coupling slots is given by

$$[T_w] = \begin{bmatrix} 0 & \exp(j\beta_{10}d) \\ \exp(-j\beta_{10}d) & 0 \end{bmatrix} \tag{7.28}$$

where β_{10} is the phase constant of the TE_{10} mode in the waveguide. The slot spacing d is chosen to produce the main beam at an angle, typically two to three beamwidths, away from broadside. The aperture distribution is chosen to produce the desired pattern shape. The parameter r' of the Nth coupling slot closest to the matched load is assigned an arbitrary value. This value is not critical since it will be changed in subsequent iterations. A larger value of r' will reduce the power dissipated in the load while increasing the reflection coefficient at the input port whereas a smaller value will produce the opposite effect. The complex phasor representing the wave going towards the load, b_{2N} is initially set at a reference value of unity. The complex phasor $a_{2N} = 0$ since there is no reflection from the matched load. The excitation of the radiating waveguide N is determined readily. From the transmission matrices $[T_w]$ and $[T_n]$ of slot N, and a_{2N} and b_{2N}, we can determine the complex phasors a_{1N} and b_{1N}, and hence a_{2N-1} and b_{2N-1}. It is then possible to choose the value r' of slot $N-1$ so that the radiating waveguide $N-1$ has the desired amplitude of excitation. We then determine the complex phasors a_{1N-1} and b_{1N-1}, and hence a_{2N-2} and b_{2N-2}. One proceeds this way towards the source end, until all the coupling slot parameters r' and hence the scattering parameter of each four-port coupler, r are known. From the values of complex phasors a_{11} and b_{11} of slot 1, the input reflection coefficient is determined. The power dissipated in the load relative to the power incident in the input port is $1/|a_{11}^2|$. The procedure is somewhat similar to the design of travelling wave arrays discussed in Chapter 6.

Subsequent iterations with greater or smaller values of r' of slot N will eventually yield a desirable solution with an optimum value of load power and input reflection coefficient. From an interpolation of the computed moment method data for the slot parameters, all the coupling slot tilt angles and lengths are obtained. From the excitations of the radiating waveguides computed in this exercise, radiating slot parameters may be obtained so as to attain the desired total normalised active slot conductances, using Elliott's procedure discussed in Section 7.2. This process will yield the offsets and lengths of radiating slots. In this discussion, all coupling slots are assumed to be resonant and hence the phases of the radiating waveguides are not constrained. The phase distribution is nominally that of the incident TE_{10} mode waveguide with 180° phase reversals due to the alternating slot offsets. With non-resonant coupling slots, an additional degree of freedom may be gained at the cost of increased complexity of the design.

7.7.2 Computed results

Computed results for the 80 × 8 element array with a travelling wave feed are shown in Figures 7.28–7.30. The aperture distribution is assumed to be uniform in amplitude and the coupling slot spacing is chosen to be $d/\lambda_g = 0.464$ where λ_g is the guide

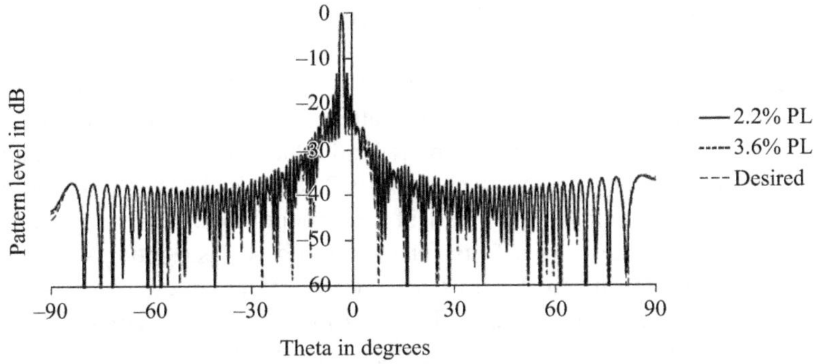

Figure 7.28 E-plane pattern of the planar array with an 80-element travelling wave array feed (PL = load power/incident power in per cent). © IEEE 2011, reprinted from [22], with permission

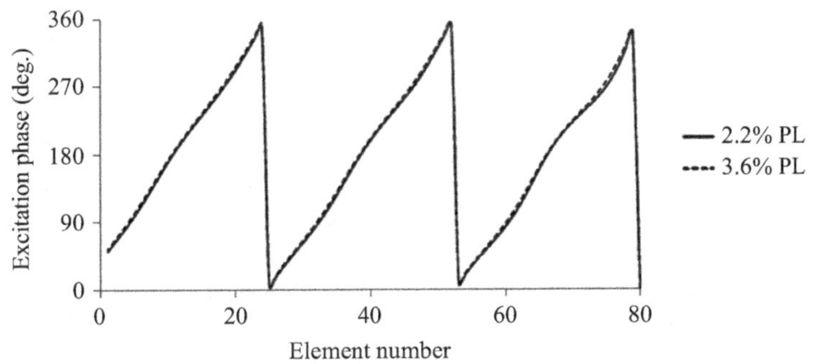

Figure 7.29 Excitation phase of the radiating waveguides (PL = relative power in the load). © IEEE 2011, reprinted from [22], with permission

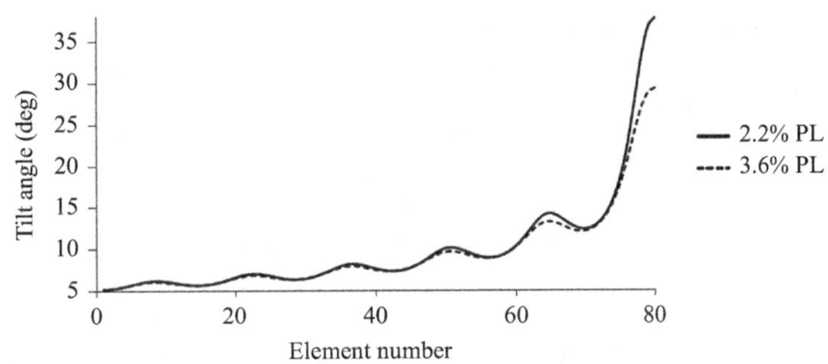

Figure 7.30 Tilt angles of coupling slots (PL = relative power in the load). © IEEE 2011, reprinted from [22], with permission

wavelength = 1.4741 λ_0 and λ_0 is the free space wavelength. In this case the main beam in the E-plane will be at $\theta = 93°$ or 3° off normal if the radiating slots in each waveguide are designed to be of standing-wave type with uniform phase aperture distribution. Figure 7.28 shows the E-plane pattern of the array shown in Figure 7.26 for two designs. The power dissipated in the load relative to that at the input port for the two cases are 2.2% and 3.6%, respectively. The desired case shows the pattern with ideal phase distribution of the incident TE_{10} mode wave but for the 180° phase reversal due to alternating slot offsets. The actual phase distribution deviates from the ideal linear phase due to the presence of reflections at each slot. The phase distributions for the two designs are illustrated in Figure 7.29. The effect of such phase deviations on the patterns is not significant. Figure 7.30 shows the tilt angles of coupling slots. Clearly the elements close to the feed have small amounts of coupling and hence relatively small values of tilt angles as discussed in Chapter 6. Slots with small values of tilt are sensitive to manufacturing tolerances. In the design of shaped patterns asymmetric amplitude distributions with larger amplitudes near the feed end are preferred to reduce the effects of manufacturing tolerances.

7.8 A review of other methods for analysis and design in the literature

In Chapters 6 and 7 we have primarily used Elliott's technique and some variations for the design of linear and planar arrays. For large arrays, the infinite array mutual coupling model is an efficient alternative to the original Elliott procedure. In this section we first present other variations of Elliott's design procedure found in the literature. In addition, some of the analysis techniques are listed. Subsequently some generalised scattering wave models used in the design and analysis will be discussed.

In the design examples we used the moment method to obtain the resonant length and scattering parameters of radiating and coupling slots. Other computational electromagnetics codes have been used to obtain slot data. Brown employed Elliott's design procedure with the scattering characteristics of isolated slots obtained from the commercial finite element analysis code HFSS [24,25]. The longitudinal radiating slot is generally modelled as a shunt element with a conductance in parallel with a susceptance.

Solbach [26] recommended a model for the longitudinal radiating slot in the form of a shunt element consisting of a resistance R in series with an open circuited stub of characteristic impedance Z_0. At the resonant frequency of the slot, R is the reciprocal of the resonant slot conductance and the stub length is a quarter wavelength. At other frequencies the stub length is inversely proportional to the wavelength. We examined this model for slots in standard height waveguides and half height waveguides in a range of ±6% over the resonant frequency. In that frequency band R varied about ±2.5% while the characteristic impedance of the stub, Z_0 varied about ±10%. There was no particular advantage of this model over Lagrangian interpolation of S parameter data and resonant lengths of slots.

Taddei *et al.* employ Elliott's design and moment method analysis with a graphical user interface [27]. Gatti *et al.* [28,29] extended Elliott's design method to complex excitations by making the active admittances complex instead of pure real values. Since resonant slots have limited phase control, the phase value is constrained to within $\pm 60^\circ$ from the resonant values. Infinite array mutual coupling model has been found to be applicable in a number of designs [30–32].

Gulick and Elliott have used the moment method solution of the integral equations for the aperture electric fields of all slots for the design of a small planar array [33]. Even for a moderate sized array the computational burden of this method will make it impractical. One may start with an initial design using Elliott's procedure and then improve that by using the moment method analysis of the slot array [34]. In Chapter 8 we will discuss the use of moment method analysis in the optimisation of a slot array initially designed by Elliott's method.

Elliott's design equations were cast in the form of an analysis for a linear array by Hamadallah [35]. His analysis employed an iterative procedure as well as a direct matrix solution. Coetzee *et al.* extended the iterative analysis to a planar array [36] whereas Casula and Mazzarella used the direct solution to a planar array by adding the coupling slot model [37]. Although the above-mentioned analysis techniques may also be used for design optimisation, we recommend two accurate analysis methods based on generalised scattering matrices, discussed in Section 7.9, with a potential for design and optimisation discussed at the end of this section.

Manuilov *et al.* [38] solve the integral equations for the slot aperture field using the Galerkin's method with spectral domain Green's functions and Chebyshev polynomials of the first and second kind for basis and test functions. Fast and accurate full wave analysis of large slot arrays is accomplished by enforcing the edge conditions. Liu and Long analyse a unit cell of an infinite uniformly slit waveguide, or periodically cut longitudinal slots or crossed slots [39]. The former supports a leaky wave mode as well as a slot mode (the latter has a propagation constant close to free space with a near end-fire beam contaminating the leaky wave radiation) while the periodic longitudinal slot array supports only the leaky wave mode. The crossed slot array supports both a leaky wave mode and a surface wave mode. RWG basis functions are used in the slot apertures [40]. The commercial finite element code HFSS has been found to provide very accurate results for the radiation pattern and the input impedance of slot arrays for radiometer applications, requiring low sidelobes [15]. While such general purpose codes are ideal for the analysis, they are not so convenient for design and optimisation. A genetic algorithm (GA) optimisation using the moment method analysis was presented in [41]. For that technique the initial design was carried out by Elliott's method and subsequently optimised. This method will be discussed in detail in Chapter 8.

7.9 The generalised scattering matrix model for analysis and design

Chiarandini and Morini have presented a generalised scattering matrix model for the analysis and design of slot arrays [42]. In this model, a planar array is modelled in the form of three blocks as shown in Figure 7.31. B1 represents the main line with four ports. It is connected to block B2 representing the branch waveguides through

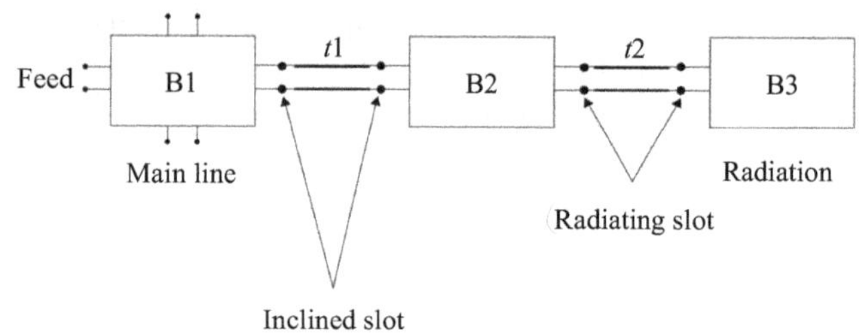

Figure 7.31 A planar array of slots modelled by three blocks. © IEEE 2001, reprinted from [42], with permission

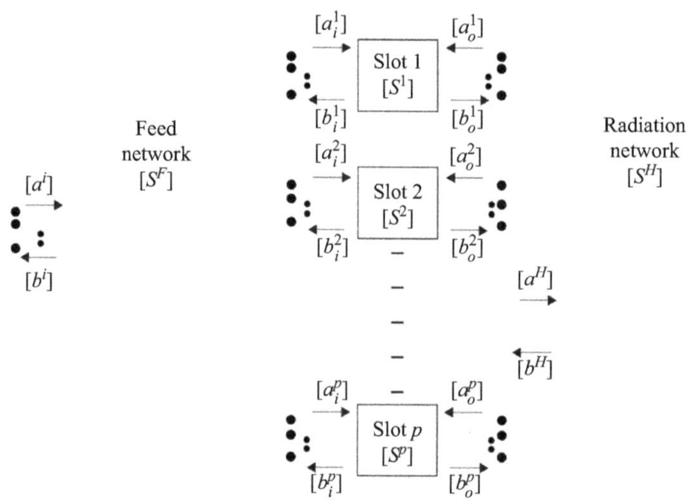

Figure 7.32 A planar array of slots modelled by generalised scattering matrices (GSM). © IEEE 2000, reprinted from [43], with permission

an inclined coupling slot of a certain thickness $t1$. The coupling slot is treated as a waveguide of length $t1$. Block B3 represents the exterior half space region. It is connected to each branch line or radiating waveguide through radiating slots of thickness $t2$. Each block is characterised by generalised scattering matrices determined by a full wave method such as the commercial code HFSS. The interconnection between blocks by slots or waveguides is modelled in terms of a few modes. Thus the generalised scattering matrix can be used for analysis as well as design and optimisation. One of the limitations of this model is that the singularity in the coupling term, if present, cannot be modelled in terms of the eigen function expansion or modes used in the generalised scattering matrix. The singularity in the Green's functions in the coupling terms is present when a coupling slot is projected on the wall containing radiating slot overlaps with the radiating slot [9]. This is discussed in detail in Chapter 8.

Figure 7.32 shows the generalised scattering matrix (GSM) model for a planar slot array developed by Enneking *et al.* [43]. The GSM of the feed network is

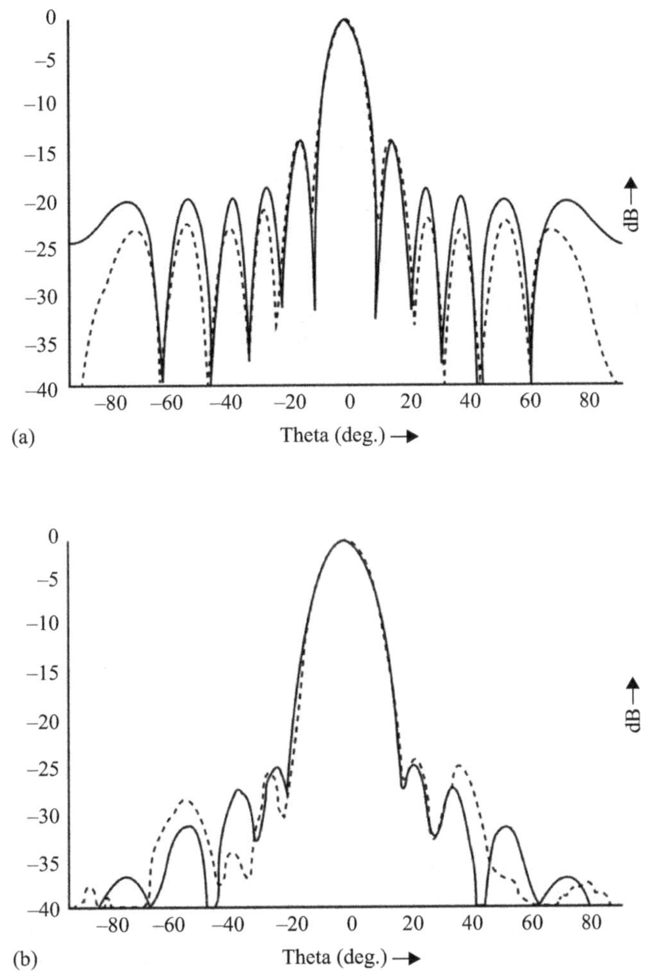

Figure 7.33 Radiation pattern of a planar array of 6 × 8 slots at 9.375 GHz computed by the GSM technique compared to measured results: (a) E-plane pattern; (b) H-plane pattern. © IEEE 2000, reprinted from [43], with permission

represented by $[S^F]$, and the GSM of the P slots are denoted by $[S^1]$ through $[S^P]$. The GSM of the external half space is $[S^H]$. The slot thickness is accounted for in the model by treating each slot as a waveguide of length equal to the slot thickness.

The GSM of the feed network is obtained by the mode matching technique. For waveguides of arbitrary cross sections the eigenvalue problem is solved by the finite element method. A scattering matrix representation is used for each slot and a scattering matrix for the external radiation network. Higher-order modes up to a certain number are accounted for. An arbitrary number of modes is used to allow coupling between modes of the same slot and modes between different slots. The exterior waveguide apertures are assumed to be located on an infinite conducting ground plane with any number of dielectric layers. The scattering matrix is computed from the admittance matrix of the slot apertures in an infinite ground plane in a manner similar to the mutual coupling calculation discussed in Section 4.5.2 in Chapter 4. However, for the GSM, each slot aperture has a number of modes, instead of the single mode used in Chapter 4. The GSM of the slots are obtained by the mode matching technique. The single GSM's are determined separately and then cascaded as shown in Figure 7.32 to obtain the overall GSM.

This analysis code is implemented in the computer programme Wasp-Net as a computer aided design and analysis package [44]. Figure 7.33 shows very good agreement between the radiation patterns computed by the GSM technique for a 6×8 array with the corresponding measured results.

References

[1] Saltzman H., Stavis G. 'A dual beam planar antenna for Janus type Doppler navigation systems'. *IRE National Convention Record.* 1958, pp. 240–247.

[2] Simmons A. J., Giddings O. M. 'A multiple-beam two-dimensional waveguide slot array'. *IRE International Convention Record.* 1963, pp. 56–69.

[3] Miller J. R., Forman R. J. 'A planar slot array with four independent beams'. *IEEE Transactions on Antennas and Propagation.* 1966;**14**(5):560–566.

[4] Elliott R., Kurtz L. 'The design of small slot arrays'. *IEEE Transactions on Antennas and Propagation.* 1978;**26**(2):214–229.

[5] Lo Y. T., Lee S. W. (eds.). *Antenna Handbook.* Van Nostrand Reinhold, New York, 1988, p. 12–1.

[6] Elliott R. S., O'Loughlin W. R. 'The design of slot arrays including internal mutual coupling'. *IEEE Transactions on Antennas and Propagation.* 1986; **34**(9):1149–1154.

[7] Derneryd A., Petersson R. 'Bandwidth characteristics of monopulse slotted waveguide antennas'. *Proceedings of the Fourth International Conference on Antennas and Propagation*, 1985, pp. 27–30.

[8] Press W. H., Teukolsky S. A., Vetterling W. T., Flannery B. P. *Numerical Recipes: The Art of Scientific Computing*, 3rd edition. Cambridge University Press, Cambridge, 2007.

[9] Rengarajan S. R., Shaw G. M. 'Accurate characterization of coupling junctions in waveguide-fed planar slot arrays'. *IEEE Transactions on Microwave Theory and Techniques.* 1994;**42**(12):2239–2248.

[10] Pozar D. M., Targonski S. D., Syrigos H. D. 'Design of millimeter wave microstrip reflectarrays'. *IEEE Transactions on Antennas and Propagation.* 1997;**45**(2):287–296.

[11] Yee H. Y. 'The design of a large waveguide arrays of shunt slots'. *IEEE Transactions on Antennas and Propagation.* 1992;**40**(7):775–781.

[12] Scharstein R. W. 'Mutual coupling in a slotted phased array, infinite *E*-plane and finite in *H*-plane'. *IEEE Transactions on Antennas and Propagation.* 1990;**38**(8):1186–1191.

[13] Mailloux R. J. *Phased Array Antenna Handbook.* Artech House, Boston, 1994.

[14] Pease M. C. 'The iterated network and its applications to differentiators'. *Proceedings of the IRE.* 1952;**40**(6):709–711.

[15] Rengarajan S. R., Zawadzki M. S., Hodges R. E. 'Design, analysis, and development of a large Ka-band slot array for digital beam-forming application'. *IEEE Transactions on Antennas and Propagation.* 2009;**57**(10): 3103–3109.

[16] Rengarajan S. R., Zawadzki M. S., Hodges R. E. 'Waveguide-slot array antenna designs for low-average-sidelobe specifications'. *IEEE Antennas and Propagation Magazine.* 2010;**52**(6):89–98.

[17] Miura Y., Hirokawa J., Ando M., Shibuya Y., Yoshida G. 'Double-layer full-corporate-feed hollow-waveguide slot array antenna in the 60-GHz band'. *IEEE Transactions on Antennas and Propagation.* 2011;**59**(8): 2844–2851.

[18] Gruenberg H. 'Second-order beams of slotted waveguide arrays'. *Canadian Journal of Physics.* 1953;**31**(1):55–69.

[19] Kurtz L. A., Yee J. S. 'Second-order beams of two-dimensional slot arrays'. *IRE Transactions on Antennas and Propagation.* 1957;**5**(4):356–362.

[20] Tang R. 'A slot with variable coupling and its application to a linear array'. *IRE Transactions on Antennas and Propagation.* 1960;**8**(1):97–101.

[21] Green J., Shnitkin H., Bertalan P. J. 'Asymmetric ridge waveguide ridge waveguide radiating element for a scanned planar array'. *IEEE Transactions on Antennas and Propagation.* 1990;**38**(8):1161–1165.

[22] Rengarajan S. R. 'Theory of a traveling wave feed for a planar slot array antenna'. *Proceedings of the XXXth URSI General Assembly and Scientific Symposium.* 2011, Istanbul, Turkey, pp. 1–4.

[23] Rengarajan S. R. 'Analysis of a centered-inclined waveguide slot coupler'. *IEEE Transactions on Microwave Theory and Techniques.* 1989;**37**(5): 884–889.

[24] Brown K. W. 'Design of waveguide slotted arrays using commercially available finite element analysis software'. *Proceedings of the IEEE International Antennas and Propagation Symposium*, 1996, pp. 1000–1003.

[25] http://www.ansys.com. Retrieved March 2018.

[26] Schrank H., Solbach K. 'Simulation of slot array antenna using standard network analysis program'. *IEEE Antennas and Propagation Magazine.* 1991;**33**(6):45–47.

[27] Taddei R., Guarnieri G., Mauriello G. 'Automated waveguide array-integrated design environment (AWA-IDE): An automated and modular tool for planar waveguide array synthesis and analysis'. *IEEE Antennas and Propagation Magazine.* 2013;**55**(4):204–216.

[28] Gatti R. V., Sorrentino R. 'Slotted waveguide antennas with arbitrary radiation pattern'. *Proceedings of the 34th European Microwave Conference*, Amsterdam, The Netherlands, 2004, pp. 821–824.

[29] http://www.swan-soft.com. Retrieved March 2018.

[30] Ando M., Zhang M., Lee J., Hirokawa J. 'Design and fabrication of millimeter wave slotted waveguide arrays'. *Proceedings European Conference on Antennas and Propagation*, 2010, pp. 1–6.

[31] Volakis J. (ed.). *Antenna Engineering Handbook*. McGraw-Hill, New York, 2007. p. 9–1.

[32] Enjiu R. K., Protoni M. B., Andre S. 'Slotted waveguide antenna design using 3D EM simulation'. *Microwave Journal*. 2013;**56**(7):72–83.

[33] Gulick J. J., Elliott R. S. 'A general design procedure for small slot arrays'. *Proceedings of the IEEE International Antennas and Propagation Symposium*, 1987, pp. 302–305.

[34] Rengarajan S. R. 'Improved design procedure for waveguide-fed slot arrays using the method-of-moments analysis'. *Electromagnetics*. 2012;**32**(4): 221–232.

[35] Hamadallah M. 'Frequency limitations on broad-band performance of shunt slot arrays'. *IEEE Transactions on Antennas and Propagation*. 1989;**37**(7): 817–823.

[36] Coetzee J. C., Joubert J., McNamara D. 'Off-center-frequency analysis of a complete planar slotted-waveguide array consisting of sub-arrays'. *IEEE Transactions on Antennas and Propagation*. 2000;**48**(11):1746–1755.

[37] Casula G. A., Mazzarella G. 'A direct computation of the frequency response of planar waveguide slot arrays'. *IEEE Transactions on Antennas and Propagation*. 2004;**52**(7):1909–1912.

[38] Manuilov M., Lerer V., Sinyavsky G. 'Fast and accurate full-wave analysis of large slotted waveguide array antennas'. *Proceedings of the 37th European Microwave Conference*, 2007, pp. 360–363.

[39] Liu J., Long Y. 'A full-wave numerical approach for analyzing rectangular waveguides with periodic slots'. *IEEE Transactions on Antennas and Propagation*. 2012;**60**(8):3754–3762.

[40] Rao S., Wilton D., Glisson A., 'Electromagnetic scattering by surfaces of arbitrary shape'. *IEEE Transactions on Antennas and Propagation*. 1982; **30**(3):409–418.

[41] Rengarajan S. R. 'Genetic algorithm optimization of a planar array using full wave method of moments analysis'. *International Journal of Computer Aided RF and Microwave Engineering*. 2013;**23**(4):430–436.

[42] Chiarandini S., Morini A. 'Full wave CAD oriented technique for the analysis of airborne flat plate arrays'. *Proceedings of the IEEE International Antennas and Propagation Symposium*, 2001, pp. 210–213.

[43] Enneking A., Beyer R., Arndt F. 'Rigorous analysis of large finite waveguide-fed slot arrays including the mutual internal and external higher order mode coupling'. *Proceedings of the IEEE International Antennas and Propagation Symposium*, 2000, pp. 74–77.

[44] http://www.wasp-net.com. Retrieved March 2018.

Chapter 8
Concepts and models for advanced designs

In this chapter we will discuss a number of concepts and models used in advanced design and optimisation of slot arrays. Models for coupling slots will be presented followed by a discussion of the edge wall slot and the compound radiating slot. Iris-excited longitudinal slot arrays, slot arrays in ridge waveguides and slot arrays covered by a dielectric layer will also be discussed. Higher-order mode coupling between adjacent coupling slots and that between a coupling slot and radiating slots in its immediate vicinity will be presented. A method of incorporating the finite ground plane effects in the design and analysis will be described. The MoM solution to the coupled integral equations for the apertures of all slots in a planar array will be discussed. Some examples employing the moment method solutions of slot arrays for improved designs will be presented.

8.1 Coupling slot models

Coupling slots are used to distribute power from a main waveguide to branch waveguides or radiating waveguides of a planar array as shown in Figure 8.1. The main waveguide or the feed waveguide is usually orthogonal to a number of stacked radiating waveguides consisting of radiating slots. Some of the commonly employed coupling slots are the centred-inclined coupling slot and the longitudinal-transverse coupling slot, discussed in Sections 8.1.1 and 8.1.2, respectively. Each of the above-mentioned coupling slots is cut in the common broad wall of two orthogonal waveguides. The centred-inclined coupling slot is centred in the common broad wall and tilted with respect the broad wall centreline. It is also referred to as a series–series slot since it may be represented as a series impedance in both the main waveguide as well as the branch waveguide. The second type of coupling slot is longitudinal with an offset from the centreline in the main waveguide and transverse and centred in the branch waveguide's broad wall. This element is called a shunt-series slot since its representation in the main and branch waveguides are shunt admittance and series impedance, respectively. In multi-layer waveguide power dividers one may use a coupling slot, discussed in Section 8.1.3, that is transverse to the common broad wall of two parallel rectangular waveguides

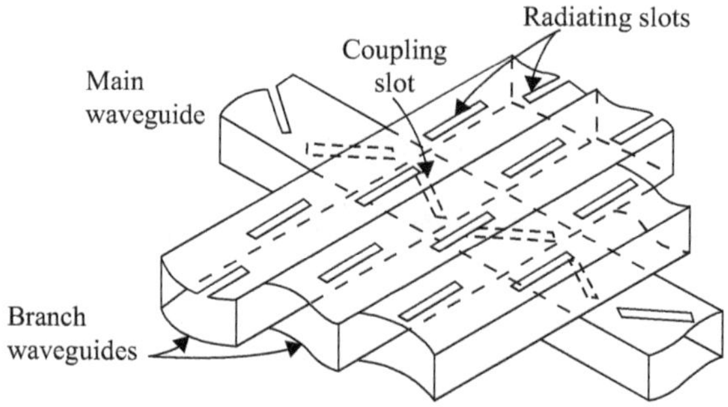

Figure 8.1 A planar array of radiating slots and coupling slots. © IEEE 1994, reprinted from [5], with permission

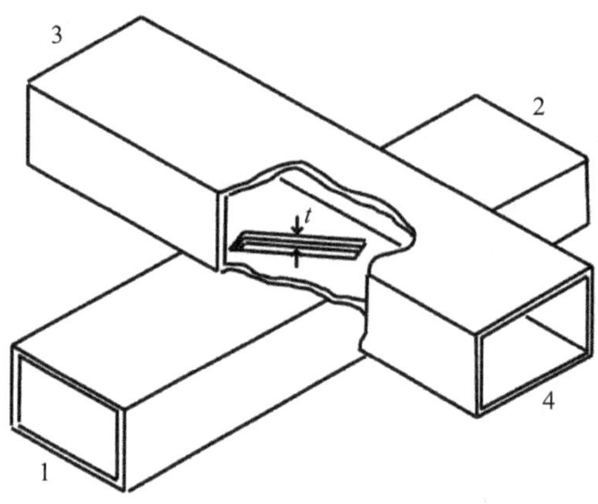

Figure 8.2 A centred-inclined slot coupler in the common broad wall of two orthogonal waveguides, with ports 1 and 2 in the main waveguide and ports 3 and 4 in the branch waveguide. © IEEE 1989, reprinted from [1], with permission

with one on top of the other. Such a slot transmits the entire power from a waveguide in one layer to the next layer. In that case the coupling slot acts as a one-to-one transformer.

8.1.1 Centred-inclined coupling slot

Centred-inclined coupling slots are commonly used to couple power from a feed waveguide into a number of radiating waveguides that are stacked orthogonally to the feed waveguide. Such a coupler may be characterised as a four-port device as shown in Figure 8.2. The coordinate system in the upper broad wall of the main waveguide is shown in Figure 8.3.

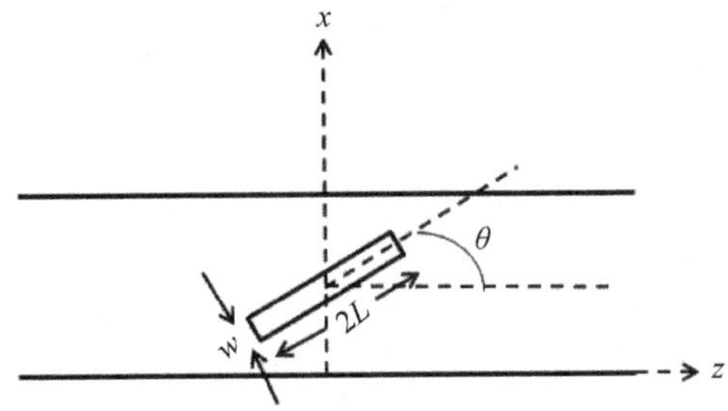

Figure 8.3 Slot coordinates at the upper broad wall of the main waveguide with ports 1 and 2. © IEEE 1989, reprinted from [1], with permission

Assuming that the reference planes of all four ports pass through the centre of the slot, a coupling slot at resonance may be characterised by a scattering matrix with only one unknown, that is, S_{11} as given in (8.1) [1].

$$S(\theta) = f(r, p, q) = \begin{bmatrix} r & p & q & -q \\ p & r & -q & q \\ q & -q & p & r \\ -q & q & r & p \end{bmatrix} \tag{8.1}$$

where $r = S_{11}(\theta)$, $p = 1 - r$ and $q = [r(1 - r)]^{1/2}$.

From symmetry considerations we can obtain the scattering matrix for complementary tilt angles and negative tilt angles as given by (8.2) and (8.3) [1].

$$S(-\theta) = f(r, p, -q) \tag{8.2}$$

$$S(\pi/2 - \theta) = f(p, r, q) \tag{8.3}$$

Thus a θ range of 0 to $\pi/4$ radians is sufficient to characterise the coupling slot.

Equation (4.68) in Chapter 4 may be used to determine the TE_{10} mode wave scattered by a centred-inclined slot in the broad wall of a rectangular waveguide. It may be shown that the forward and backward scattered waves are equal in magnitude but 180° out of phase. A similar odd-symmetric scattering behaviour is found when a wave is incident in a transmission line loaded by a series impedance. Therefore a centred-inclined slot in the broad wall of a rectangular waveguide may be modelled as a series impedance in an equivalent transmission line. The scattered TE_{10} mode waves in the backward and forward direction are in phase and 180° out of phase, respectively, with respect to the incident TE_{10} mode wave at resonance. Analysis of the centred-inclined slot coupler using the MoM shows that the slot may be modelled as a series impedance in the main line as well as in the branch line with excellent accuracy. If the branch line ports 3 and 4 are match-terminated, the series resistance normalised to the characteristic impedance of the TE_{10} mode wave in the main line may be expressed in terms of S_{11} as

$$\bar{R} = 2S_{11}/(1 - S_{11}) \tag{8.4}$$

The coupling coefficient used in (7.2) in Chapter 7 is one-half of the normalised resistance of a resonant coupling slot given in (8.4) and is repeated next.

$$\chi^2 = S_{11}/(1 - S_{11}) \tag{8.5}$$

This can be easily derived from the scattering matrix (8.1) if the load admittances in ports 3 and 4 are known.

If ports 3 and 4 are not matched, the series impedance seen in the main line at the location of the coupling slot may be derived easily from (8.6) using the scattering matrix (8.1) with the impedance terminations in ports 3 and 4.

$$[b] = [S][a] \tag{8.6}$$

where $[b]$ is a column matrix containing the TE_{10} mode wave amplitudes leaving the slot from each of the four ports while $[a]$ contains the corresponding quantities of the waves incident at the slot. The normalised series impedance $\bar{Z}$ in the main line is given by (8.7).

$$\bar{Z} = (\bar{Z}_3 + \bar{Z}_4)\chi^2 \tag{8.7}$$

where $\bar{Z}_3$ and $\bar{Z}_4$ are the normalised impedances seen by the slot at ports 3 and 4, respectively. The bars on top imply normalised impedances whereas the un-normalised impedances are denoted without bars. Characteristics of this type of coupler have been analysed by the moment method solutions to the coupled integral equations for the electric fields of apertures of a thick slot [1].

Resonant coupling slots are employed in planar standing wave arrays to realise the aperture distribution with uniform phase. Data on the resonant length and $|S_{11}|$ at resonance for a range of tilt angles, generated by a computational electromagnetics (CEM) code are needed in such designs. The variation of the resonant length with the tilt angle of the slot is not significant for standard height waveguides whereas for reduced height waveguides it is not negligible.

Figure 8.4 shows a circuit model for a non-resonant series–series slot [2]. The transformer ratio n, the inductance L and the capacitance C needed to provide the susceptance in the main waveguide are functions of frequency. A simplified model has been proposed by Mazzarella and Montisci [2]. In that model, the transformer

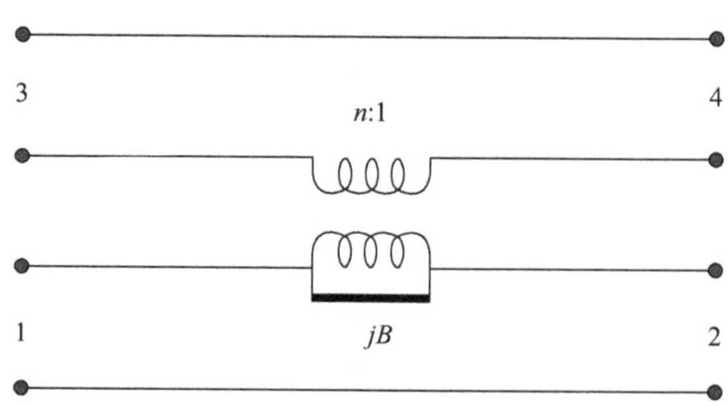

Figure 8.4 Circuit model for a non-resonant series–series coupling slot

ratio n is determined at the resonant frequency. Over a frequency range of 6.6% below the resonant frequency $L = L_1$ and $C = C_1$ were found to produce accurate values of the susceptance. Similarly $L = L_2$ and $C = C_2$ yielded accurate susceptances over a 6.6% frequency band above resonance. By choosing the values of L and C between the two sets they were able to produce an accurate model over a 10% frequency band [2]. The value of n at resonance is the coupling coefficient χ defined in (8.5).

8.1.2 Shunt-series coupling slot

Figure 8.5 shows the shunt-series coupling slot in the common broad wall of two orthogonal waveguides. The slot is longitudinal and offset from the broad wall centreline of the main waveguide with ports 1 and 2. In the branch waveguide with ports 3 and 4, the slot is centred in the transverse direction. Slot length, $2L$, width w and offset δ from the centreline are shown. Representation of the slot as a series element in the branch waveguide is excellent while the shunt representation in the main line becomes poor for large offsets and reduced height waveguides [3]. However, we refer to it as a shunt-series coupling slot based on its approximate representation. When the shunt model fails, one may work with the computed scattering parameters.

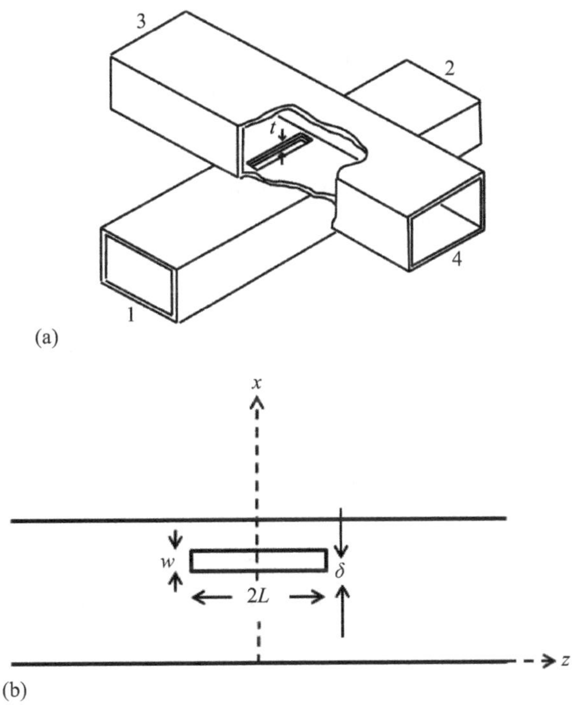

Figure 8.5 Shunt-series coupling slot.: (a) cut-away view of the slot; (b) slot parameters and localised slot coordinates. © IEEE 1989, reprinted from [3], with permission

The scattering matrix of the four-port resonant coupler with all ports defined to be planes passing through the centre of the slot is given by

$$S = \begin{bmatrix} -r & p & jq & -jq \\ p & -r & jq & -jq \\ jq & jq & p & r \\ -jq & -jq & r & p \end{bmatrix} \tag{8.8}$$

where $-r = S_{11}$. If the shunt model is acceptable $p = 1 - r$, and $q = [r(1 - r)]^{1/2}$. For a negative offset, q will be replaced by $-q$ in (8.8). A resonant shunt-series coupling slot was used (see Figure 7.12 in Chapter 7) at the centre of a feed waveguide of a planar array as a transformer to match the total resistance seen by the feed waveguide into a match at the input port [4]. In that case the normalised input admittance is given by (8.9) with the coupling coefficient determined by (8.10) using S_{11} at resonance.

$$\bar{Y} = (\bar{Z}_3 + \bar{Z}_4)\chi^2 \tag{8.9}$$

where

$$\chi^2 = -S_{11}/(1 + S_{11}) \tag{8.10}$$

For the case of a non-resonant shunt-series coupling slot, although a circuit model may be derived, it is suggested that the scattering matrix representation be used. Because of symmetry, reciprocity and power balance considerations the scattering matrix gets simplified.

The shunt-series coupling slot employed in a planar array (see Figure 8.6) has the advantage that the higher-order mode coupling between such a coupling slot and the radiating slots in its immediate vicinity is negligible whereas such coupling

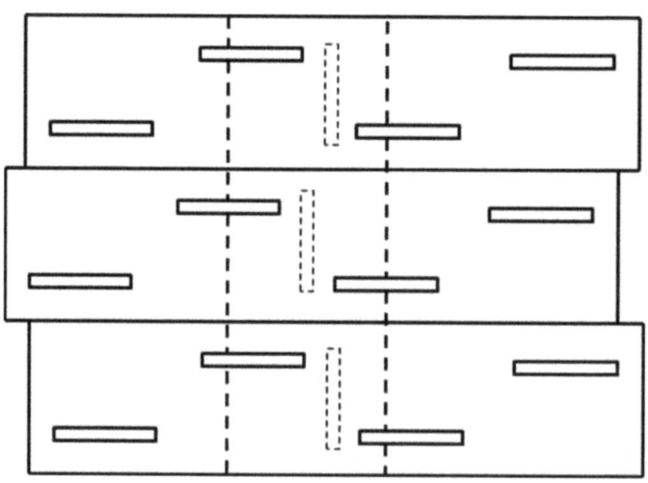

Figure 8.6 A planar slot array with shunt-series coupling slots

in arrays employing centred inclined or series–series coupling slots must be accounted for under certain conditions (see Section 8.7) [5,6].

8.1.3 Transverse/transverse coupling slot

Consider two identical rectangular waveguides sharing a common broad wall as shown in Figure 8.7(a). A centred transverse resonant slot cut in the common broad wall behaves as a series element in both waveguides [7]. If the slot is resonant at the operating frequency and it is centred in both waveguides, the scattering matrix of this four-port device may be derived from reciprocity, symmetry and power balance considerations as follows.

$$S = \begin{bmatrix} 0.5 & 0.5 & 0.5 & -0.5 \\ 0.5 & 0.5 & -0.5 & 0.5 \\ 0.5 & -0.5 & 0.5 & 0.5 \\ -0.5 & 0.5 & 0.5 & 0.5 \end{bmatrix} \tag{8.11}$$

Figure 8.7(b) shows short circuits placed in ports 2 and 3 at a distance s from the centre of the slot. The reflection coefficients at ports 2 and 3, referenced to the centre of the slot, are equal to $e^{j\psi}$, where $\psi = \pi - 2\beta_{10}s$. In this case the two-port

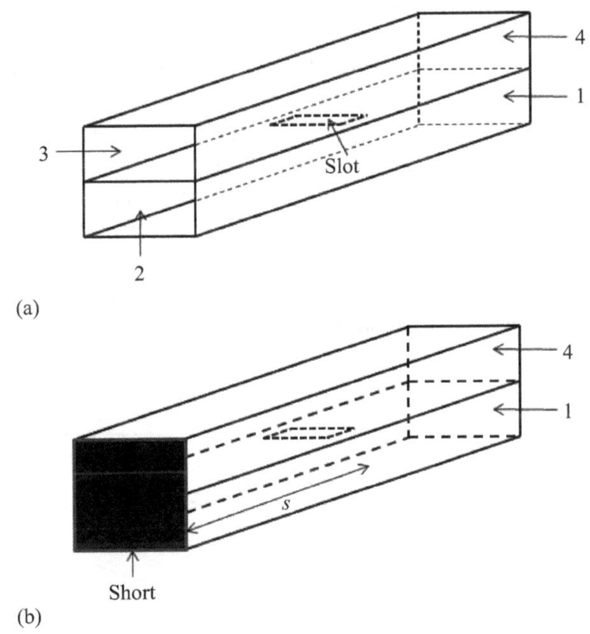

Figure 8.7 Transverse/transverse coupling slot in the common broad wall of parallel waveguides: (a) four-port coupler; (b) ports 2 and 3 shorted at a distance s from the slot centre

scattering matrix of the junction with ports 1 and 4 referenced to the centre of the slot is

$$S = 0.5\begin{bmatrix} 1+e^{j\psi} & -1+e^{j\psi} \\ -1+e^{j\psi} & 1+e^{j\psi} \end{bmatrix} \tag{8.12}$$

For the special case of $s = 0$, the slot with half the original width will be in front of the short. The previous result may be shown to be valid for this special case by considering the image of the equivalent magnetic current of the slot. In this case the two-port device has zero reflection at ports 1 and 4 and the transmission coefficient between ports 1 and 4 is unit amplitude and 180° phase. Such a device is conveniently used in multi-layer power dividers for total transmission from one layer to another and also to introduce the required amounts of phase shifts through extra lengths of waveguides between ports 1 and 4 and the slot centre [8].

8.2 Edge wall slot

Figure 8.8 shows an edge wall slot cut in a rectangular waveguide. Since wall currents in the edge wall are vertical (see Figure 1.1), in order to disturb the wall currents, and thus to radiate from the waveguide, such a slot should have an inclination with respect to the vertical. Greater values of the tilt angle with respect to the vertical will provide stronger excitations of the slot. In order to provide an equi-phase excitation from an array of such slots, a slot spacing of half a guide wavelength is used with adjacent slots having alternating inclinations of positive and negative angles with respect to the vertical. An array of edge wall slots with alternating tilts will radiate in the broadside direction with a principal polarisation along the axial direction of the waveguide (see Figure 8.9). Thus the longitudinal offset slots and edge wall slots offer two orthogonal polarisations. Another advantage of using edge wall slots is that adjacent waveguides with edge wall slots may be placed with a spacing in the order of half wavelength in free space. Therefore it is possible to scan such arrays electronically over a wide angular range using phase shifters between adjacent waveguides. In the case of longitudinal offset

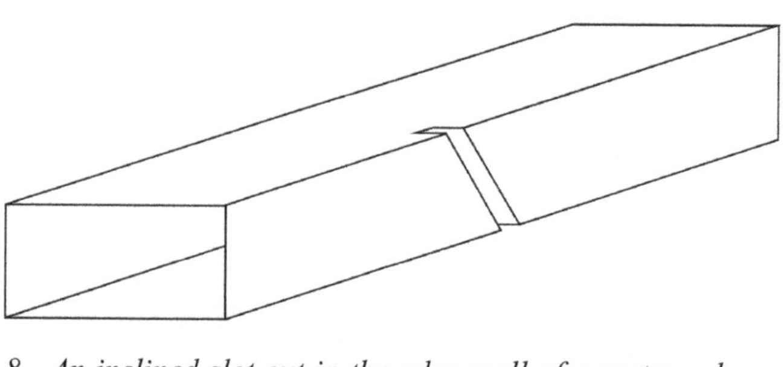

Figure 8.8 An inclined slot cut in the edge wall of a rectangular waveguide

slots in the broad walls, the scanning range is limited because of the broad wall dimension being greater than half wavelength in free space.

The edge wall slot may be modelled as a shunt element since it scatters symmetrically. Since the edge wall dimension is less than a half wavelength in free space, a slot cut in that wall may have to be wrapped around both broad walls in order to accommodate the required resonant slot length. This presents a problem in stacking waveguides one on top of another. The tilted slots introduce a certain amount of cross-polarisation. Untilted edge wall slots excited by probes, wires or irises do not radiate cross-polarisation. Figure 8.10 shows a concept for producing dual polarisation. One set of waveguides with longitudinal offset slots cut in the broad walls produce linear polarisation perpendicular to the waveguide axes. The second set of waveguides with edge wall slots will produce the orthogonal polarisation along the waveguide axes. See also Chapter 9.

In the design of edge wall slot arrays, mutual coupling can be incorporated by using an infinite array model [9,10]. It is possible to compute the mutual coupling by the so-called element by element method by considering a pair of slots at a time using the spectrum of two-dimensional solutions [11,12].

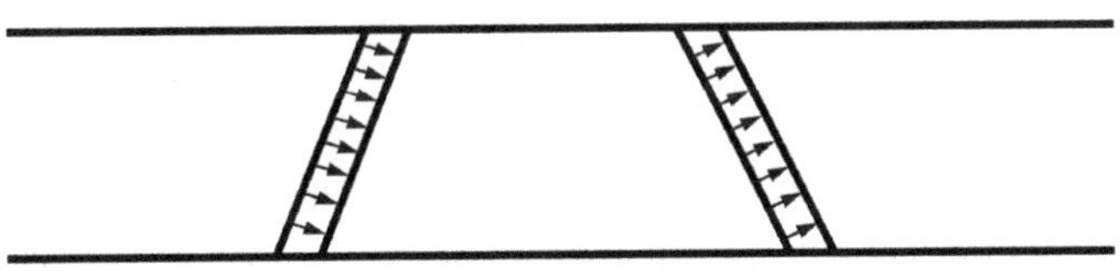

Figure 8.9 The aperture electric field in two edge wall slots with a half guide wavelength spacing

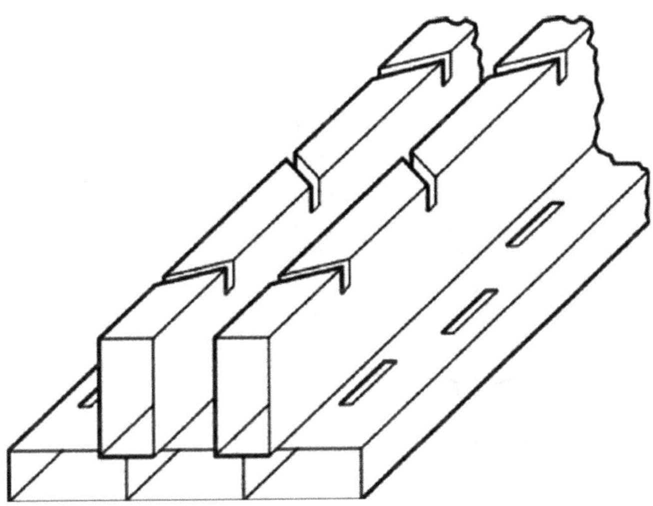

Figure 8.10 Planar array of slots for dual polarisation

8.3 Radiating compound broad wall slot

A broad wall slot offset from the centreline and tilted with respect to the waveguide axis, called the compound slot, can be designed to be resonant and still produce any excitation amplitude and phase. Figure 8.11 shows different broad wall radiating slots. The aperture electric field of a resonant longitudinal offset slot is 90° with respect to the TE_{10} waveguide mode amplitude at the centre of the slot for positive offsets whereas the phase of the aperture field becomes 270° if the offset is negative. Similarly the phase becomes 0° and 180° for a centred-inclined resonant slot for positive and negative tilts, respectively. For a resonant compound slot, the phase of the aperture electric field with respect to the incident TE_{10} mode wave may be anywhere in the range of 0°–360° depending on the sign of the offset and tilt as shown in Figure 8.12(a)–(d) [13]. Figure 8.12(e) shows the range of possible values of offsets and tilts bounded by the four curves in the four quadrants. A standard X-band waveguide with a slot length equal to one-half free space wavelength at 9.3 GHz and a slot width equal to 1/10 of the length were used to generate this figure. The horizontal axis is the tilt in degrees and the vertical is the offset normalised to $a/2$, where a is the waveguide inner width. For a longitudinal slot ($\theta = 0°$) the maximum possible offset is $\pm(a - w)/2 \approx \pm 0.92\, a/2$, where w is the slot width. For a transverse slot ($\theta = 90°$) the maximum possible offset is $\pm(a/2 - L) \approx \pm 0.289\, a/2$, where L is one-half of the slot length.

For a compound slot, resonance occurs when the forward scattered TE_{10} mode wave is 180° out of phase with the incident TE_{10} mode wave, both referenced to the centre of the slot. The back scattered TE_{10} mode wave phase for a resonant slot is twice that of the aperture electric field.

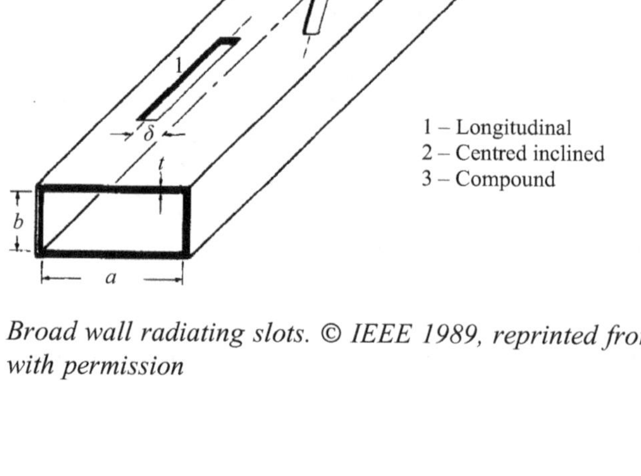

Figure 8.11 Broad wall radiating slots. © IEEE 1989, reprinted from [13], with permission

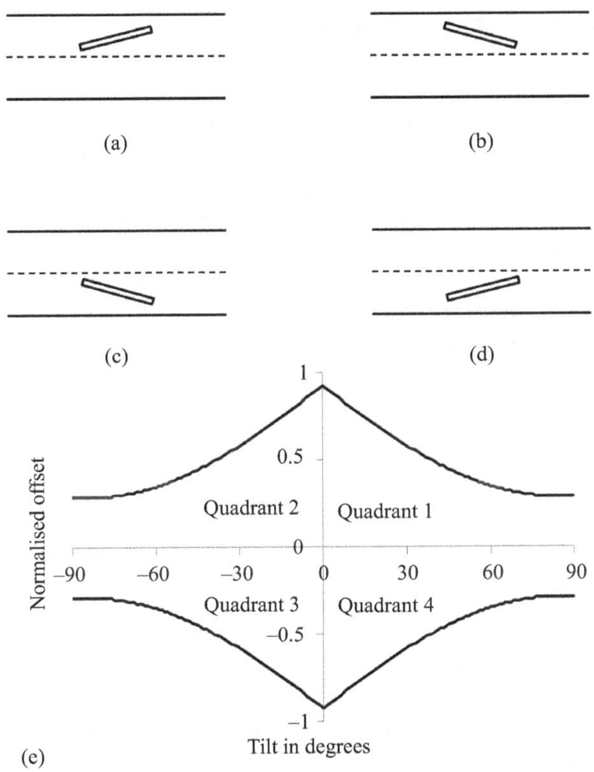

Figure 8.12 Four ranges of phase for the aperture electric field and possible range of offsets and tilts for compound slots. (a) $\delta > 0$, $0 \leq \theta \leq 90°$, $0 \leq \arg E \leq 90°$; *(b)* $\delta > 0$, $-90° \leq \theta \leq 0°$, $90° \leq \arg E \leq 180°$; *(c)* $\delta < 0$, $-90° \leq \theta \leq 0$, $180° \leq \arg E \leq 270°$; *(d)* $\delta < 0$, $0 \leq \theta \leq 90°$, $270° \leq \arg E \leq 360°$; *(e) possible range of normalised slot offsets (vertical axis) and tilts (horizontal axis) for a compound slot*

Although the compound radiating slot was introduced in the 1940s [7], it was not used in an antenna application until 1960 [14], when a linear travelling wave array of compound radiating slots was designed, and built. It is ideally suited for travelling wave arrays such as the fuze antenna of a missile system in which the beam is at an angle with respect to the missile axis. Compound slots are also useful as coupling elements. Although an isolated resonant compound slot can be designed to produce any value of the aperture amplitude and phase, the amount of realisable phase values will be limited by the reflection coefficient at the output port, with a greater range of phase values for smaller reflection coefficients. This result was shown for a compound coupling slot in [15] but it is true for compound radiating slots also. Compound slot arrays may be designed using the procedure described in Section 6.8 in Chapter 6 in terms of scattering wave parameters. An array of compound slots all with a tilt of 45° but different offsets has been used in an automotive collision avoidance radar [16].

8.4 Iris-excited longitudinal slot

Longitudinal slot arrays with alternating slot offsets introduce grating lobes called butterfly lobes [17]. These lobes may be suppressed by splitting the array in the *H*-plane in two halves with even symmetry. Such an arrangement was discussed in Section 7.6.2 and Figure 7.23 in Chapter 7. It is also possible to eliminate the butterfly lobes by letting the longitudinal offset slots radiate into an exterior region consisting of baffles [18,19].

No butterfly lobes appear if the slots are collinear in a waveguide. This is possible with centred longitudinal slots excited by asymmetric ridges (see Section 9.9 in Chapter 9), probes, wires or irises. In this section we will discuss iris-excited slots. The compound iris, shown in Figure 8.13, consisting of a combination of an inductive iris of width I and a capacitive iris of width C, is particularly attractive since the resonant characteristics of the iris and that of the slot are complementary and therefore the combination of iris and slot exhibits greater bandwidth than the conventional offset slot [20]. MoM analysis and experimental measurements show that the longitudinal slot generally behaves as a shunt element in the form of a parallel resonant circuit whereas a compound iris behaves also as a shunt element but in the form of a series resonant circuit. Tang [20] demonstrated that it is possible to excite the collinear slots with appropriate magnitude and in phase by different sized irises.

A systematic experimental investigation of iris-excited centred longitudinal slots in the broad wall of a rectangular waveguides was undertaken by Nyström [21]. Pairs of values of I and C were determined such that the compound iris would be resonant at the design frequency of 5.3 GHz in a WR 159 waveguide of inner dimensions 40.39 mm × 20.19 mm. Marcuvitz's expressions [22] for the inductive and capacitive reactances were initially used to determine the approximate values of I and C needed to have a resonant compound iris at 5.3 GHz. Later the C values were changed slightly so that the measured resonances were at 5.3 GHz for each pair of I and C values. Table 8.1 shows the values of C needed for each value of I in order to produce the desired resonance. Smaller values of I and C will produce smaller amounts of excitation of a centred longitudinal slot while larger values of I and C will produce larger excitations.

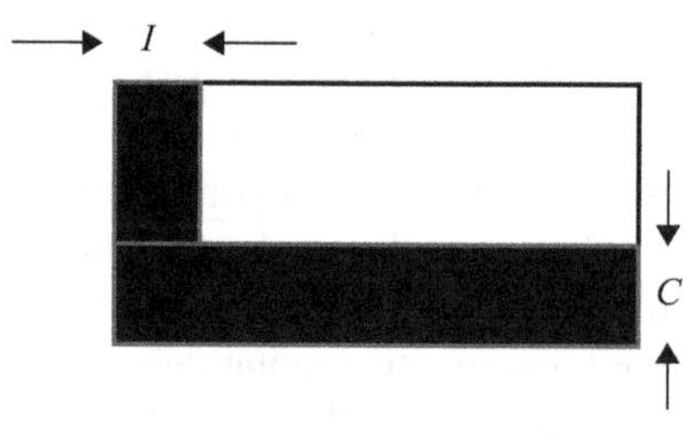

Figure 8.13 The cross section of a compound iris inside a rectangular waveguide

Table 8.1 Values of I and C for the iris to produce resonance at 5.3 GHz

I (mm)	*C* (mm)
5.0	4.7
6.4	6.1
7.5	7.5
8.8	9.1
10.0	10.8

Source: ©1991, I. Landmark/Nyström. Modified from [21], with permission.

Table 8.2 Values of I, C and the length of a centred slot to produce resonance at 5.3 GHz compared to an equivalent longitudinal offset slot producing the same resonant conductance

Case	Value of *I* (mm)	Value of *C* (mm)	Centred slot length (mm)	Equivalent longitudinal slot offset (mm)	Equivalent longitudinal slot length (mm)
A	5.0	4.7	26.5	1.3	26.6
B	6.4	6.1	27.0	2.7	26.7
C	7.5	7.5	27.3	4.1	26.8
D	8.8	9.1	27.7	6.5	27.1
E	10.0	10.8	28.0	10.4	27.6

Subsequently a centred longitudinal slot of appropriate length was cut in the waveguide such that the iris and the slot combination would be resonant at 5.3 GHz. In the absence of the iris the incident TE_{10} mode will not excite a centred slot whereas the asymmetry produced by the iris will excite it.

Table 8.2 shows the resonant lengths of centred slots for different iris widths. All slot widths were chosen to be 2.0 mm. Each combination of iris widths and the length of the centred longitudinal radiating slot produces resonance at 5.3 GHz. For each case the offset and the length of an equivalent longitudinal offset slot producing the same resonant conductance at 5.3 GHz are also shown in Table 8.2.

Figure 8.14 shows measured values of the slot conductance and susceptance normalised to the resonant conductance of the iris-excited slot A in Table 8.2. Normalised conductance and susceptance of an equivalent offset longitudinal slot whose length and offset are shown for case A in Table 8.2 are also compared. The iris-excited slot exhibits greater bandwidth. Similarly Figure 8.15 compares the iris-excited slot and an equivalent conventional offset longitudinal slot for case C. Once again the iris-excited slot is found to exhibit greater bandwidth. Adjacent iris-excited slots should have the inductive section on the opposite sides of the broad

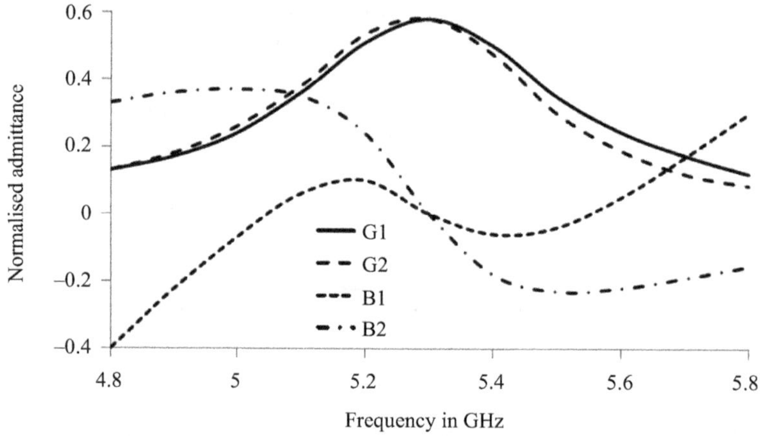

Figure 8.14 Susceptance (B1) and conductance (G1) versus frequency for an iris-excited centred slot and the susceptance (B2) and conductance (G2) of an equivalent conventional offset longitudinal slot for case A in Table 8.2. ©1991, I. Landmark/ Nyström. Modified from [21] with permission

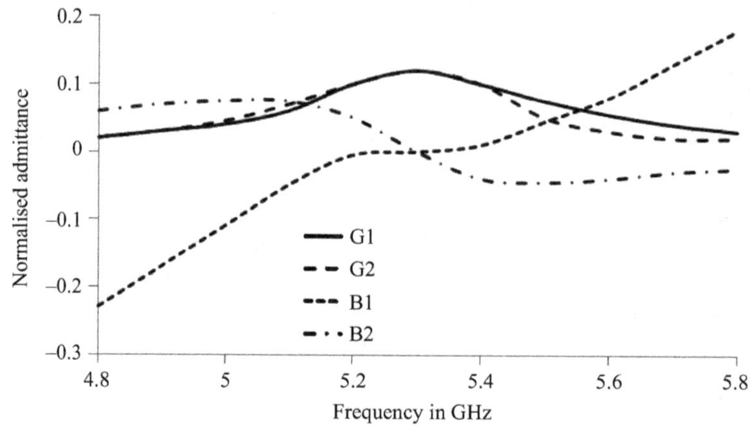

Figure 8.15 Susceptance (B1) and conductance (G1) versus frequency for an iris-excited centred slot and the susceptance (B2) and conductance (G2) of an equivalent conventional offset longitudinal slot for case C in Table 8.2. ©1991, I. Landmark/ Nyström. Modified from [21] with permission

wall centreline for a standing wave application so as to provide in phase excitation of slots. In a travelling wave array the inductive sections may be on the same side of the centreline or alternate, corresponding to the two cases discussed in Section 6.6 in Chapter 6.

8.4.1 *Elliott's design procedure for iris-excited slotted waveguide arrays*

For the conventional longitudinal slot, we obtain resonant lengths for different offsets and then obtain the admittance data and the two-port scattering parameters of the slot. The slot voltage is usually determined from the scattering parameters or the admittance data. In the case of an iris-excited slot, we treat the iris dimensions I and C as a couplet for a range of values as shown in Table 8.1. For each couplet, the resonant length and resonant conductance may be obtained from a CEM code. We then obtain the admittance data as a function of the slot length normalised to its resonant length for a range of slot lengths. In addition, for each of these cases we can obtain the slot aperture field from the same CEM code, assuming a half cosinusoidal equiphase distribution for the slot aperture. The relationship between the backscattered and forward scattered TE_{10} mode amplitude and the voltage of the nth slot can be expressed as follows.

$$B_{10} = C_{10} = f_{in} V_n^s \tag{8.13}$$

In (8.13) f_{in} is a function of iris dimensions, slot length and width. Similar to the first design equation (6.10) in Chapter 6 and in [23] we derive

$$\frac{Y_n^a}{G_0} = \frac{V_n^s}{V_n} f_{in} K_i \tag{8.14}$$

where

$$K_i = \frac{1}{\pi/a}\sqrt{\frac{2\omega\mu_0\beta_{10}ab}{G_0}} \tag{8.15}$$

The second design equation (6.15) derived in Chapter 6 is valid for the iris-excited slot also and it is reproduced as (8.16).

$$\boxed{\frac{Y^a}{G_0} = \frac{2}{\dfrac{2}{Y_s/G_0} + \dfrac{1}{\Gamma}\cdot\dfrac{V_{\text{scoupl}}}{V_s}}} \tag{8.16}$$

However, the voltage coupled to the slot from the external mutual coupling, given in (6.17), gets modified for iris-excited slots as

$$V_{\text{scoupl}} = V_{s\ \text{ext}} = \frac{j\pi}{2(k_0 b)\beta_{10}a^3(\omega\mu_0)^2 f_i^2}\frac{1}{Y_s/G_0 + 2}\frac{Y_s/G_0}{}\sum_{m=1}^{N}{}' V_m^s g_{mn}(L_m, L_n, X_{mn}, Z_{mn}) \tag{8.17}$$

Using the design equations given here, one can design linear and planar arrays of iris-excited slots using computed values of resonant lengths and admittances as functions of normalised slot lengths for different couplets I and C.

8.5 Slot arrays in ridge waveguides

Planar arrays of longitudinal radiating slots cut in the broad walls of rectangular waveguides may be electronically scanned in the *E*-plane by including phase shifters between adjacent waveguides. Conventional rectangular waveguides typically have an '*a*' dimension in the order of 0.7 λ_0 at the operating frequency. In this case, it is possible to scan to a maximum angle of 25° off broadside before grating lobes start appearing in the visible space. In order to scan to larger angles the spacing between waveguides should be closer to one-half wavelength. One way to accomplish this is by reducing the waveguide width which can be realised in ridge waveguides. Figure 8.16 shows cross sections of ridge waveguides.

Single-ridge waveguides are convenient for slot array applications since longitudinal radiating slots may be cut in the broad wall opposite to the ridge over a wide range of offsets (see Figure 8.17). Kim and Elliott presented a design procedure for slot arrays in ridge waveguides using experimental data on isolated slot admittances [24]. For analysis one can use a model for ridge waveguides presented by Falk [25,26] or use commercial CEM software. In this section we will briefly discuss the design procedure developed in [24].

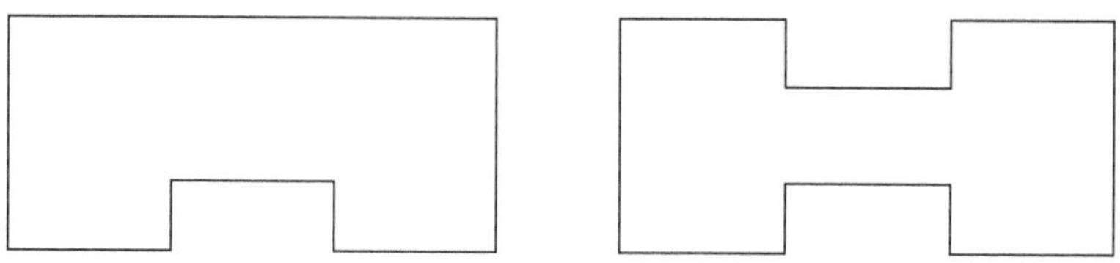

(a) (b)

Figure 8.16 Ridge waveguide cross sections: (a) single-ridge waveguide; (b) double-ridge waveguide

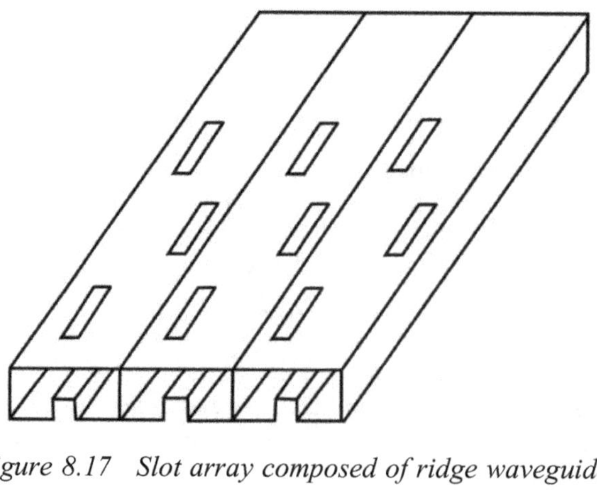

Figure 8.17 Slot array composed of ridge waveguides

The longitudinal slot cut in the broad wall of a ridge waveguide scatters symmetrically, just like the longitudinal slot in a rectangular waveguide and may be modelled in the form of a shunt admittance on an equivalent transmission line. Using Montgomery's expressions [27] for the TE_{10} mode electric field, the first design equation similar to (6.10) of the longitudinal slot in the conventional rectangular waveguide is obtained next.

$$\frac{Y_n^a}{G_0} = K_1 f_n \frac{V_n^s}{V_n} \tag{8.18}$$

where

$$K_1 = \frac{2k_t}{k_0}\sqrt{\frac{k_t^2}{\omega\mu_0\beta_{10}G_0}} \tag{8.19}$$

and k_t is the transverse wavenumber of the TE_{10} mode given by [27].

The second design equation (6.15) derived in Chapter 6 is valid for the ridge waveguide slot as well, and it is reproduced in (8.20).

$$\boxed{\frac{Y^a}{G_0} = \frac{2}{\dfrac{2}{Y_s/G_0} + \dfrac{1}{\Gamma}\cdot\dfrac{V_{\text{scoupl}}}{V_s}}} \tag{8.20}$$

However, the voltage coupled to the slot from the external mutual coupling, given in (6.17) gets modified for the ridge waveguide slot as [24]

$$V_{\text{scoupl}} = V_{s\ \text{ext}} = \frac{-j\beta_{10}k_0^2}{4\pi k_t^4}\frac{1}{f_n^2}\frac{Y_s/G_0}{Y_s/G_0 + 2}\sum_{m=1}^{N}{}' V_m^s g_{mn}(L_m, L_n, X_{mn}, Z_{mn}) \tag{8.21}$$

A planar array consisting of two ridge waveguides with eight radiating slots in each was designed using (8.18) and (8.20), and it was built and tested [24]. The measured results for the radiation pattern and the voltage standing wave ratio (VSWR) compared well with the theory.

8.6 Slot arrays with a dielectric cover

In high-power radar antennas, waveguides are pressurised to prevent electrical breakdown. Slotted waveguide arrays in such applications have often a dielectric layer on top of the slot apertures to seal the pressurised waveguides. The presence of the dielectric affects the resonant length and the resonant conductance of the slots. The problem of dielectric covered waveguide slots was first studied by Bailey using the variational approach [28,29]. Rexberg and Katehi studied the properties of dielectric covered slots using the MoM solution to the integral equations with the appropriate Green's functions in the waveguide and the exterior region [30,31]. The dielectric cover generally reduces the resonant length of the slot as shown in Figure 8.18 since the effective dielectric constant seen by the slot increases when a

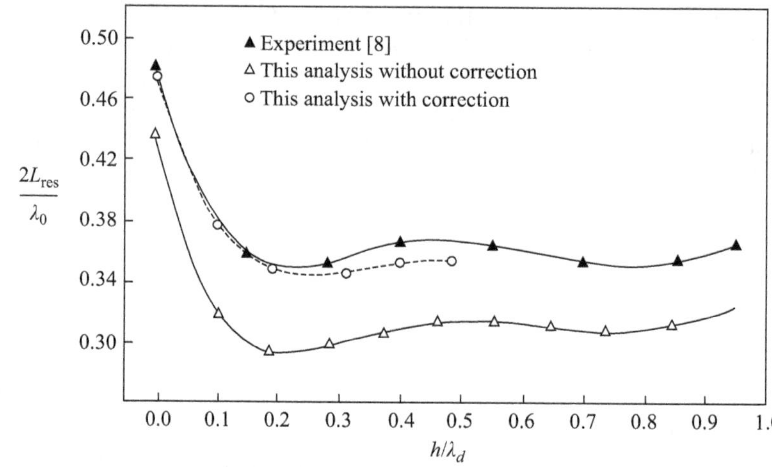

Figure 8.18 The effect of the dielectric cover thickness on the resonant length of the slot. © IEEE 1990, reprinted from [31], with permission

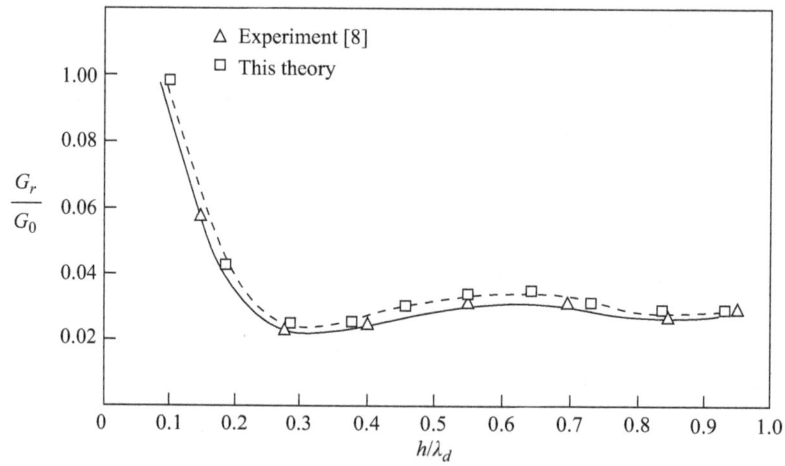

Figure 8.19 The effect of the dielectric cover thickness of the resonant conductance of the slot. © IEEE 1990, reprinted from [31], with permission

dielectric cover is used. The resonant slot conductance also decreases in presence of the dielectric cover (see Figure 8.19). For both of these plots, the slot length is 12.065 mm, and the slot offset is 2.36 mm. The dielectric constant is 3.31. The waveguide inner dimensions are 22.86 mm × 10.16 mm and the slot width is 1.59 mm.

Casula and Montisci have modified the Elliott's design procedure for a slot array with a dielectric cover [32]. The modification is needed only in the second design

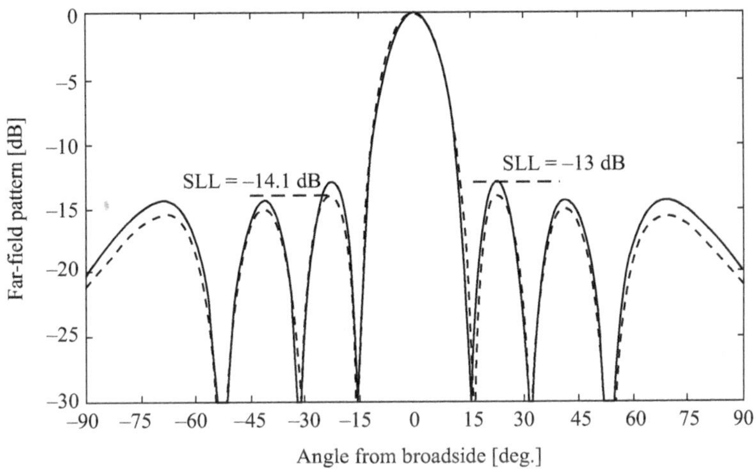

Figure 8.20 Simulated radiation pattern in the E-plane. Solid line (with dielectric); dashed line (without dielectric). © IEEE 2009, reprinted from [32], with permission

equation involving the external mutual coupling. The external mutual coupling is evaluated in the spectral domain in the form of numerical double integrals in polar coordinates. The radial integral exhibits highly oscillating terms and has poles due to surface waves. For a thin dielectric cover, only the TM_0 surface wave pole is present. The residue contribution of the pole has to be considered in addition to its principal value [33] or the external mutual coupling integral in spectral domain may be deformed to avoid the pole singularity [30]. The highly oscillating integral is evaluated by the method of weighted averaging and by the use of Shank's transforms [34,35].

Casula and Montisci demonstrate the significance of a dielectric cover of thickness as small as 0.012 free space wavelength and a dielectric constant equal to 2.2 by comparing the radiation pattern and input reflection coefficient of a 10×5 array (five waveguides with ten slots in each) simulated with and without the dielectric [32]. Those results are reproduced in Figures 8.20 and 8.21.

8.7 Higher-order mode coupling between centred-inclined coupling slots

Elliott's design procedure discussed in Chapters 6 and 7 does not incorporate higher-order mode coupling between coupling slots of a planar slot array (see Figure 8.22). Higher-order mode coupling between a coupling slot and a pair of straddling radiating slots in each coupling junction, also not accounted for in the design procedure, will be discussed in Section 8.8. It has been shown that the higher-order mode coupling between coupling slots is significant for small tilt angles or small slot offsets from the centreline [36]. The TE_{20} mode coupling accounts for most of this effect in reduced-height waveguides whereas in standard

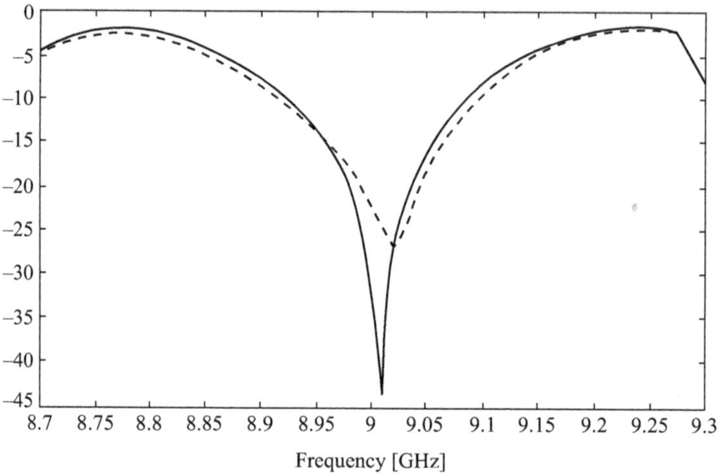

Figure 8.21 Simulated reflection coefficient. Solid line (with dielectric); dashed line (without dielectric). © IEEE 2009, reprinted from [32], with permission

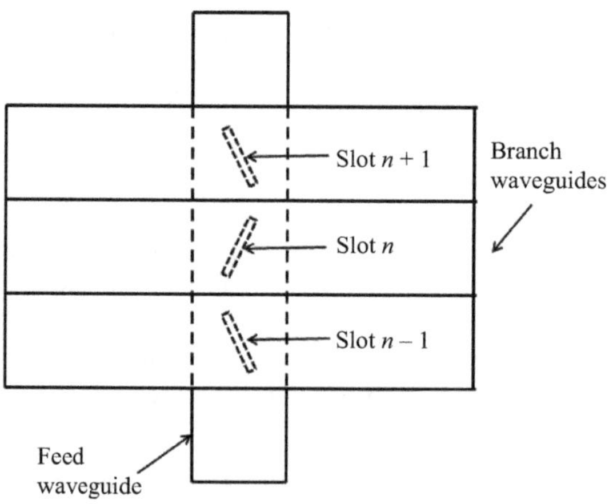

Figure 8.22 Geometry of the higher-order mode coupling between slot n and slots n – 1 and n + 1

height waveguides, in addition to the TE_{20} mode coupling, there may be some significant TE_{01} mode coupling as well. The higher-order mode coupling between adjacent shunt-series coupling slots may be accounted for by using a procedure similar to that for such coupling between adjacent radiating slots [37]. In this section we will discuss a method for incorporating the higher-order mode coupling between centred-inclined coupling slots [38].

8.7.1 Theory for the higher-order mode coupling between centred-inclined coupling slots

Using (4.68) in Chapter 4, we can derive an expression for the backward and forward scattered TE_{10} mode waves in the main waveguide by a centred-inclined coupling slot [see (8.22)].

$$B_{10} = -C_{10} = K_2 C_n V_n^s \tag{8.22}$$

Here V_n^s is the voltage of the nth slot, $K_2 = -\dfrac{(\pi/a)^2}{\omega\mu_0 w\beta_{10}ab}$, and C_n is the reaction integral (8.23) carried out over the axis of slot n with a tilt angle θ_n with respect to the longitudinal centreline of the broad wall.

$$C_n = w\int_{-L_n}^{L_n}\left[-j\sin\{\pi/a(\zeta\sin\theta)\} - \frac{\beta_{10}}{\pi/a}\cos\{\pi/a(\zeta\sin\theta)\}\right]$$
$$\times\exp(-j\beta_{10}\zeta\cos\theta)\cos\left(\frac{\pi}{2L_n}\zeta\right)d\zeta \tag{8.23}$$

The centred-inclined coupling slot may be modelled as a series impedance, as mentioned in Section 8.1.1, in an equivalent transmission line having the phase constant β_{10} and the characteristic impedance, Z_0 equal to the wave impedance of the TE_{10} mode wave. We can define an active impedance, Z_n^a for the coupling slot similar to the active admittance of a longitudinal radiating slot specified by (6.10) and (6.16) in Chapter 6 as follows.

$$Z_n^a = -\frac{C_n(\pi/a)}{w\beta_{10}[ab/2]^{1/2}}\frac{V_n^s}{I_n} \tag{8.24}$$

The term 'active impedance' includes mutual coupling from higher-order modes between adjacent coupling slots. In the absence of higher-order mode coupling, the active impedance becomes self-impedance. The term I_n in (8.24) is the mode current of the equivalent transmission line at the location of the active impedance. The derivation of mutual coupling expression follows the procedure used in deriving (6.15) in Chapter 6 for radiating slots.

The normalised active impedance is expressed in terms of the self-impedance and the isolated slot reflection coefficient as follows.

$$\frac{Z_n^a}{Z_0} = \frac{2}{\frac{2Z_0}{Z_n} - \frac{1}{\Gamma_n}V_{nh}^s/V_n^s} \tag{8.25}$$

where Γ_n is the reflection coefficient of the slot without accounting for higher-order mode coupling, Z_n is the impedance of the coupling slot in the absence of higher-order mode coupling and V_{nh}^s is the slot voltage induced by the higher-order mode coupling from adjacent slots. Using (6.36) in Chapter 6 and (8.22) and (8.23) we obtain an expression for V_{nh}^s as

$$V_{nh}^s = \frac{V_{n1}^s}{I_{10}}I_h = \frac{V_{n1}^s I_h}{A_{10}C_n} = \frac{\Gamma_n}{K_2 C_n^2}I_h \tag{8.26}$$

where I_h is given by the reaction integral (8.27) with $H_{h\zeta}$ the magnetic field along the axis of slot n due to higher-order modes scattered from adjacent slots. V^s_{n1} is the slot voltage induced by a TE_{10} mode wave of amplitude A_{10}.

$$I_h = \int_{-L_n}^{L_n} H_{h\zeta} \cos\left(\frac{\pi}{2L_n}\zeta\right) d\zeta \tag{8.27}$$

Combining (8.25) and (8.26) we obtain

$$\frac{Z^a_n}{Z_0} = \frac{1}{Z_0/Z_n - \frac{1}{2K_2C^2_n}\left(I_h/V^s_n\right)} = \frac{1}{Z_0/Z_n - \frac{1}{2K_2C^2_n}\left[I_1\frac{V^s_{n-1}}{V^s_n} + I_2\frac{V^s_{n+1}}{V^s_n}\right]} \tag{8.28}$$

The dominant contribution to I_h is expected to be from the TE_{20} mode but it is easy to account for a few other modes to get an accurate evaluation. Using the dyadic Green's function for the magnetic field of the magnetic current given in (2.62) in Chapter 2 we can determine I_1 for the coupling between slots $n-1$ and n and I_2 between slots $n+1$ and n, shown in Figure 8.22 using (8.29). The term I_1 is non-existent for the first coupling slot and likewise I_2 is not relevant for the last one.

$$\begin{aligned} I_1 = \int_{-L_n}^{L_n} \cos\left(\frac{\pi}{2L_n}\zeta\right) \int_{-L_{n-1}}^{L_{n-1}} \cos\left(\frac{\pi}{2L_{n-1}}\zeta'\right) \\ \times\big[G_{1xx}(x,x',z,z')\sin(\theta_n)\sin(\theta_{n-1}) \\ + G_{1xz}(x,x',z,z')\sin(\theta_n)\cos(\theta_{n-1}) \\ + G_{1zx}(x,x',z,z')\cos(\theta_n)\sin(\theta_{n-1}) \\ + G_{1zz}(x,x',z,z')\cos(\theta_n)\cos(\theta_{n-1})\big]d\zeta' d\zeta \end{aligned} \tag{8.29}$$

where $x = a/2 + \zeta \sin\theta_n$, $x' = a/2 + \zeta' \sin\theta_{n-1}$, $z = \zeta\cos\theta_n$, $z' = \zeta'\cos\theta_{n-1} - D$, and D is the slot spacing. The tilt angles of slots n and $n-1$ are θ_n, and θ_{n-1}, respectively. The expression for I_2 is obtained readily from I_1 given in (8.29) by replacing subscripts $n-1$ by $n+1$ and $-D$ by D in z' in (8.29). The terms G_{1xx}, G_{1xz}, G_{1zx} and G_{1zz} are the $\hat{x}\hat{x}, \hat{x}\hat{z}, \hat{z}\hat{x}$ and $\hat{z}\hat{z}$ components respectively of the dyadic Green function $\overline{\overline{G}}_1$ given in (2.63) and (2.64) in Chapter 2.

8.7.2 Procedure to incorporate the higher-order mode coupling in the design of centred-inclined coupling slots

Initially coupling slots are designed using Elliott's procedure described in Chapter 7. It is assumed that coupling slots are resonant at the design frequency and that they are spaced a half guide wavelength apart at the design frequency for a standing wave array. The procedure to be described is applicable even if the slots are not resonant, and for travelling wave type feeds as well. From the scattering parameter S_{11} of each coupling slot, treated as a four-port device, we determine its self-impedance. The mode current appearing in Equation (8.24) may be assigned an arbitrary value of unity for the first coupling slot. For a standing wave array all mode currents are equal in magnitude with alternating positive and negative signs between adjacent slots. For coupling slots in a travelling wave feed, mode currents are obtained using transmission line equations in a manner similar to the mode voltages of a travelling wave array of longitudinal slots in Section 6.6 in Chapter 6. In (8.24) we initially set the values of active impedances to self-impedances and obtain the values of slot voltages. Then the tilt angles and lengths of coupling slots are perturbed so as to make the active impedance of each coupling slot equal to its self-impedance in (8.28) and (8.24) while the slot voltages remain the same as the original values. This process converges rapidly since higher-order mode effects are usually small. Including coupling from a few higher-order modes such as TE_{20}, TE_{30}, TE_{11}, TM_{11} and TE_{01} will provide very accurate results.

In a study of a planar slot array with spacings between adjacent coupling slots of 0.6 λ_0 it was found that there was a significant amount of higher-order mode coupling effect on the reflection coefficient, whereas the radiations patterns were not affected much [38]. An optimisation procedure that perturbed the coupling slots by not more than 3% reduction in slot lengths and less than 0.3° reduction in the tilt angle produced excellent results with very good match at the centre frequency.

8.8 Higher-order mode coupling between a centred-inclined coupling slot and longitudinal radiating slots

Figure 8.1 shows the geometry of a planar array consisting of coupling junctions, each with a centred-inclined coupling slot and a pair of straddling longitudinal slots. In Figure 8.23, the coupling slot is projected on the broad wall containing the radiating slots. As shown in Figure 8.23 we define two cases of coupling called hard coupling and soft coupling. Experimental results show that higher-order mode coupling is greater in the case of hard coupling than in the soft coupling geometry. Each radiating slot is at a distance of a quarter guide wavelength from the coupling slot. Therefore higher-order mode coupling is expected to be significant in this type of coupling junctions. In some cases there is even an overlap between the coupling slot and the radiating slot when the former is projected on the wall containing the radiating slot, thereby requiring the use of the singularity in the Green's function (when $\vec{z} = \vec{z}'$ in Equation (2.60) in Section 2.8 in Chapter 2) to evaluate the coupling.

An extensive investigation of higher-order mode coupling in coupling junctions was carried out [5] using the MoM solution to the integral equations of the aperture fields of the three slots in a coupling junction. That study showed that the higher-order mode coupling can be neglected for half-height branch waveguides ($a \sim b/4$) whereas it is significant for standard height waveguides ($a \sim b/2$) as well as quarter height waveguides ($a \sim b/8$) and it is greater for the hard coupling geometry compared to the soft coupling. In a large array or sub-array consisting of many radiating slots in each radiating waveguide, the errors in the slot excitations

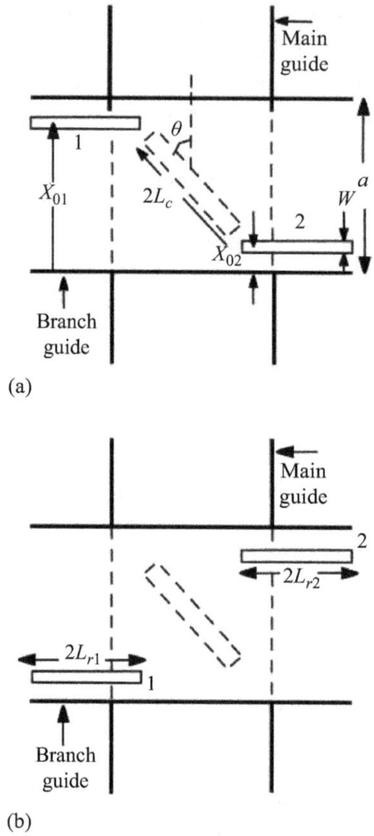

Figure 8.23 Geometry of the coupling junction: (a) hard coupling; (b) soft coupling. © IEEE 1994, reprinted from [5], with permission

are primarily confined to the coupling and radiating slots in the junction regions. If the vast majority of the radiating slots elsewhere have close to the desired excitations, the radiation pattern is expected to be of acceptable accuracy in the H-plane. However, the errors in the coupling slots may affect the input match and affect the pattern in the E-plane. In a small array such errors may become substantial since the percentage of total number of radiating slots in the coupling junction region is greater.

There is a simple method to include the higher-order mode coupling discussed earlier in the design of planar arrays. Let us consider the problem of three slots in the junction region using an MoM solution to the coupled integral equations of the aperture fields of slots. The calculation can be made for two sets of excitations, with and without higher-order mode coupling in the junction region. Then one can perturb the coupling slot's tilt and length and the offsets and lengths of the radiating slots, such that the excitations of the perturbed slots with higher-order mode coupling are the same as the excitations of the original slot parameters without higher-order mode coupling. We then replace the original slots in Elliott's design by the perturbed slots. The rationale for this design is that higher-order mode coupling is a localised phenomenon and therefore we can account for it by considering only the three slots in the junction region, without accounting for the presence of other slots. In some cases, however, this method does not produce satisfactory results and one may need to use a full wave method to analyse the entire slot array to account for the coupling effects. Such a technique is discussed in Section 8.10 [39].

8.9 Finite ground plane effects

In the design and analysis of slot arrays, it is usually assumed that the radiating slots are embedded in an infinite perfectly conducting ground plane. While the perfectly conducting ground plane assumption generally is very good, the effect of a finite ground plane on the pattern characteristics can be significant. Edge diffraction effects influence the excitation of slots close to the edge, in addition to affecting the radiation patterns in angular regions far from broadside. The mutual coupling expression for the infinite ground plane case can be improved by an additional term to include the edge diffraction effects [40], especially for slots that are close to the edges of the ground plane. A simple model using Keller's geometrical theory of diffraction (GTD) was introduced by Josefsson [41] to account for edge effects in mutual coupling calculations, and its accuracy to predict the mutual coupling between elements of a linear array of slots aligned in the *H*-plane has been validated with experimental results. Josefsson's model is easy to implement in the design procedure, not only for the mutual coupling terms but for the self-coupling also, as explained later.

Figure 8.24 shows two slots *m* and *n* and one edge of the ground plane. The ray incident from slot *m* gets diffracted at the edge and couples a voltage at slot *n*. The expression g_{mn} for the external mutual coupling between slots *m* and *n* in (5.66) in Chapter 5 is proportional to the integral of the magnetic field produced by a unit magnetic current in slot *m* in the aperture of slot *n* weighted by the cosine function of the aperture electric field of slot *n* [see (5.40) and (5.41) in Chapter 5]. There is

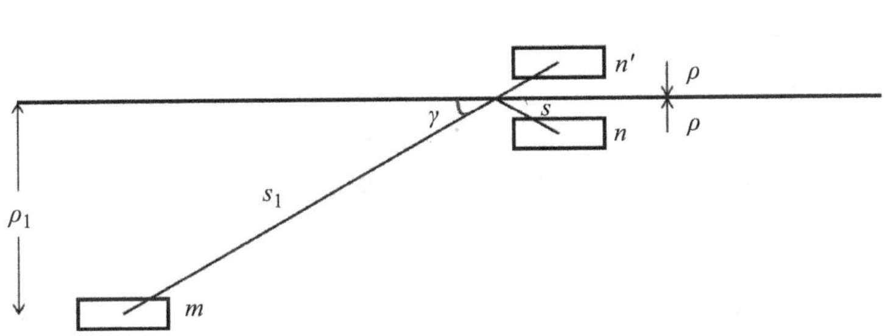

Figure 8.24 The geometry of the edge diffraction effect

an additional term for the magnetic field in the region of slot n due to a unit magnetic current in slot n, due to edge diffraction, derived later.

The magnetic field incident at the edge is given by

$$H^{\text{inc}} = H_0 \exp(-jks_1)/s_1 \tag{8.30}$$

where H_0 is a constant.

The edge-diffracted field is given by

$$H^d = H^{\text{inc}} D_{\text{h}} \exp(-jks)\sqrt{\frac{s_1}{s(s+s_1)}} \tag{8.31}$$

D_{h} is the hard diffraction coefficient given by [42]

$$D_{\text{h}} = \frac{\exp(-j\pi/4)\,\sin(\pi/n)}{n\sqrt{2\pi k}\,\sin(\gamma)}\left[\frac{1}{\cos\frac{\pi}{n} - \cos\frac{\phi-\phi'}{n}} + \frac{1}{\cos\frac{\pi}{n} - \cos\frac{\phi+\phi'}{n}}\right] \tag{8.32}$$

The parameter n is given by the interior wedge angle $(2 - n)\pi$. For a ground plane of zero thickness, $n = 2$ whereas for a thick edge, the interior wedge angle is 90° and $n = 0.5$. φ and φ' are the angles that the incident and diffracted rays make with respect to the plane surface, and they are zero in this case. Because of grazing incidence, a factor ½ has been included in the diffracted field since the total incident field at grazing incidence is the sum of an incident and a reflected field [43]. Thus we obtain the diffracted field as

$$H^d = -H_0 \frac{\exp[-jk(s+s_1)]}{(s+s_1)} \frac{(1-j)}{\sqrt{16\pi}} \sqrt{\frac{(s+s_1)}{k\rho\rho_1}} \tag{8.33}$$

The diffracted field at slot n can be written as the incident field at the image slot n' multiplied by the edge diffraction parameters. Therefore the original mutual coupling term g_{mn} discussed in Section 6.4 in Chapter 6 is augmented by adding an extra term containing the integral between slots m and n' (image of n) as follows.

$$g_{mn} \to g_{mn} + H_0\, g_{mn'} \tag{8.34}$$

where

$$H_0 = \frac{-1+j}{\sqrt{16\pi}} \sqrt{\frac{s+s_1}{k\rho\rho_1}} \tag{8.35}$$

The field of slot n, incident at the edge, after getting diffracted may couple to itself. This term can be added to the summation in (6.46b) in Chapter 6 as

$$V_n^s g_{nn'}(L_n, L_n) \tag{8.36}$$

The $g_{nn'}$ integral will be carried out using the procedure discussed in Section 6.4 in Chapter 6 for slots n and n' of half-length L_n, separated by a distance 2ρ in the E-plane direction and aligned in the H-plane direction, as shown in Figure 8.24.

8.10 MoM solution to the integral equations of a planar array and applications

Section 5.2 in Chapter 5 discusses the MoM solution to the integral equation of the aperture field of a single slot. This is the most efficient technique to compute the characteristics of a slot since the unknowns are the equivalent magnetic currents in the slot apertures, as opposed to techniques such as the finite element method and finite difference method where the unknowns are in a volume. In addition, when one uses entire domain sinusoidal basis functions for the longitudinal variations of the magnetic current, the first term, a half cosinusoidal distribution, is an excellent approximation to the aperture electric field of a slot near resonance. Therefore a few unknowns are usually adequate to obtain good accuracy.

8.10.1 Formulation of the coupled integral equations

Several simplifying assumptions are made in deriving the integral equations of a planar slot array. For narrow slots the longitudinal component of the aperture electric field is ignored. Consistent with this narrow slot approximation, the boundary conditions are enforced for the transverse component of the electric field and the longitudinal component of the magnetic field only. The slots are assumed to be square-ended and embedded in a perfectly conducting infinite ground plane.

The milling process in manufacturing usually yields round-ended slots. Round-ended slots are not easy to work with in an analysis model. Therefore one uses an approximate equivalence between round-ended slots in the experimental work with the square-ended model used in the analysis based on an equal area criterion. The equal area criterion yields the following relationship between round-ended and square-ended slots [see (5.54)].

$$2L_r = 2L_s + 0.215w \tag{8.37}$$

where $2L_r$ is the length of the round-ended slot, $2L_s$ is the length of the square-ended slot and w is the width.

In a study of round-ended slots using the MoM/finite element method by Zhang *et al.* [44], the aperture equivalent magnetic current was assumed to be even symmetric and essentially longitudinal to the slot except near the round ends where an edge based finite element model was used. The authors show good comparison between computed and measured S_{21} values. It is not clear how well such a model works for S_{11} especially for slots with small scattering. Based on their work the authors recommend the following expression for square-ended to round-ended slot length conversion.

$$2L_r = 2L_s + 0.245w \tag{8.38}$$

For typical narrow slots employed in many practical waveguide array antennas with a width to length ratio of 1/10, the two formulas are within 0.3% of the slot lengths. For very wide slots with a width to length ratio of ¼, the deviation between the lengths in the two formulations is 0.7%.

The slot thickness is accounted for by considering the two (inner and outer) apertures of the slot. The formulation of the coupled integral equations starts with shorting out each slot aperture by an infinitesimally thin perfectly conducting sheet. By invoking Schelkunoff's equivalence principle (see Section 2.7), we place equivalent magnetic currents on each such conducting sheet. Enforcement of the continuity of the tangential electric fields across each aperture of the original problem requires that the magnetic current on either side of the conducting sheet shall be equal in magnitude but opposite in direction. Figure 8.25 shows the cross section of a thick radiating slot which we assume is the *m*th slot.

By enforcing the continuity of the longitudinal component of the aperture magnetic field for the *m*th slot we obtain (8.39) and (8.40).

$$H_{\xi m}^{\text{ext}+}\left(\vec{M}_m^{\text{ext}}\right) - H_{\xi m}^{\text{ext}-}\left(\vec{M}_m^{\text{ext}}\right) - H_{\xi m}^{\text{ext}-}\left(\vec{M}_m^{\text{int}}\right) + \sum_q H_{\xi m}^{\text{ext}+}\left(\vec{M}_q^{\text{int}}\right) = 0 \tag{8.39}$$

$$H_{\xi m}^{\text{int}+}\left(\vec{M}_m^{\text{int}}\right) + H_{\xi m}^{\text{int}+}\left(\vec{M}_m^{\text{ext}}\right) - H_{\xi m}^{\text{int}-}\left(\vec{M}_m^{\text{int}}\right) - \sum_{n \neq m} H_{\xi m}^{\text{int}-}\left(\vec{M}_n^{\text{int}}\right) = H_{\xi m}^{\text{inc}} \tag{8.40}$$

Thus, there are two coupled integral equations for slot *m*, one for the exterior aperture facing the exterior half space and the other for the interior aperture facing the waveguide. There are two unknowns for slot *m*, $\vec{M}_m^{\text{ext}}$ and $\vec{M}_m^{\text{int}}$, the magnetic currents in the exterior (ext) and interior (int) aperture, respectively. The slot *m* is coupled to all the radiating slots in the entire array through external mutual coupling in the summation over *q*. Similarly the slot *m* is coupled to all the other radiating slots and the coupling slot in that radiating waveguide, in the summation over *n*, through TE_{10} mode scattering. In addition, slot *m* is coupled to adjacent radiating slots and to the coupling slot, if it is in the immediate vicinity, through higher-order modes. The magnetic field of the magnetic current in each of the three regions (waveguide interior, exterior half space and the cavity region to account for the slot thickness) is expressed in terms of integrals over the magnetic current with the appropriate Green's function for the waveguide, half space and cavity, respectively.

For a linear array of standing wave or travelling wave type directly fed by an incident TE_{10} mode wave, an additional contribution on the right side of (8.40) exists. This source term is absent in planar standing wave arrays where all the

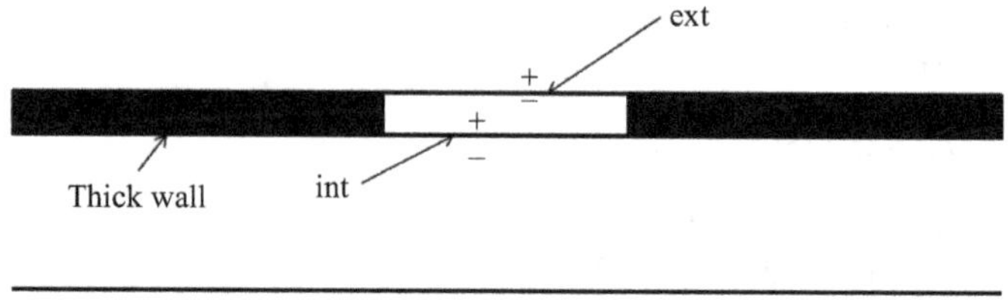

Figure 8.25 The internal and external apertures of the mth radiating slot

radiating waveguides are supposed to be shorted at both ends and excited by coupling slots. A similar pair of coupled integral equations results when the continuity of the longitudinal component of the magnetic field is enforced across the apertures of each coupling slot. The incident field term is absent if the feed waveguide containing the coupling slots is shorted at both ends and excited by an input slot coupling to another waveguide layer. An example of such an input shunt-series coupling slot is found in Figure 7.12. The source term then appears in the integral equation for this slot.

8.10.2 MoM solution

The coupled integral equations (8.39) and (8.40) are reduced to a system of simultaneous equations in the MoM where we expand each unknown magnetic current in terms of a set of expansion functions with unknown coefficients and form the inner product of each integral equation with a set of weighting functions. The resulting matrix equation is of order $2N \times M$ where N is the total number of slots in the array and M is the number of unknown coefficients in each. M need not be the same for all slots. Entire domain sinusoidal expansion is a good choice for longitudinal variation of the aperture magnetic current. The transverse variation is assumed to be uniform. For narrow slots there is no need to use the edge conditions discussed in Section 5.3. The weighting functions may be chosen same as the expansion functions, sinusoidal and pulse or sinusoidal and delta. In the latter case we use point matching in the transverse direction along the centreline of the slot.

In deriving the integral equations one faces a difficulty in obtaining the magnetic field in the waveguide cavity region for a standing wave array. Figure 8.26 shows a waveguide with several slots with shorts at both ends. In the coupling terms between slots m and n the contribution from the forward scattered TE_{10} mode will become $C_{10}/[1-\Gamma_L\Gamma_R \exp(-j2\beta_{10}D)]$ in place of just C_{10}. Γ_L and Γ_R are the reflection coefficients of the left and right shorts, β_{10} is the phase constant of the TE_{10} mode and D is the spacing between the shorts at the two ends. This expression results from the infinite number of reflected waves from both ends. D is an integer multiple of a half guide wavelength at the design frequency. For a lossless system, the denominator becomes zero since Γ_L and Γ_R are both -1 for the short. One could avoid this problem by assuming a small amount of loss in the metallic waveguide, so that the resulting round trip attenuation will make the sum of the infinite number of reflected waves finite.

Once the matrix equations are solved, we determine the expansion coefficients of the magnetic equivalent currents of all slots. The aperture distribution is then

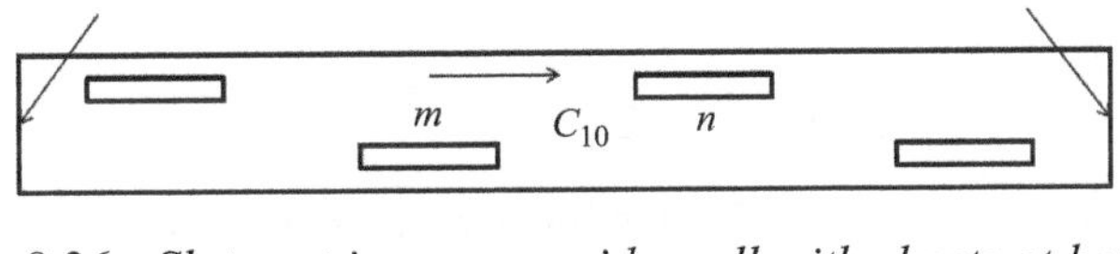

Figure 8.26 Slots cut in a waveguide wall with shorts at both ends

known. The radiation pattern may be computed by summing the radiated far fields of the magnetic currents of all slots. The reflection coefficient at the input port is found from the scattering from the coupling slots. Directivity is readily found since the incident wave power and reflected power are known and the power density in the main beam peak direction can easily be calculated.

8.10.3 Improved design procedure using the MoM solution

In this section we discuss the use of the MoM solution to improve the design of a planar slot array. We will start with the radiating slots.

The design of a planar array using Elliott's technique yields offsets and lengths of all radiating slots and lengths and tilt angles of coupling slots as well (see Section 7.2 in Chapter 7). A radiating slot with a strong excitation coefficient in the middle of the array is chosen as a reference element. Let the offset and the half-length of the nth slot be x_0 and L, respectively. The complex voltage of the nth slot normalised to the reference slot voltage is V_{sn} $(=u+jv)$ while the desired value of the normalised voltage of the nth slot is V_{in}. The difference $(V_{sn}-V_{in})=\delta u+j\delta v$ is the error. The mean square error for all radiating slots in the aperture is $\frac{1}{N}\sum_n |V_{sn}-V_{in}|^2$ where N is the number of radiating slots in the array.

By using the MoM analysis program three times, first for the original values of each slot parameter and then by incrementing the offset plus half the waveguide width, x_0 alone, and then by incrementing the half-length, L alone, we determine the numerical values of the partial derivatives for slot n. The perturbations for the parameters of slot n to eliminate the errors are then given by (8.41).

$$\begin{bmatrix} \Delta x_0 \\ \Delta L \end{bmatrix} = \begin{bmatrix} \dfrac{\partial u}{\partial x_0} & \dfrac{\partial u}{\partial L} \\ \dfrac{\partial v}{\partial x_0} & \dfrac{\partial v}{\partial L} \end{bmatrix}^{-1} \begin{bmatrix} -\delta u \\ -\delta v \end{bmatrix} \tag{8.41}$$

The partial derivatives obtained numerically are not robust for very small values of the derivatives. In that case the slot parameters are left unchanged. Since radiating slot perturbations affect the return loss, coupling slots have to be perturbed also, as discussed next.

It is assumed that the feed waveguide is excited at one end. The reflection coefficient Γ at this input port, consists of the incident TE_{10} wave that is reflected at the end of the feed waveguide short with reflection r, and the TE_{10} mode scattered from all coupling slots, s. The initial design based on Elliott's technique will produce $s\approx -r$.

$$\Gamma = r+s \tag{8.42}$$

$$\text{Let } \alpha = \arg(s)+\arg(r) \tag{8.43}$$

For a match α is zero, assuming that the magnitudes of r and s are nearly equal. If α is not equal to zero, we need to delay s by α to compensate for the phase error.

Therefore all coupling slots need to be detuned so that the input reflection coefficient will be equal to that given by (8.44).

$$\Gamma' = r + s\exp(-j\alpha) \tag{8.44}$$

The normalised input impedance, z, then would become

$$Z = (1 + \Gamma')/(1 - \Gamma') \tag{8.45}$$

Z can be made approximately equal to 1 by introducing a factor by which coupling coefficients of all slots are scaled so that the new values of the mth coupling coefficients is given next.

$$\chi_m^2 = |Z^{-1}|S_{11}/(1 - S_{11}) \tag{8.46}$$

S_{11} in (8.46) is the original scattering parameter of the mth four port coupling slot with match terminations in all ports with the ports referenced to planes passing through the centre of the slot. Then the perturbed scattering parameter S_{11m} of the same will become

$$S_{11m} = \chi_m^2 \exp(-j\alpha)/\left(1 + \chi_m^2\right) \tag{8.47}$$

The magnitude is perturbed by the scale factor introduced by the coupling coefficient in (8.46) and the phase change is due to perturbation for s in (8.44). Since the coupling coefficients of all coupling slots are scaled by the same factor and the phase of all S_{11} are perturbed by the same amount there is very little additional error introduced in the aperture distribution of the radiating slots. However, because of the approximation in the assumption that the phase of the scattering parameters of all coupling slots track over a small frequency range near resonance, the perturbation produces only an approximate match condition and further iterations are required. Equations (8.46) and (8.47) can be modified such that the coupling slot perturbations produce the desired value of the total normalised resistance if there is a matching input slot with centred feed as in some arrays [4,45]. Such a case is not discussed here.

The radiating slot perturbations are carried out one slot at a time, first for all radiating slots, followed by the perturbations of all coupling slots. We then iterate the process 3–4 times until the errors in the aperture distribution are small enough with an acceptable match at the design frequency. Criteria for stopping the iteration may use the maximum allowable error or the maximum mean absolute error for the aperture distribution and the desired value of the reflection coefficient. It is possible to perturb only slots exhibiting large errors in the aperture distribution thereby speeding up the process.

The improved design process was demonstrated in a theoretical study of a 5 × 4 array, shown in Figure 8.27 for a radiometer type application designed at 13.285 GHz [39]. The design parameters are shown in Table 8.3. This desired distribution produces E-plane sidelobes in the range of 35.5–32 dB below the peak and H-plane sidelobes in the range of 21.5–31.7 dB below the peak.

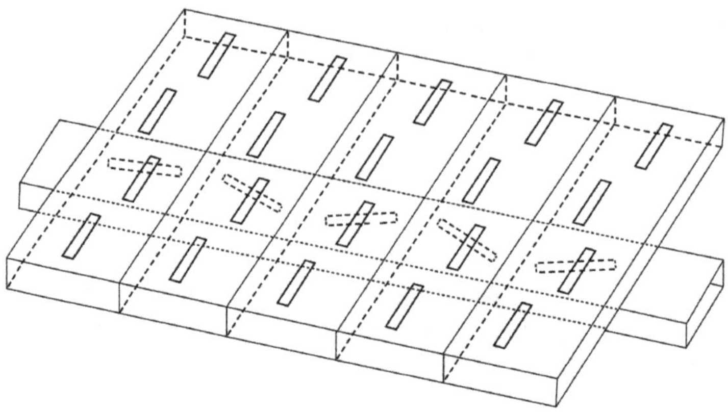

Figure 8.27 A 5 × 4 slotted waveguide array

Table 8.3 Design parameters of the 5 × 4 array

Feed waveguide	Radiating waveguide	Slot width (mm)	Wall thickness (mm)	Aperture distribution (*E*-plane)	Aperture distribution (*H*-plane)
16.33 mm × 8.16 mm	15.10 mm × 7.55 mm	1.27 mm	0.508 mm	0.28:0.75:1:0.75:0.28	0.606:1:1:0.606

In the initial design using the Elliott's procedure, the total normalised slot conductance of all radiating slots in each radiating waveguide was assumed to be 2 with an input match at the feed port enforced. Figure 8.28 shows the *E*-plane pattern of this array with the legend 'original'. A substantial amount of higher-order mode coupling is present in the slots in the five coupling junctions, thereby affecting the return loss and the *E*-plane pattern. It was concluded that the major source of error in the original design was the higher-order mode coupling in the five coupling junctions.

Figure 8.28 shows also the *E*-plane patterns computed with an approximate method of accounting for the higher-order mode coupling in the junction slots in the original design. In this method for which results are shown by the legend 'approximate', one perturbs each coupling slot and a pair of straddling radiating slots by considering one junction at a time. The perturbation compensates for the errors introduced by the higher-order mode coupling in the original design. This is a major simplification considering only six coupled integral equations of the three slots for each junction. However, the method is approximate and, as shown by Figure 8.28, the improvement produced by this method is marginal. Obviously the iterative method of perturbing all slots using the moment method analysis of the entire array antenna, shown by the legend 'improved', produces very good results.

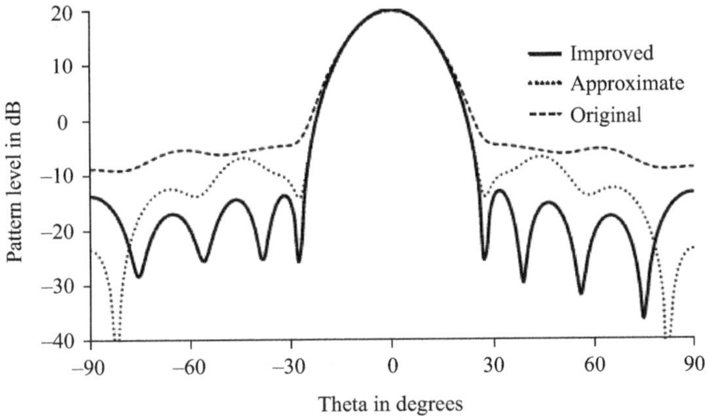

Figure 8.28 E-plane patterns of the 5 × 4 array using the original, improved and approximate methods. Reprinted from [39], with permission

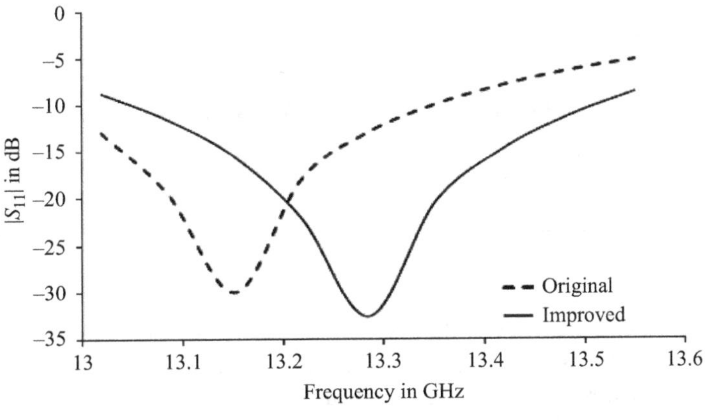

Figure 8.29 Magnitude of the reflection coefficient of the original and the improved designs. Reprinted from [39], with permission

For arrays made up of sub-arrays, the proposed method can be applied by considering one sub-array at a time while treating all other sub-arrays as additional source terms or external excitations.

The improved design procedure using MOM applied to this antenna yielded radiation patterns close to the ideal patterns. Since the higher-order mode coupling in the junctions did not have a significant effect in the *H*-plane patterns, they are not shown here. Figure 8.29 shows the reflection coefficient in dB over a range of frequencies. The original design has the best match at 1% below the centre of the desired frequency band. The improved design places the best performance at the

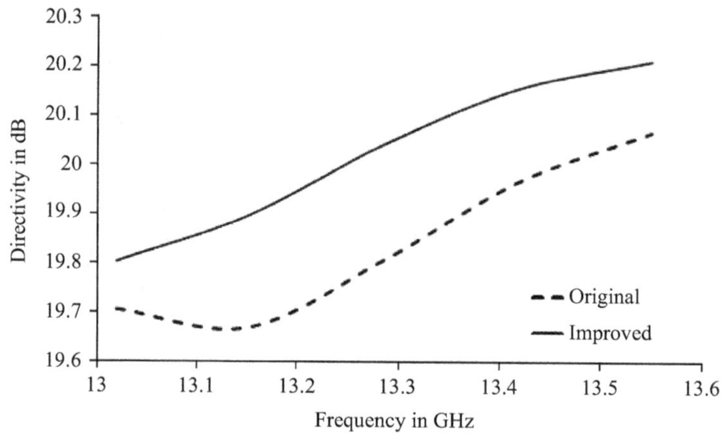

Figure 8.30 Directivity of the original and the improved designs. Reprinted from [39], with permission

Table 8.4 Coupling slot parameters in the original and improved designs

Original design		**Improved design**	
Tilt angle (°)	**Length (mm)**	**Tilt angle (°)**	**Length (mm)**
6.40	10.53	6.92	10.49
−17.45	10.55	−19.25	10.51
22.62	10.56	24.96	10.53

Source: Reprinted from [39], with permission.

centre frequency and achieves the desired 10 dB return loss over 4% frequency band. Similarly, the improved design achieves between 0.1 and 0.235 dB increase in the values of directivity within the required 4% frequency band compared to the original design. Figure 8.30 compares the directivity of the two designs. E-plane patterns of the improved design at band edges, not shown here, were closer to the desired patterns.

Table 8.4 shows the lengths and tilt angles with respect to the feed waveguide axis of the coupling slots for the original and improved designs. Since coupling slots exhibit symmetry with respect to the centre only the three elements closest to the feed port are specified. The perturbations are found to be small. Table 8.5 shows similar results for the offsets and lengths of radiating slots in the three radiating waveguides starting from the one nearest to the feed port. In the original design radiating slot parameters exhibit symmetry with respect to the diagonal, but for the sign of the offset. In the improved design such symmetry is present in an approximate sense. Radiating slot perturbations are significant compared to those of coupling slots.

Table 8.5 Radiating slot parameters in the original and improved designs

Original design		Improved design	
Offset (mm)	**Length (mm)**	**Offset (mm)**	**Length (mm)**
1.240	10.943	1.669	10.997
−2.689	11.092	−2.553	10.859
2.130	11.074	2.506	11.143
−1.661	10.896	−1.297	10.744
1.490	10.849	1.192	10.705
−2.422	10.923	−2.026	10.665
2.483	11.050	1.948	10.709
−1.433	10.755	−1.316	10.639
1.453	10.774	1.423	10.744
−2.372	10.945	−2.372	10.945
2.372	10.945	2.358	10.939
−1.453	10.774	−1.412	10.746

Reprinted from [39], with permission.

8.11 Global optimisations for improved designs

Biology-inspired global optimisations such as genetic algorithms (GA) [46–48] and particle swarm optimisations [49] have been applied to many antenna problems. The design procedure discussed previously for linear and planar slot arrays is generally carried out at the centre frequency of the desired band. Such a design usually results in performance degradation at band edges. In a monopulse array, the slot excitations are different for sum and difference patterns. Therefore, the Elliott design procedure is not readily applicable if the sum mode and difference mode patterns are to be realised. In a previous work a design was implemented by averaging the values of offsets and lengths obtained for each of three patterns, sum, elevation and azimuth difference patterns [50]. Multi-objective and multi-frequency designs can be carried out as an optimisation exercise using an analysis program. In this section we show a couple of examples of GA optimisations of slot arrays. GA is a powerful technique to find optimum solutions without getting stuck in local maxima or minima. It is a robust process to search a large solution space efficiently since it uses probabilistic rules.

8.11.1 GA optimisation of a travelling wave linear array

The GA optimisation was carried out for a travelling wave array [51], based on a previous design [52], also studied in Section 6.6, using an analysis model in terms of equivalent circuits on a transmission line. The analysis was facilitated by rearranging Elliott's equations for an N-element array by Hamadallah [53]. The main analysis equations shown in (6.51)–(6.53) are reproduced here without the TE_{20}

mode internal higher-order mode coupling terms. The slot voltages in the left column matrix are given by

$$[G][V^s] = \frac{k_2}{k_1} V_1 [E] \tag{8.48}$$

where

$$G_{mn} = g_{mn} \quad \text{if } n \geq m \tag{8.49}$$

$$G_{mn} = g_{mn} - jk_2 f_m f_n \sin\{(m-n)\beta_{10} d\} \quad \text{if } n < m \tag{8.50}$$

In (8.48) $[G]$ is an $N \times N$ square matrix whereas $[V^s]$ and $[E]$ are $N \times 1$ column matrices where N is the number of slots. Column matrix elements $[E]$ are specified in (6.55). The constants k_1 and k_2 are specified in (6.55a) and (6.55b) in Chapter 6. Using (6.49) and (6.50) we can obtain the input admittance and then the VSWR.

The length and offset of each slot are quantised into 64 values, 6 bits or genes. For the 21-element slot array a set of values of slot lengths and offsets is represented by a chromosome of 252 genes. A figure of merit is assigned to each array design (organism) according to the following performance function.

$$D^{\alpha}\left(1.0 - \frac{P_{\text{load}}}{P_{\text{in}}}\right)^{\beta} \cdot \frac{1}{\text{VSWR}^{\gamma}} \cdot \frac{1}{\text{SLL}^{\tau}} \tag{8.51}$$

where D is the directivity, SLL the sidelobe level and $P_{\text{load}}/P_{\text{in}}$ is the ratio of power dissipated in the load to the input power. The directivity is calculated readily from the power radiated, $P_{\text{in}} - P_{\text{ref}} - P_{\text{load}}$ and the power density in the maximum far-field direction. The population size was kept at 50. The values of α, β, γ and τ were varied depending on the parameters that were optimised. When bandwidth was optimised the performance function was averaged over a number of frequencies within the band.

The parameters of the studied travelling wave array were as follows: $a = 22.86$ mm, $b = 10.16$ mm, wall thickness = 1.27 mm and slot width 1.5875 mm. The design frequency is 9.375 GHz, the slot spacing 17.4 mm and the beam peak at 45° with respect to the waveguide axis in the H-plane. The slots are all offset on the same side of the centreline. The result of this optimisation exercise indicates possible performance improvements using GA. Hamadallah's analysis equations [53] with MoM data for Elliott's travelling wave array gave the following initial results: VSWR = 1.05, $P_{\text{load}}/P_{\text{in}} = 16.7\%$, SLL = –20.1 dB. The results compare well with the experimental results in Elliott's paper [52]: VSWR = 1.05 and SLL = –22 dB. The slight discrepancy in the SLL may be attributed to the edge diffraction of the finite ground plane, especially in the E-plane. The value of $P_{\text{load}}/P_{\text{in}}$ could not be measured but it was computed by Elliott to be 12.3%. The computed results in [51] are closer to the experimental results. In subsequent optimisations we varied the values of slot lengths and offsets to optimise the figure of merit.

When VSWR and $P_{\text{load}}/P_{\text{in}}$ are minimised, values of 1.011% and 2.1%, respectively, were obtained. The SLL improved to −22.1 dB. If the sidelobe level alone was optimised, it reduced to a low value of −25.2 dB, yielding VSWR of 1.016 and $P_{\text{load}}/P_{\text{in}}$

of 4.6%. When the design was optimised for 5% bandwidth, the SLL was better than −21.9 dB, while P_{load}/P_{in} was below 6.2% within the band. The VSWR was better than 1.02 in the frequency range of interest. For 10% bandwidth optimisation, SLL was better than −20.9 dB and P_{load}/P_{in} was lower than 7.1%. VSWR values are higher than those for the 5% bandwidth case but still below 1.04 in the entire band.

The optimisation exercises show the power of the GA when combined with an antenna analysis technique. However, this theoretical exercise did not include the internal higher-order mode coupling and has the limitations of the Elliott's design procedure discussed in Chapter 6.

8.11.2 GA optimisation of a planar standing wave array using the MoM analysis

In this section we describe a GA optimisation of a planar array to maximise the directivity and return loss over 5% bandwidth [54]. With GA the moment method solution of the array is computed thousands of times making the procedure prohibitively time consuming. The computational effort can be reduced substantially, however, by using only one expansion and test function for each slot aperture in the MoM solution. The basis function used is a half sine wave and it is a good approximation to the aperture distribution of a slot near resonance. However, the accuracy of the solution with only one basis function is not adequate. We will establish the equivalence between a slot with one basis function and a slot using many basis functions so as to obtain better results later.

In order to match slotted waveguide arrays reactive elements can be inserted in the input waveguide port. Irises are easy to design and manufacture and are very useful as matching devices in waveguides. The inductive irises are preferred over the capacitive irises (see Figure 8.31) because of their greater power handling capability. Figure 8.32 shows a waveguide section consisting of two inductive irises. There are

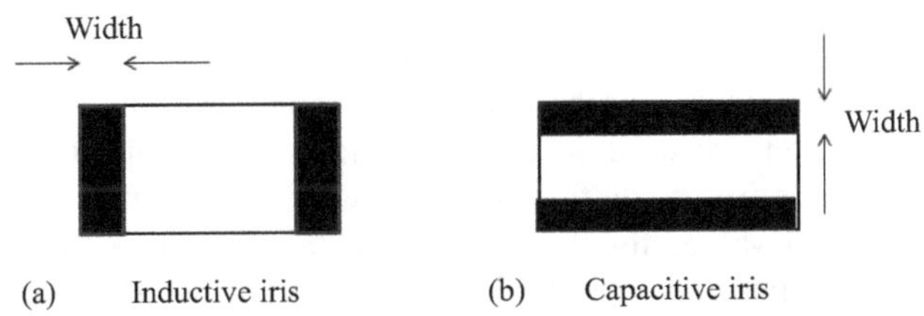

Figure 8.31 Cross sections of inductive and capacitive irises in rectangular waveguides

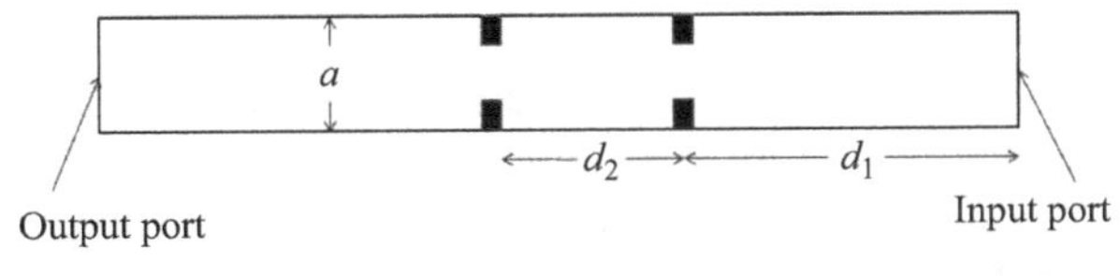

Figure 8.32 A waveguide H-plane section with two thick inductive irises

four parameters, widths of the two irises in the *H*-plane, the spacing between the irises, d_2, and the distance between the input port and the nearest iris, d_1, that may be optimised to maximise the return loss over a frequency band. Iris thickness is usually fixed. The input port is connected to the slot array to be matched.

Mode matching (MM) is a fast and accurate technique for solving boundary value problems for waveguide discontinuities [55]. It is ideally suited to the analysis of waveguides with thick irises. GA optimisation in conjunction with MM may be used to improve the impedance match over the required frequency bandwidth.

We start with the design of an 8×10 array consisting of eight radiating waveguides, each containing ten radiating longitudinal slots, using Elliott's procedure. The aperture distribution is assumed to be uniform. The dimensions of the input and radiating waveguides are 5.70 mm × 2.85 mm while those of the feed waveguide are 5.673 mm × 2.836 mm. The slot widths and thicknesses are 0.508 mm and the centre frequency is 35.75 GHz. Diagonal symmetry is assumed in the array for the radiating slots whereas centred-inclined coupling slots are assumed to be symmetric with respect to the centre but for alternating signs for tilt angles. In the GA design there were 85 parameters, offset and length for each of the 40 radiating slots, tilt angle for each of the four centred-inclined coupling slots and offset from the centreline for the input shunt-series coupling slot. Other radiating and coupling slot data are obtained from symmetry. Coupling slots are all assumed to be resonant. Binary GA with 7 bits for each parameter is employed so that the length of each chromosome is 595. This allows a resolution in the variable size to be less than the machine tolerance, assumed to be 1 mil (~25 μ).

For each element of the population (chromosome) GA performs the moment method analysis of the slot array at nine equally spaced frequency points within a 4% frequency band centred at 35.75 GHz. The lowest values of the broadside gain and the return loss are weighted to specify the fitness parameter in GA. Thus we optimise the worst case values of gain and return loss within the bandwidth for a nominally uniform aperture distribution. In the GA the fitness parameter is not allowed to increase if the return loss is greater than 12 dB. A population size of 10, tournament selection, mutation probability of 0.1, and probability of crossover of 0.5 with uniform crossover are used with the best individual replicated into the next generation [46–48]. Optimum solutions are achieved typically after 600 generations. The range of allowed values for the offset from the centreline for radiating slots and the input shunt-series coupling slot is about 1/10th of the waveguide '*a*' dimension. The slot length and the distance between the short and the nearest slot centre are allowed to have a range of about 5% of their nominal values while the tilt angle of each centred-inclined coupling slot is allowed to vary ±5° with respect to the initial value. The computer time for each GA execution for about 600 generations is about one week in a dedicated PC with an Intel Core Duo processor T9600 and a clock speed of 2.8 GHz with 4GB of RAM. The computer time may be drastically reduced by using many parallel processors or using resources such as cloud computing.

In the case of matching sections with irises, the range of values for the spacing between irises and the spacing between the array feed port and the nearest iris is a maximum of half a guide wavelength at the centre frequency. The irises are assumed to be 20 mil (~0.5 mm) thick and their widths are allowed to have a range of 1/3 of the waveguide '*a*' dimension. There are four variables and each is 10 bits long so that the length of the chromosome is 40. Other parameters of GA in this case are identical to the previous GA MOM optimisation. The fitness parameter is the lowest value of the return loss computed at nine equally spaced frequency points within 4% bandwidth centred at 35.75 GHz. GA mode matching (MM) is an extremely fast computational process.

8.11.3 Equivalence between a slot with one basis function and a slot with many basis functions

Prior experience shows that for the radiating slots and for the input shunt-series coupling slot nine basis functions are needed for good accuracy while three basis functions are sufficient for centred-inclined coupling slots. If the slot parameters obtained by GA are used in a moment method analysis program with nine (or three) basis functions, the results for gain and input reflection coefficient would deviate substantially from the corresponding GA results with one basis function for each slot. Scattering data were computed for a range of values of slot offsets and lengths for the longitudinal radiating slots and the shunt-series input slot, and for a range of values of lengths and tilt angles for the coupling slots using the moment method solution of an isolated slot for the pertinent integral equations. Notice that for the nth slot the two slot parameters, p_{1n} and p_{2n}, are the offset and the length, respectively, for a radiating slot and the input shunt-series coupling slot whereas they are the tilt angle and the length, respectively, for a centred-inclined coupling slot. The desired slot parameters p_{1n} and p_{2n} (m basis functions) are obtained from the parameters q_{1n} and q_{2n} (one basis function) of the corresponding slot in the optimum GA design such that (8.52) is satisfied.

$$S_{11}(p_{1n}, p_{2n}, m) = S_{11}(q_{1n}, q_{2n}, 1) \tag{8.52}$$

In (8.52), the scattering parameter S_{11} is the reflection coefficient at the input port of an isolated slot where all the other ports are matched. The right side of (8.52) is computed with one basis function with slot parameters q_{1n}, and q_{2n}. In order to determine slot parameters p_{1n}, and p_{2n}, we perform a two-dimensional interpolation of the computed data with m (nine or three) basis functions over a range of values of the magnitude and angle of S_{11} for radiating slots. For resonant centred-inclined coupling slots and input shunt-series coupling slot, one-dimensional interpolation is carried out. For slots near resonance, all scattering parameters can be expressed in terms of S_{11}. Therefore (8.52) provides equivalence. The rationale for the equivalence stated here is further justified by the fact that the array analysis employing nine or three basis functions, with parameters p_{1n} and p_{2n} for the nth slot, n ranging from 1 through 89, yields results close to those produced by the GA design using one basis function for each slot aperture. Ando *et al.* employ a similar equivalence between a slot with one basis function and a corresponding slot with nine basis functions where the length is the only parameter used as a variable [56]. We found better results with two variables in (8.52).

Figure 8.33 shows gain values for three cases. The original design using Elliott's procedure has a gain plot with a wide variation in the frequency range. A GA solution obtained with one basis function for each slot aperture has nearly a constant gain over a 4% bandwidth. For each radiating slot and the input shunt-series coupling slot in this GA optimisation, we obtain an equivalent slot with nine basis functions while for each coupling slot the equivalent slot with three basis functions is obtained from Equation (8.52). The gain values calculated for an array with these equivalent slots is shown using the legend 'nine bases GA'. The latter two plots are very similar. Figure 8.34 shows results for the reflection coefficients for the three cases. These results further justify the equivalence established by (8.52).

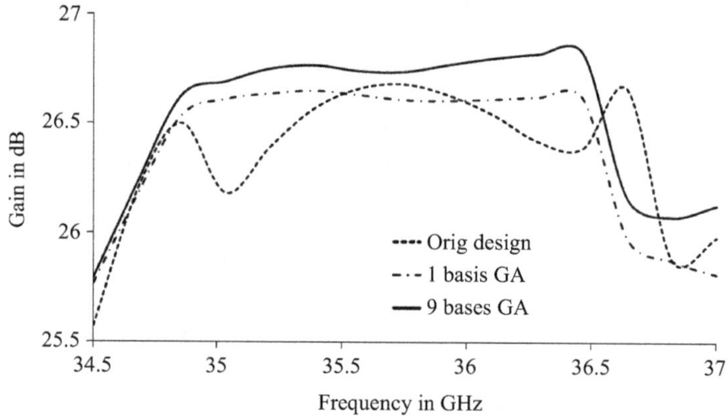

Figure 8.33 Gain plots of GA designs with one basis function and an equivalent array with nine basis functions. The legend 'orig design' used Elliott's procedure. Reprinted from [54], with permission

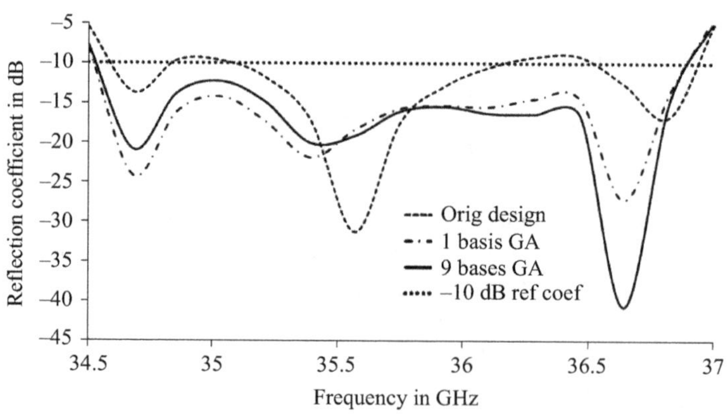

Figure 8.34 Reflection coefficient plots of GA designs with one basis function and an equivalent array with nine basis functions. The legend 'orig design' used Elliott's procedure. Reprinted from [54], with permission

8.11.4 GA optimised results

Figure 8.35 illustrates the computed values of gain as a function of frequency for the GA optimisation. The GA yields nearly uniform gain over 5% frequency range. In the GA optimisation the fitness parameter was first specified as a weighted sum of the return loss and gain. The reflection coefficient was not allowed to exceed –12 dB. The GA produces significantly better gain values. Figure 8.36 shows the reflection coefficient as a function of frequency. The reflection coefficient value of the original design is somewhat higher than the nominal –10 dB

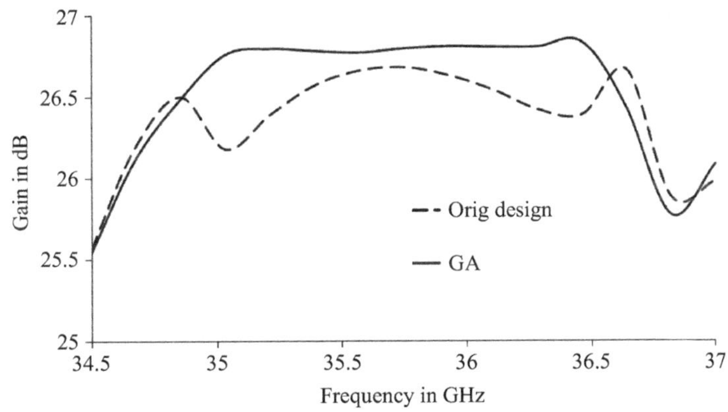

Figure 8.35 Gain as a function of frequency of different array designs; 'orig design' used Elliott's procedure and in the GA design the gain was optimised over a 4% bandwidth. Reprinted from [54], with permission

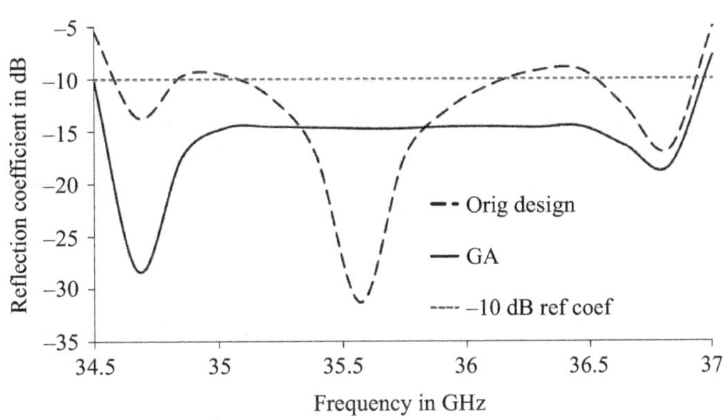

Figure 8.36 Reflection coefficient as a function of frequency for different array designs; 'orig design' used Elliott's procedure and GA optimised the gain and the reflection coefficient. Reprinted from [54], with permission

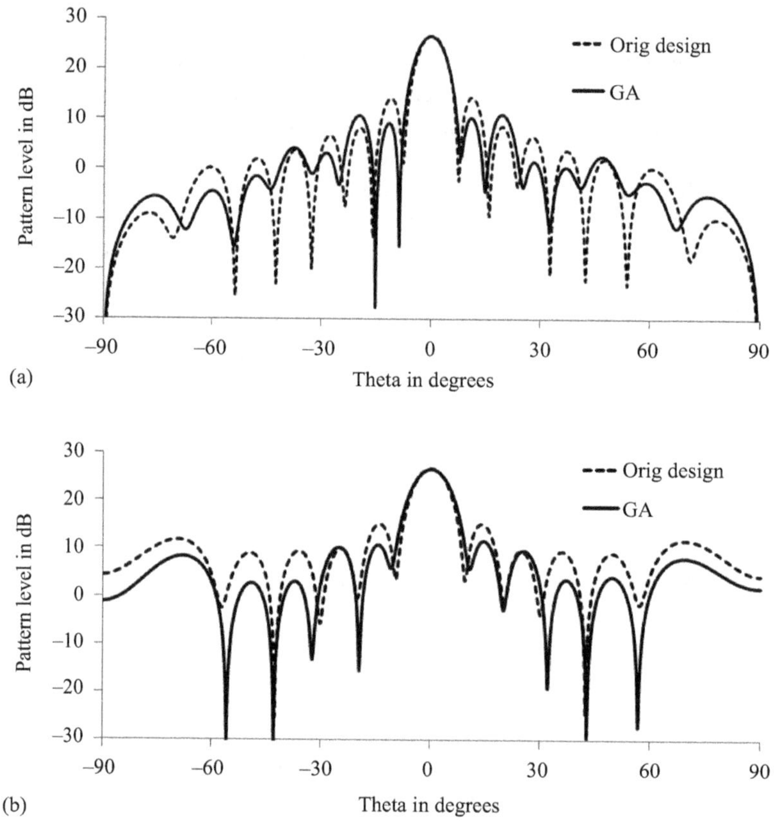

Figure 8.37 Radiation patterns at 35.75 GHz; 'orig design' used Elliott's procedure and GA optimised the gain and the reflection coefficient: (a) H-plane pattern; (b) E-plane pattern. Reprinted from [54], with permission

within the 5% frequency band. The initial GA optimisation that produced no higher than −12 dB reflection coefficient is further optimised by GA with a separate design of two irises. Thus Figure 8.36 shows a nearly flat reflection coefficient of −15 dB and exhibits greater than 6% bandwidth for input match with better than −10 dB reflection coefficient.

Typical radiation patterns computed from the GA optimised design are shown in Figure 8.37 for the centre frequency. The pattern peaks are normalised to 0 dB for easy comparison of sidelobe levels. There is no cross-polarised radiation in the principal planes. GA patterns generally exhibit lower levels of sidelobe radiation. The GA optimised the gain over a 5% bandwidth. The pattern with the legend 'orig design' was obtained from the array designed by the Elliott procedure. The GA design produced lower sidelobes compared to the original design using the Elliott procedure.

References

[1] Rengarajan S. R. 'Analysis of a centered-inclined waveguide slot coupler'. *IEEE Transactions on Microwave Theory and Techniques.* 1989;**37**(5): 884–889.

[2] Mazzarella G., Montisci G. 'Wideband equivalent circuit of a centered-inclined waveguide slot coupler'. *Journal of Electromagnetics and Applications.* 2000;**14**:133–151.

[3] Rengarajan S. R. 'Characteristics of a longitudinal/transverse coupling slot in crossed rectangular waveguides'. *IEEE Transactions on Microwave Theory and Techniques.* 1989;**37**(8):1171–1177.

[4] Rengarajan S. R., Zawadzki M. S., Hodges R. E. 'Waveguide-slot array antenna designs for low-average-sidelobe specifications'. *IEEE Antennas and Propagation Magazine.* 2010;**52**(6):89–98.

[5] Rengarajan S. R., Shaw G. M. 'Accurate characterization of coupling junctions in waveguide-fed planar slot arrays'. *IEEE Transactions on Microwave Theory and Techniques.* 1994;**42**(12):2239–2248.

[6] Senior D. C. *Higher-order mode coupling effects in a shunt-series coupling junction of a planar slot array.* Ph.D. dissertation, University of California, Los Angeles, CA, 1986.

[7] Watson W. H. 'Resonant slots'. *IEE Journal.* 1946;**93**(Part 3A):747–777.

[8] Zawadzki M. S., Rengarajan S. R., Hodges R. E. 'The design of H- and V-pol waveguide slot array feeds for a scanned offset dual-polarized reflectarray'. *IEEE Antennas and Propagation Society International Symposium.* 2005;**2B**:417–420.

[9] Silver S. *Microwave Antenna Theory and Design.* MIT Radiation Laboratory Series, vol. 12, McGraw-Hill, New York, 1949.

[10] Yee H. Y. 'Slot antenna arrays' in Johnson R. C., Jasik H. (eds.). *Antenna Engineering Handbook.* McGraw-Hill, New York, 1984.

[11] Kildal P.-S., Rengarajan S. R., Moldsvor A. 'Analysis of nearly cylindrical antennas and scattering problems using a spectrum of two-dimensional solutions'. *IEEE Transactions on Antennas and Propagation.* 1996;**44**(8): 1183–1192.

[12] Rengarajan S. R. 'Mutual coupling between slots cut in rectangular cylindrical structures: spectral domain technique'. *Radio Science.* 1996;**31**(6): 1651–1661.

[13] Rengarajan S. R. 'Compound radiating slots in a broad wall of a rectangular waveguide'. *IEEE Transactions on Antennas and Propagation.* 1989;**37**(9): 1116–1123.

[14] Maxum B. J. 'Resonant slots with independent control of amplitude and phase'. *IRE Transactions on Antennas and Propagation.* 1960;**8**(4):384–389.

[15] Derneryd A., Petersson R. 'Bandwidth characteristics of monopulse slotted waveguide antennas'. *Proceedings of the Fourth International Conference on Antennas and Propagation*, 1985, pp. 27–30.

[16] Montesinos-Ortego I., Zhang M., Sierra-Perez M., Hirokawa J., Ando M. 'Systematic design methodology for one-dimensional compound slot-arrays combining method of moments, equivalent circuit model and forward matching procedure'. *IEEE Transactions on Antennas Propagation.* 2013; **61**(1):453–458.

[17] Gruenberg H. 'Second order beams of slotted waveguide arrays'. *Canadian Journal of Physics.* 1953;**31**(1):55–59.

[18] Gruenberg H. 'Theory of waveguide-fed slots radiating into parallel plate regions'. *Journal of Applied Physics.* 1952;**23**(7):733–737.

[19] Forooraghi K., Kildal P.-S., Rengarajan S. R. 'Admittance of an isolated waveguide-fed slot radiating between baffles using a spectrum of two-dimensional solutions'. *IEEE Transactions on Antennas and Propagation.* 1993;**41**(4):422–428.

[20] Tang R. 'A slot with variable coupling and its application to a linear array'. *IRE Transactions on Antennas and Propagation.* 1960;**8**(1):97–101.

[21] Nystrom I. *Excitering av centrerad slits i vågledare.* Report R/AG-91:188, Ericsson Radar Electronics AB, Sweden, 1991.

[22] Marcuvitz N. *Waveguide Handbook,* MIT Radiation Laboratory Series, Vol. 10, Boston Technical Publishers, Inc., 1964.

[23] Elliott R. S. 'An improved design procedure for small arrays of shunt slots'. *IEEE Transactions on Antennas and Propagation.* 1983;**31**(1):48–53.

[24] Kim D., Elliott R. S. 'A design procedure for slot arrays fed by a ridge waveguide'. *IEEE Transactions on Antennas and Propagation.* 1988;**36**(11): 1531–1536.

[25] Falk K. 'Conductance of a longitudinal resonant slot in a ridge waveguide'. *IEE Proceedings, part H.* 1987;**134**(1):98–100.

[26] Falk K. 'Admittance of a longitudinal slot in a ridge waveguide'. *IEE Proceedings, part H.* 1988;**135**(4):263–268.

[27] Montgomery J. P. 'On the complete eigenvalue solution of ridge waveguide'. *IEEE Transactions on Microwave Theory and Techniques.* 1971;**19**(6):547–555.

[28] Bailey M. C. 'Design of dielectric-covered rectangular resonant slots in a rectangular waveguide'. *IEEE Transactions on Antennas and Propagation.* 1967;**15**(5):594–598.

[29] Bailey M. C. 'The impedance properties of dielectric-covered narrow radiating slots in the broad face of a rectangular waveguide'. *IEEE Transactions on Antennas and Propagation.* 1970;**18**(5):596–603.

[30] Rexberg L. 'Vector Fourier transform analysis of dielectric-covered slot in the broad wall of a waveguide'. *Microwave and Optical Technology Letters.* 1988;**1**(10):360–363.

[31] Katehi P. B. 'Dielectric-covered waveguide longitudinal slots with finite wall thickness'. *IEEE Transactions on Antennas and Propagation.* 1990; **38**(7):1039–1045.

[32] Casula G. A., Montisci G. 'Design of dielectric-covered planar arrays of longitudinal slots'. *IEEE Antennas and Wireless Propagation Letters.* 2009;**8**:752–755.

[33] Mosig J. R. 'Integral equation techniques' in *Numerical Techniques for Microwave and Millimeter-Wave Passive Structures*, Itoh, I. (ed.), John Wiley & Sons, New York, 1989.

[34] Mosig J. R., Gardiol F. E. 'A dynamical radiation model for microstrip structures'. *Advances in Electronics and Electron Physics*. 1982;**59**:139–237.

[35] Alaylioglu A., Evans G., Hyslop J. 'The evaluation of oscillatory integrals with infinite limits'. *Journal of Computational Physics*. 1972;**13**:433–438.

[36] Rengarajan S. R. 'Higher-order mode coupling between coupling slots in a planar slot array'. *IEEE Transactions on Microwave Theory and Techniques*. 1991;**39**(7):1219–1223.

[37] Elliott R. S., O'Loughlin W. R. 'The design of slot arrays including internal mutual coupling'. *IEEE Transactions on Antennas and Propagation*. 1986;**34**(9):1149–1154.

[38] Rengarajan S. R. 'An optimization procedure for including the higher order mode coupling between coupling slots in the design of a planar slot array'. *IEEE Antennas and Wireless Propagation Letters*. 2008;**7**:785–787.

[39] Rengarajan S. R. 'Improved design procedure for waveguide-fed slot arrays using the method-of-moments analysis'. *Electromagnetics*. 2012;**32**(4): 221–232.

[40] Mazzarella G., Panariello G. 'Evaluation of edge effects in slot arrays using the geometrical theory of diffraction'. *IEEE Transactions on Antennas and Propagation*. 1989;**37**(3):392–395.

[41] Josefsson L. 'Mutual coupling calculations including edge effects'. *Electronics Letters*. 1994;**30**(25):2087–2088.

[42] Keller J. B. 'Geometrical theory of diffraction'. *Journal of Optical Society of America*. 1962;**52**(2):116–130.

[43] Kouyoumjian R. G. 'The geometrical theory of diffraction and its application'. *Numerical and Asymptotic Techniques in Electromagnetics*, in Mittra, R. (ed.). Springer-Verlag, Berlin, 1975.

[44] Zhang M., Hirano T., Hirokawa J., Ando M. 'Analysis of a waveguide with a round-ended wide straight slot by the method of moments using numerical-eigenmode basis functions'. *IEICE Transactions on Communications*. 2004; **E-87B**(8):2319–2326.

[45] Rengarajan S. R., Zawadzki M. S., Hodges R. E. 'Design, analysis, and development of a large Ka-band slot array for digital beam-forming application'. *IEEE Transactions on Antennas and Propagation*. 2009;**57**(10): 3103–3109.

[46] Goldberg D. *Genetic Algorithms in Search, Optimization and Machine Learning*, Addison-Wesley, Reading, MA, 1989.

[47] Haupt R., Haupt S. E. *Practical Genetic Algorithms*, 2nd edn., Wiley, Hoboken, NJ, 2004.

[48] Haupt R., Werner D. H. *Genetic Algorithms in Electromagnetics*. Wiley-IEEE Press, Hoboken, NJ, 2007.

[49] Kennedy J., Eberhart R. 'Particle swarm optimization'. *Proceedings of the IEEE International Conference on Neural Networks IV*, 1995, pp. 1942–1948.

[50] Schaffner J., Kim D., Elliott R. S. 'Compromises among optimum sum and difference patterns for planar waveguide-fed slot arrays'. *Alta Frequenza.* 1981;**L**(6):312–319.

[51] Jensen A., Rengarajan S. R. 'Genetic algorithm optimization of a traveling wave array of longitudinal slots in a rectangular waveguide'. *Applied Computational Electromagnetics Society Journal.* 2006;**21**(3):337–341.

[52] Elliott R. S. 'On the design of traveling-wave-fed longitudinal shunt slot arrays'. *IEEE Transactions on Antennas and Propagation.* 1979;**27**(5): 717–720.

[53] Hamadallah M. 'Frequency limitations on broad-band performance of shunt slot arrays'. *IEEE Transactions on Antennas and Propagation.* 1989;**37**(7): 817–823.

[54] Rengarajan S. R. 'Genetic algorithm optimization of a planar array using full wave method of moments analysis'. *International Journal of Computer Aided RF and Microwave Engineering.* 2013;**23**(4):430–436.

[55] Wexler A. 'Solution of waveguide discontinuities by modal analysis'. *IEEE Transactions on Microwave Theory and Techniques.* 1967;**17**(9):508–517.

[56] Ando M., Zhang M., Lee J., Hirokawa J. 'Design and fabrication of millimeter wave slotted waveguide arrays'. *Proceedings of the European Conference on Antennas and Propagation*, 2010, pp. 1–6.

Chapter 9

Antenna systems and special requirements

So far we have mainly discussed flat slotted waveguide array antennas consisting of rectangular waveguides, typically with slots in the broad wall, radiating a fixed narrow beam. Many other configurations are possible, however, and even more suitable in special cases.

In this chapter we will investigate how slotted waveguide arrays are used in system applications where more advanced antenna functions or special configurations are required. As this is a very broad subject we will concentrate on cases where advantages and perhaps difficulties are of particular interest.

9.1 Phase scanned slotted waveguide arrays

Slotted waveguide arrays have been used extensively in phase steered antenna systems, in particular for beam steering in one dimension. In this case each slotted waveguide acts as one element in a one-dimensional scanning array. This may seem as a severe limitation, unable to do full beam steering in all directions. However, there are many situations where one-dimensional electronic scan is sufficient. The principal layout of a phase steered slotted waveguide array is shown in Figure 9.1.

The element size $\approx$ element spacing (d in Figure 9.1) leads to a theoretical maximum scan angle θ_{max} according to the formula

$$\sin\theta_{max} = \lambda/d - 1 \tag{9.1}$$

in order to prevent grating lobes and array mismatch. A typical slotted waveguide array with longitudinal slots in the broad wall has a waveguide width of roughly 0.7 λ. When used as an element in a phased array this 'element size' leads to a maximum scan angle of about 25° according to (9.1). However, at this angle the grating lobe already enters visible space and there is a considerable mismatch due to mutual coupling [1,2]. Fifteen degrees can therefore be a more realistic maximum scan angle, cf. Figure 9.11 (see also Section 4.5.1). An example is shown in Figure 9.2 for the centre element of a small array of rectangular aperture elements (E-plane slots) with different spacings.

For scan angles up to ±60° it is necessary to use waveguide structures with a much reduced width for the highest frequency used. Possible alternatives include dielectric filled rectangular waveguides [3–5], ridge-loaded waveguides [6–9] and

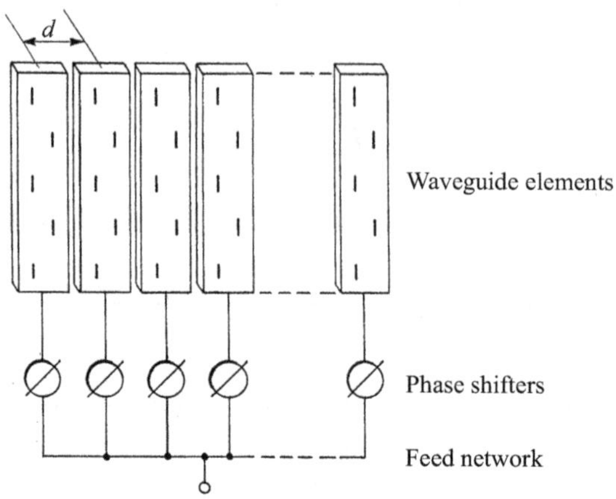

Figure 9.1 Phase steered array with slotted waveguides as radiating elements

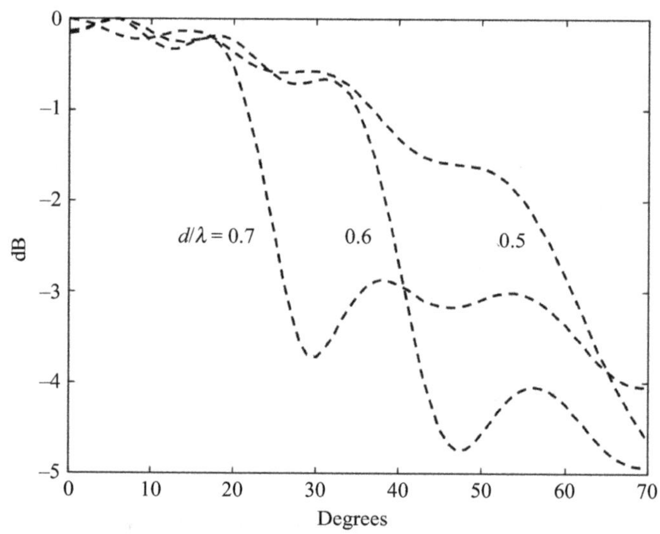

Figure 9.2 Embedded element patterns for the centre element of a 15-element array with different element spacings d/λ

edge slotted waveguides [10–12]. Figure 9.3 shows a typical size reduction using ridge-loaded waveguides. The width of the ridge guide is in this example 35% less than the width of the corresponding rectangular guide for the same guide wavelength.

Shnitkin [9] describes an interesting ridge waveguide configuration (see Figure 9.4), where the waveguide heights alternate on the two sides of the ridge. The longitudinal slots remain on the physical centreline, and no 'butterfly lobes' (see Section 9.9) are generated.

Figure 9.3 Cross sections of a rectangular waveguide (left) and a ridge-loaded waveguide (right), with the same cut-off wavelength

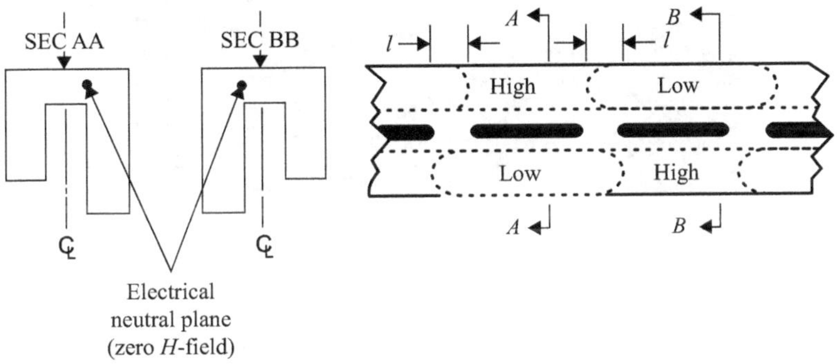

Figure 9.4 Asymmetric ridge waveguide with slots along the centreline. © 1990 IEEE, reprinted from [9], with permission

Numerical methods have been developed for the analysis of ridge waveguide characteristics; see, for example, Conciauro *et al.* [13]. Also slots in the ridge waveguide can be analysed [14,15]. Slot data for array design can of course also be obtained from measurements. Commercial codes using finite elements (e.g. HFSS) are used as well.

Systems with phase steering in elevation combined with mechanical rotation in azimuth is a common way of covering all coordinates, for example, in a 3D radar. The antenna may be composed of several slotted ridge waveguides with individual phase control; see Figure 11.7 in Chapter 11. Another interesting solution is to combine phase steering with frequency scanning [16,17].

A large S-band radar with wide angle phase steering in azimuth is used in an airborne early warning (AEW) system [18,19]. About 200 transmit/receive modules feed two slot arrays on each side of the unit carried by the aircraft. The vertical slotted ridge waveguides allow almost 360° of azimuth coverage (Figure 9.5).

9.2 Frequency scanned slotted waveguide arrays

A linear waveguide array with, say, longitudinal slots in the broad wall evenly spaced along the waveguide is a building block of many array systems. When fed in one end and with a matched termination in the other it will radiate in a direction which is determined by the frequency, the waveguide wavelength and the spacing

(a)

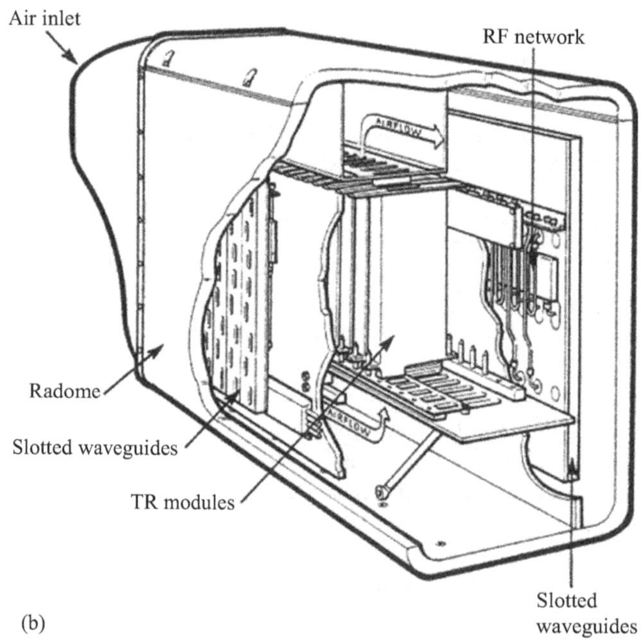

(b)

Figure 9.5 AEW radar with a phase steered slotted ridge waveguide array antenna. (a) The dorsal unit in a near field test chamber. (b) AEW demonstrator cross section showing TR modules in centre with slotted waveguide array panels on each side of the dorsal unit. Courtesy of Saab AB

between the radiating slots. Thus, the beam direction can be steered by changing the frequency. This type of electronic beam steering is fairly simple and economical; it requires no electronic phase shifters, and the antenna can be fabricated with relative ease. Several linear arrays can of course be combined to form a planar array. Frequency scanned slotted waveguide arrays have been developed for frequencies as high as at least 180 GHz [20].

The general problem of frequency scanning has been treated by many authors [21–24]. There are several basic types of frequency scanned linear arrays; the most important are

1. *The travelling wave array*
   with matched termination and no phase reversal
   (no alternating slot offsets or slot inclination angles)
2. *The travelling wave array*
   with matched termination and phase reversal
3. *The resonant series-fed array*
   short instead of a matched termination

9.2.1 Travelling wave linear array

This case is illustrated by the single slotted waveguide array in Figure 9.6. See also Section 6.6, where a detailed design of a travelling wave array is presented. The slots can be longitudinal, displaced in the same direction (not alternating offsets) or inclined (in the same direction). The waveguide is fed in one end with a matched load in the other end.

Assume the slot spacing along the waveguide to be d. Thus, the phase delay from one slot to the next one is

$$\Delta\phi = \frac{2\pi}{\lambda_g} d \tag{9.2}$$

where λ_g is the guide wavelength.

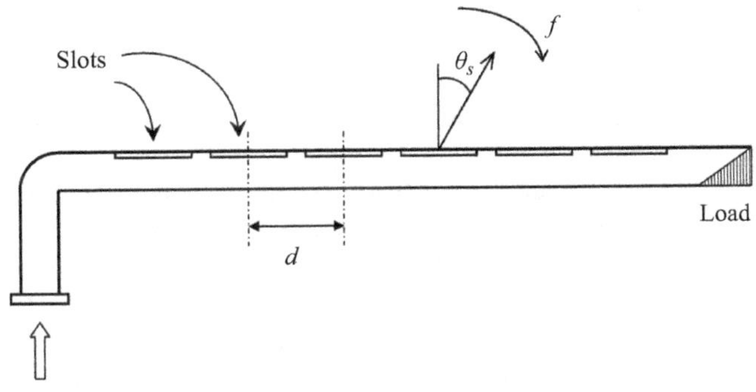

Figure 9.6 A travelling wave linear slotted waveguide array (leaky wave antenna)

The beam direction from the normal θ_s is related to the phase delay:

$$\Delta\phi = \frac{2\pi}{\lambda} d \sin\theta_s \tag{9.3}$$

yielding the beam direction

$$\sin\theta_s = \lambda/\lambda_g \tag{9.4}$$

Thus, the beam direction is independent of the spacing d, but it depends on the frequency. In a typical case with $\lambda_g = 1.5\lambda$ we obtain $\theta_s = 42°$. The antenna is called a leaky wave antenna or fast wave antenna. It is interesting to note that the beam has the same angle versus the y-direction as the constituent TE_{10} plane waves versus the x-direction (the waveguide mode can be seen as plane waves inside the waveguide bouncing between the waveguide side walls).

9.2.2 Phase reversal travelling array

Here we assume the same general configuration as before, but with the slots having alternating offsets; the same theory applies to a linear array with inclined slots with alternating inclination angles. The waveguide is terminated in a matched load as before. The phase delay from one slot to the next becomes

$$\Delta\phi = \frac{2\pi}{\lambda_g} d - \pi \tag{9.5}$$

where the minus π comes from the phase reversals. The beam direction from the normal is

$$\sin\theta_s = \frac{\lambda}{\lambda_g}\left(1 - \frac{\lambda_g}{2d}\right) \tag{9.6}$$

hence depending on both the frequency and the spacing d. Differentiating the last expression gives the approximate scan versus frequency sensitivity:

$$[\Delta\theta]_{\text{degrees}} \approx \left[\frac{\Delta f}{f_0}\right]_{\%} \tag{9.7}$$

that is, about $\pm 1°$ scan for $\pm 1\%$ frequency variation. The closer to waveguide cut-off the larger is the beam squint. An example is shown in Figure 9.7 for $\pm 5\%$ frequency scan yielding about $\pm 5°$ of beam scanning. It is here assumed that the beam shall be directed normal to the array at the frequency $f = f_0$.

The useful angular regions for this type of array are indicated in Figure 9.8. It scans backwards (towards the feed input) when the slot spacing is less than $\lambda_g/2$ and forward when the spacing is larger than $\lambda_g/2$. See also Figure 6.15. Three 'forbidden' regions are indicated: (1) slot spacing less than the slot length is assumed not possible, (2) slot spacing $d = \lambda_g/2$ implies that the reflections from all slots add in phase with a resulting high VSWR at the input and (3) too large

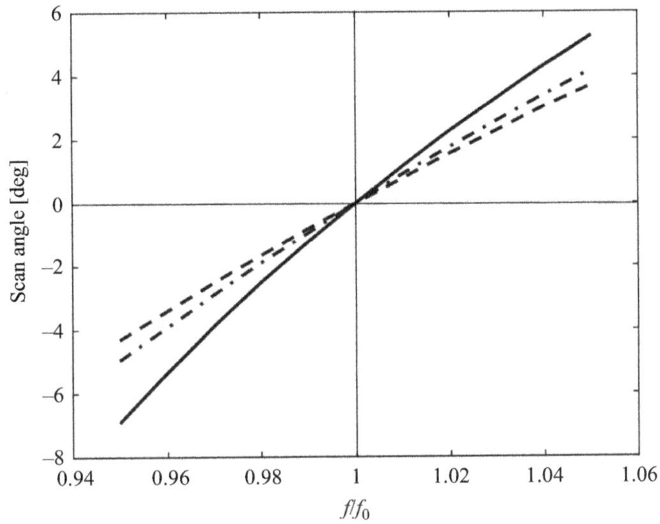

Figure 9.7 Frequency scanning of a travelling wave array antenna with phase reversal. Standard waveguide with a dimension = 22.86 mm. Centre frequencies are: solid line 7.5 GHz, dash-dotted 8.5 GHz and dashed line 9.5 GHz

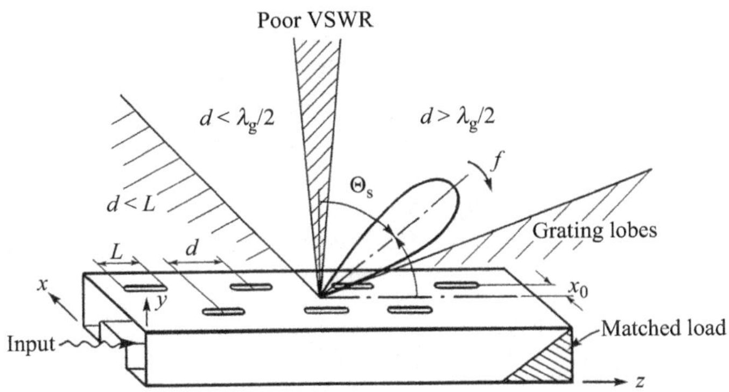

Figure 9.8 Scan limitations of a frequency scanned slotted waveguide array. Travelling wave with alternating slot offsets for phase reversal. © 1990 Microwave Exhibitions and Publishers Ltd. Reproduced by permission of Microwave Exhibitions and Publishers [2]

spacing causes a grating lobe to radiate. To prevent grating lobe radiation the requirement is

$$d/\lambda < \frac{1.5}{1 + \lambda/\lambda_g} \tag{9.8}$$

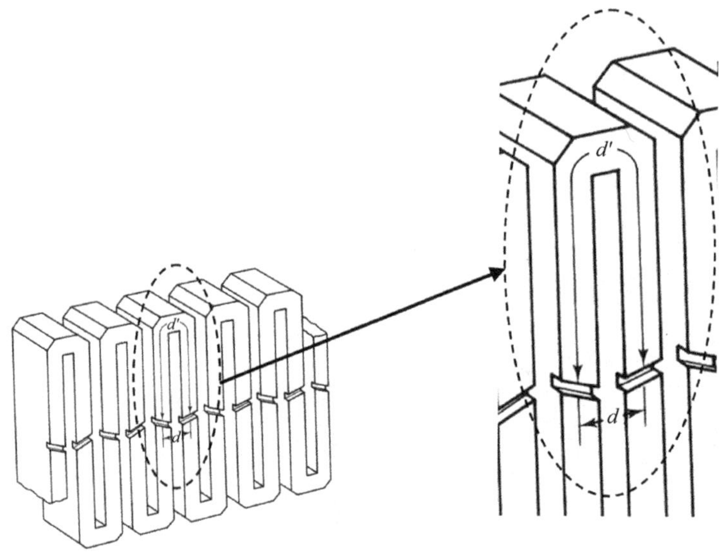

Figure 9.9 A serpentine feed. Note the increased path d′ compared to the slot spacing d. © 1981 J. H. Elliott. Adapted from [24], with permission

An example of a phased array antenna system using the phase reversal slotted travelling wave arrays as the phase steered elements was illustrated in Chapter 3 (Figure 3.5). In this case two-dimensional electronic scan was obtained: frequency scan in elevation and phase scan in azimuth. If the phase delay between slots is increased a larger scan range is obtained for the same frequency variation. This can be achieved with a folded, serpentine feed (Figure 9.9). In the figure the serpentine waveguide has radiating edge slots. It could also feed a second set of slotted waveguides to form a planar array; an example is shown in Figure 9.10. More than 90° of scan can be obtained with practical serpentines for 10% frequency variation.

With a serpentine (sinuous) feed of length d' the beam direction can be written

$$\sin\theta_s = \frac{d'\lambda}{d}\left(\frac{1}{\lambda_g} - \frac{1}{\lambda_{g0}}\right) \tag{9.9}$$

where it is assumed that the beam direction $\theta_s = 0$ when $\lambda_g = \lambda_{g0}$. From (9.9) it appears that $\sin\theta_s$ might exceed unity for large phase delays. However, phase is ambiguous and we need to pick the right branch, that is inside $[0,2\pi]$ or $[-\pi, +\pi]$. Compare the real versus imaginary space for array patterns in Figure 9.11. The necessary modification to (9.9) for any beam direction is to include an integer parameter m as shown in (9.10):

$$\sin\theta_s = \frac{\lambda}{d}\left(\frac{d'}{\lambda_g} - m - \frac{1}{2}\right) \tag{9.10}$$

The scanning range with serpentine lengths of 0.5, 3 and 5 guide wavelengths is illustrated in Figure 9.12.

Figure 9.10 The ITT Gilfillan AN/SPS-48 3D radar antenna. Note serpentine on the left. Photo courtesy of Troy Prince [25]

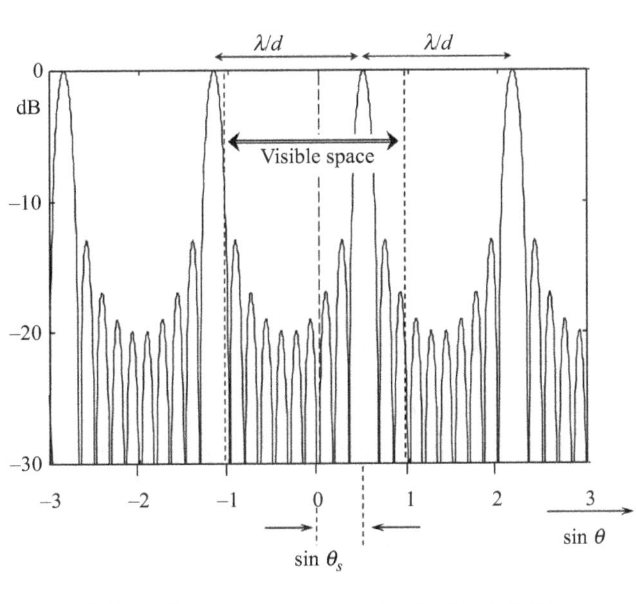

Figure 9.11 The real beam must be inside 'visible space'

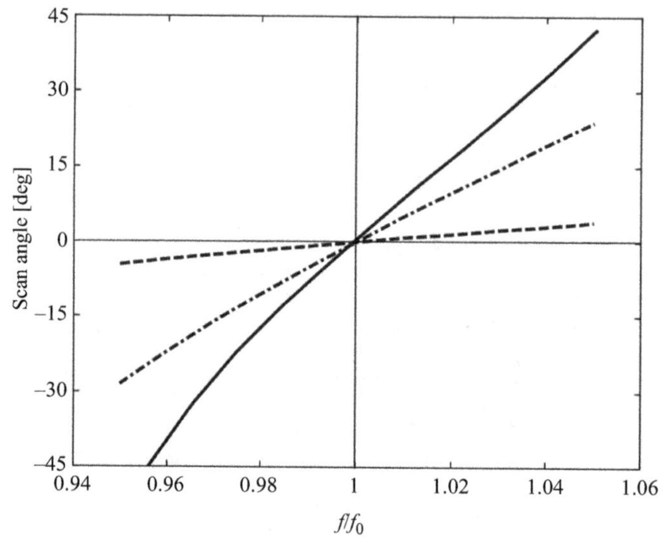

Figure 9.12 Frequency scan with different serpentine lengths d'. Waveguide a dimension 22.86 mm, centre frequency $f_0 = 9.0$ GHz. Solid line $d' = 5\lambda_g$, dash-dotted line $d' = 3\lambda_g$ and dashed line $d' = 0.5\lambda_g$

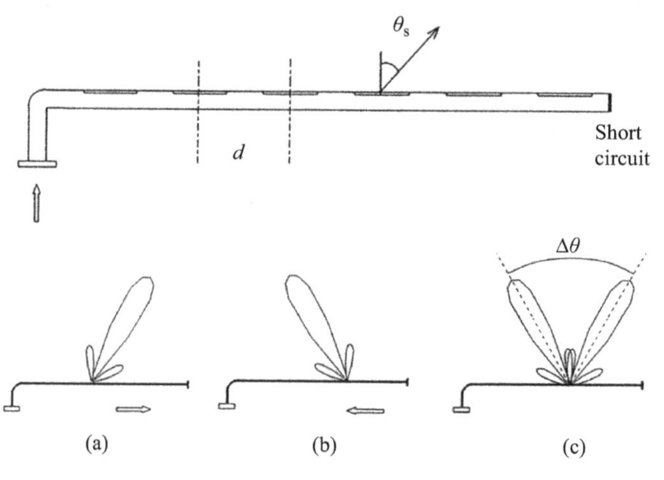

Figure 9.13 The resonant series feed array

9.2.3 Resonant series feed array

In this array the matched load is replaced by a short circuit. There will be a forward travelling wave (Figure 9.13(a)) and a reflected backward travelling wave (b) in the waveguide. Each of the waves will excite one beam and the two beams (c) will coincide only if the slot spacing is equal to half the guide wavelength. In the example shown $d > \lambda_g/2$.

The array will produce one single beam only in a narrow frequency range where the two beams overlap satisfactorily. Assuming a maximum beam separation of half the three dB beamwidth we get a useful bandwidth BW of about

$$\mathrm{BW} = \frac{50}{N}\% \tag{9.11}$$

where N is the number of slots. The number 50 is approximately valid for no taper and will be slightly larger in an amplitude tapered array. We find that the bandwidth in per cent approximately equals the beam width in degrees.

We remarked before in relation to Figure 9.7 that scanning in the broadside direction would cause prohibitively large reflections, since all slot reflections add up in this case. Especially in long waveguides this could be problematic. However, certain slot configurations have been found to reduce the reflections considerably. In one case an inductive post adjacent to the slot is used to tune out the slot reflection [26]. Another possibility is to use pairs of transverse slots (closely spaced) that both radiate but the respective reflections are out of phase [27,28]. A variation still is the offset crossed slot radiating circular polarisation which scatters only in the forward direction [29,30].

9.2.4 Centre-fed travelling array

Returning to the travelling wave array antenna we observed that it is relatively simple to build and therefore attractive. The beam squint, however, is sometimes not acceptable. One such case is in frequency agile radar systems where fast frequency changes would make the beam jump leaving blank sectors in the search pattern. A more stable beam can be obtained, however, by feeding in the centre of a travelling wave array with matched loads at both ends [2,31]. This gives two partial beams, one from each half of the antenna. The slot separations are different in the two antenna halves, but such that the partial main beams overlap. This means that one beam is forward fired, the other backward fired. The scheme is illustrated in Figure 9.14.

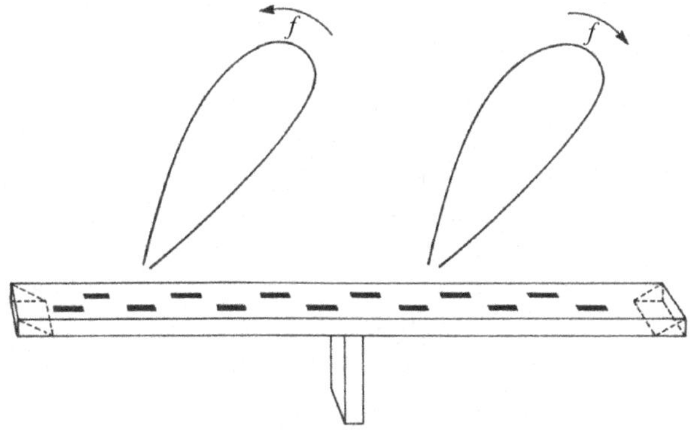

Figure 9.14 Principle for the squintless antenna. © DSTO Australia (1990), reprinted from Radarcon 90 [31], with permission

When the frequency is increased both beams scan towards their respective waveguide loads with the result that the composite beam remains relatively stable in the far field. The bandwidth realised before the beam splits depends on the array length, but can in a practical case be around 3–4%.

This antenna has several interesting features from a theoretical viewpoint. Since the slot spacing is different in the two halves the usual design methods are not directly applicable. The problem has been addressed by O'Loughlin *et al.* [32]. See also Figure 11.7 (Chapter 11).

9.3 Dual polarised slotted waveguide arrays

Dual polarised antennas can significantly improve system performance by extracting extra information from the incoming signals. Typical examples include remote sensing systems such as synthetic aperture radars and Multiple Input Multiple Output (MIMO) systems. Circular polarisation is effective in rain clutter rejection for radar. Polarisation diversity can improve the suppression of jamming and interfering signals, etc.

Many special configurations using waveguide slots combined with a second element type (dipoles, microstrip patches, stripline-fed slots) have been proposed for dual polarised arrays. One example is a flat plate waveguide array antenna that has a microstrip dipole array with feed lines bonded to the waveguide array surface, thus creating a dual polarised array sharing the same aperture [33]. However, we will in this section concentrate on solutions that employ only waveguide-fed slots for achieving the dual polarised function.

The slotted waveguide array antenna is primarily a single polarisation device as are the narrow slot radiators themselves. However, the slots in a rectangular waveguide can be oriented in many ways [34]; see Figure 1.2 in Chapter 1. A combination of two sets of arrays with different slot types and feeding the two arrays from a switching/phasing circuit would enable a variable polarisation antenna system. The question is of course how to make this in a practical and efficient way.

A waveguide with square cross section can support both the TE_{10} mode and the TE_{01} mode independently of each other. Crossed slots centred on one of the waveguide walls would therefore radiate either longitudinal polarisation or transverse polarisation depending on the waveguide mode that is excited [35]. By control of the mode amplitudes and phases any polarisation state can be radiated. A major drawback, however, of this scheme is the size of the square waveguide which does not allow for the tight spacing often required in a phase steered array. Also, spacing of slots at one waveguide wavelength intervals for in-phase excitation would cause grating lobes to radiate. Some sort of waveguide loading, for example, dielectric [5] or corrugated [36] could remedy this.

A solution which enables a relatively tight spacing was proposed by Ajioka *et al.* [37,38] using a bifurcated waveguide with crossed slots radiating from the narrow wall (Figure 9.15).

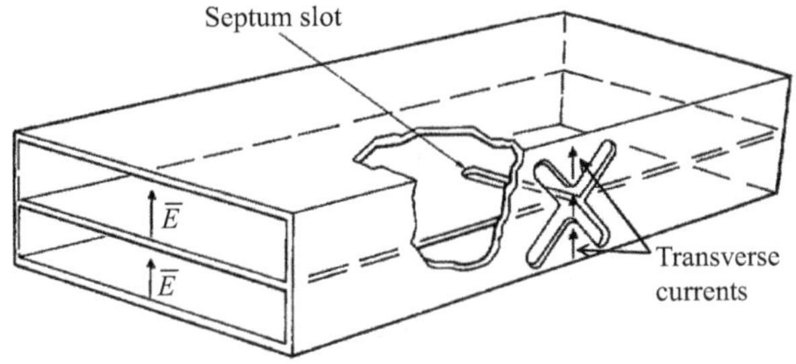

Figure 9.15 The bifurcated waveguide. With even excitation (shown) vertical polarisation is created. Odd excitation creates horizontal polarisation. © IEEE, adapted from [38], with permission

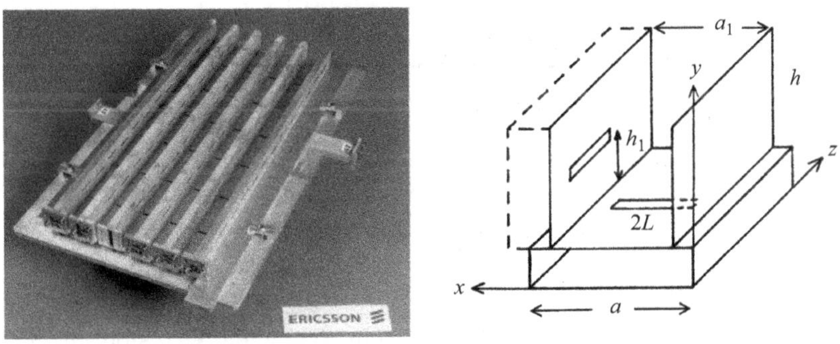

Figure 9.16 (a) Breadboard sub-array for polarimetric SAR. Courtesy of RUAG Space AB. (b) Building block with main dimensions indicated. © URSI 1992. Adapted from [40], with permission

Odd and even excitations of the dual waveguide provide the means for controlling the polarisation. The waveguide can be low height so that arraying in a phase scanned antenna is possible. Related schemes, also based on bifurcated waveguides, have been proposed by Sangster *et al.* [39]. It remains, however, to find a solution to the grating lobe problem caused by the λ_g spacing of the septum slots for in-phase excitation.

A combination of two separate waveguides with respectively longitudinal slots and transverse slots could provide the two orthogonal polarisation components required for polarisation agility. In order to accommodate the waveguide pair in an array with sufficiently dense spacing several solutions have been proposed. Figure 9.16 shows one breadboard C-band array studied for a polarimetric SAR [40].

The building block is shown to the right in Figure 9.16. Both polarisations excite the region between the vertical baffles that are formed by the waveguides standing on the narrow side. Inside the baffle region the fundamental propagating

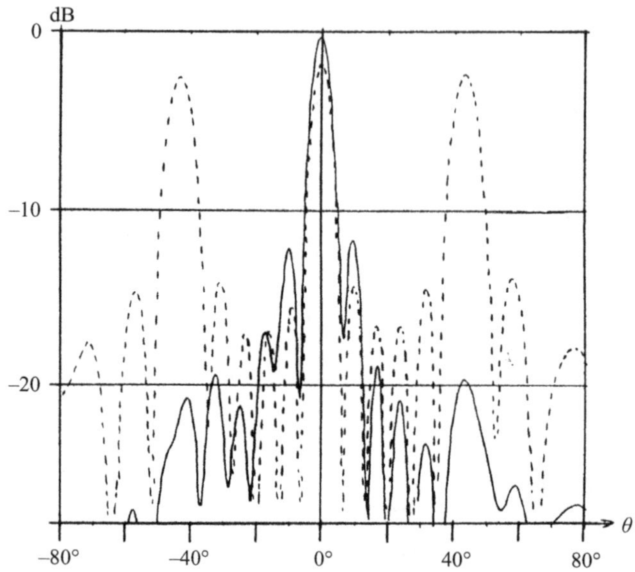

Figure 9.17 Measured patterns for a six-element array with transverse slots spaced 1.4 λ_0. Broken curve without baffle, and solid curve with baffle. © IEEE 1991, reprinted from [41], with permission

modes are a TEM parallel plate mode excited by the longitudinal slots and a TE(1) parallel plate mode excited by the transverse slots. The latter mode has a wavelength λ_1 larger than the wavelength λ_g of the feeding waveguide (since $a_1 < a$, Figure 9.16). The slots are resonantly spaced at λ_g distance. Hence the grating lobes inside the baffle region are evanescent and will be attenuated provided that $a > a_1 > \lambda_0/2$, with λ_0 the free space wavelength. In a typical case a baffle height of one-half wavelength gives about 20 dB grating lobe attenuation; see Figure 9.17 [41,42].

The baffles work as a spatial filter, with a pass band centred at normal incidence. Other spatial filter structures could be selected for the same purpose, based on, for example, dielectric layers or wire grid layers [43].

We give one more example of a proposed dual polarisation waveguide element using baffles (Figure 9.18). Here λ_g spaced transverse slots have the grating lobes suppressed as before using baffles. Longitudinal slots at $\lambda_g/2$ spacing on the waveguide centreline would normally not be excited. However, in this case the lower array, also with centred longitudinal slots *with post excitation* radiates through the upper longitudinal slots [44].

Similar to Figure 9.16 two waveguides, one for each polarisation could be used with the lower waveguide having longitudinal slots and the other waveguide standing on its side having edge slots. This simple principle is illustrated in Figure 9.19 [45].

Unfortunately, the inclined edge slots will radiate some cross-polarisation due to the inclination angle. However, this problem can be circumvented by exciting

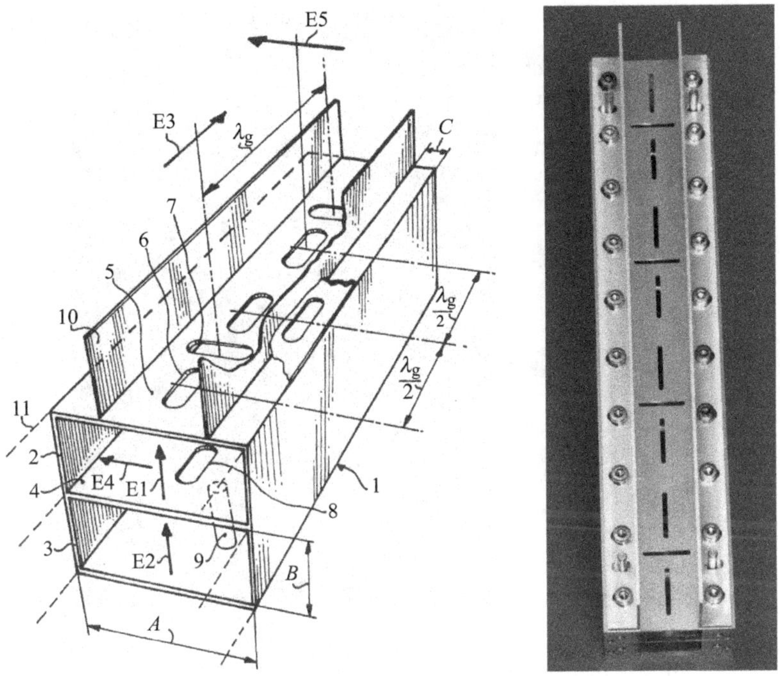

(a) (b)

Figure 9.18 Dual polarised slotted waveguide using baffle for grating lobe suppression. A breadboard sub-array is shown on the right. Courtesy of RUAG Space AB

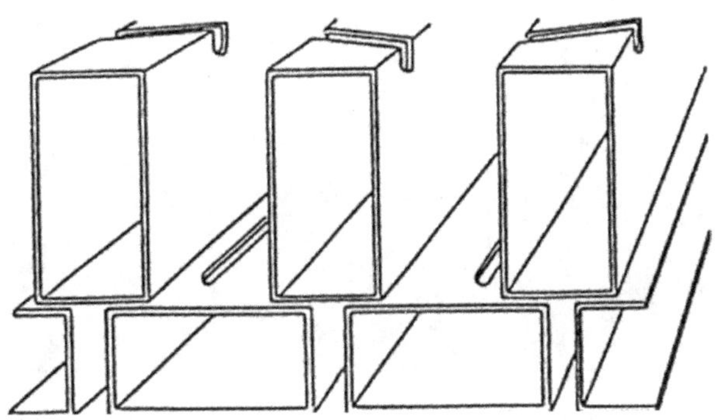

Figure 9.19 Longitudinal and inclined edge slot radiators in a dual polarised array. © IEEE 1996, reprinted from [45], with permission

non-inclined edge slots by a pair of wires inside the waveguide. Furthermore, to allow a large scan range in the transverse plane (vertical) the lower waveguide can be ridge loaded. A breadboard panel using this modified concept [46,47] is shown in Figure 9.20. This design allows about $\pm 20°$ vertical scan range.

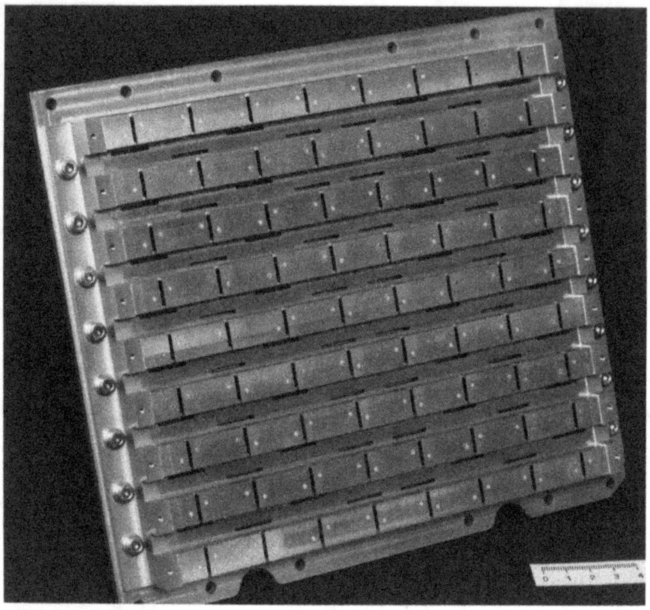

Figure 9.20 Breadboard panel of a dual polarised array. Courtesy of RUAG Space AB

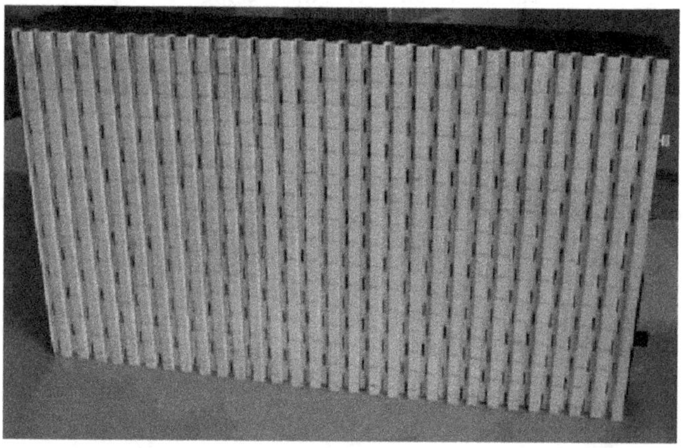

Figure 9.21 The DESA X-band SAR subpanel demonstrator. Courtesy of RUAG Space AB

The technology has been implemented in an X-band SAR Instrument Demonstrator (DESA) developed by Dornier/Astrium [48,49]. A sub-array panel of 30 dual polarised waveguide elements is shown in Figure 9.21.

9.4 Multiple beam slotted waveguide arrays

A system using two arrays with edge slots, one behind the other, has been proposed for a dual beam array [50] (Figure 9.22). The rear array radiates through the spaces between the waveguides of the front array. Two orthogonal linear polarisations are available. Each of the arrays can be phase steered in one dimension while the combined antenna is mechanically steered. This allows two independently steered beams from the same aperture, however with limited polarisation flexibility.

The aforementioned example is not strictly a multiple beam solution. With multiple beam arrays we usually mean systems that generate a number of beams in space from one common aperture. The feed system can be a Butler matrix [51], a Blass matrix [52] or equivalent. Typically, each beam is associated with a physical beam port in the feeding system. For the beam ports to carry independent signals they must be isolated from each other, in fact the isolation requirement is equivalent to a requirement on the orthogonality of the corresponding radiation patterns. For two beams with pattern functions $\overline{E}_1$ and $\overline{E}_2$ the orthogonality requirement [53] can be written as

$$\int_0^{2\pi} d\phi \int_0^{\pi} \overline{E}_1(\theta,\phi) \cdot \overline{E}_2^*(\theta,\phi) \sin\theta d\theta = 0 \tag{9.12}$$

A waveguide system for generating five independent beams is shown in Figure 9.23(a). It is shown here as a receive system but works in principle for both receive and transmit. A large receive only system was once meant to be used in a height finding air traffic control radar with more than 100 elevation beams [54,55]. The concept, usually termed Blass matrix [52], has been studied for future spaceborne SAR applications with two and three beams. A two-beam array, one

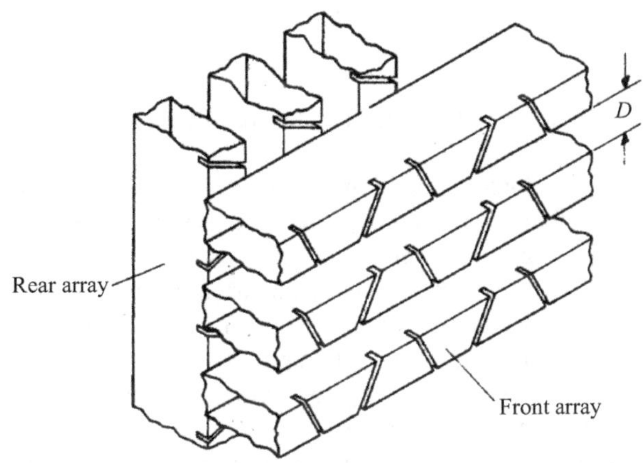

Figure 9.22 Crossed edge slot arrays according to Alexander and Sichelstiel. Reprinted from [50]

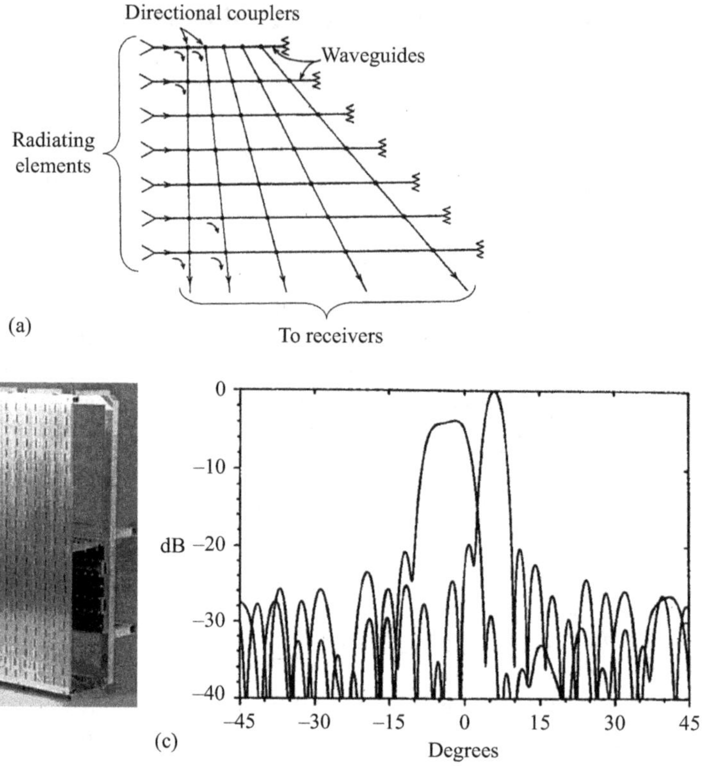

Figure 9.23 The Blass concept, example with five beam ports. © McGraw-Hill Education 1962, reprinted from Skolnik [54] with permission. (b) Dual beam slotted waveguide array. Courtesy of RUAG Space AB. (c) Measured dual beam patterns. © IEEE 1960, reprinted from [56], with permission

narrow beam and one shaped broad beam were successfully tested [56,57]; see Figure 9.23(b) and (c).

In Section 9.2 on resonant travelling wave-type slotted waveguide arrays it was mentioned that the reflection from the shorted termination causes a second beam to radiate. Thus, by feeding the two ends of a non-resonant array we can construct a dual beam antenna with the two beams symmetrically directed relative to broadside. However, a high-efficiency symmetrical taper will be difficult to arrange [23].

Classical navigational systems for aircraft normally include an X-band Doppler radar with four beams pointing towards ground in the directions forward right and left, and rear right and left. By measuring the Doppler shift of the four returns the speed and direction of the aircraft can be determined with high precision. Multi-beam arrays using non-resonant slotted waveguide arrays are suitable for this, considering their inherent stability and the high precision

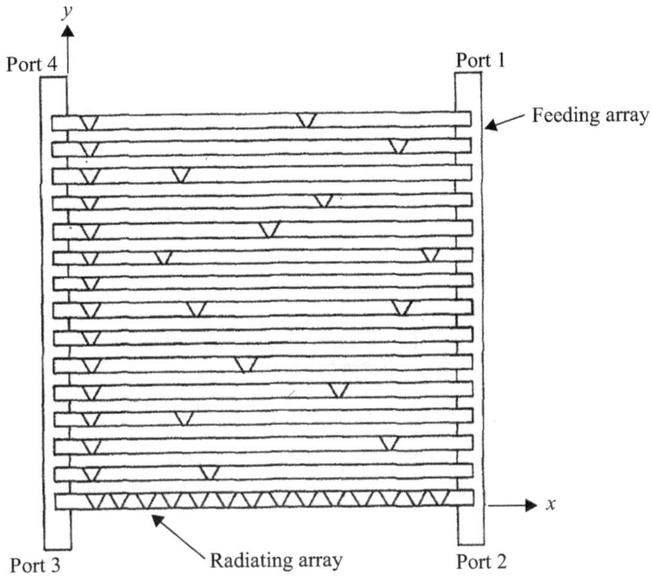

Figure 9.24 Layout of four-beam Doppler navigation array. © 1963 IEEE, reprinted from [58], with permission

manufacturing possible. By feeding a planar array in the four corners as illustrated in Figure 9.24 the four beams are realised [58].

Switching between two sets of waveguide arrays was proposed by Saltzman and Stavis [59]. Similar designs have been presented by Miller and Forman [60], who used square waveguides with TE_{10} and TE_{01} modes for better isolation between the beam ports.

9.5 Conformal slotted waveguide arrays

A conformal antenna has a shape that is dictated not by electromagnetic considerations but by, for example, aerodynamic requirements such as in aircraft structures (the IEEE Standard Definition of Terms for Antennas, IEEE Std 145-1993). Thus the antenna conforms to a *given* shape which often is not optimum for the antenna performance. In many cases, however, a particular curved shape is chosen in order to realise a specific requirement, for example, a spherical shape for a very wide angular coverage. Such an array is also said to be conformal. Straight slotted waveguides are not easily used in doubly curved array antennas. However, they fit quite well to cylindrical geometries, as shown by the example in Figure 9.25 [61].

The scattering from a cylindrical slotted waveguide array antenna was studied by Fan and Jin [62] (see Figure 9.26). See also Section 9.8.

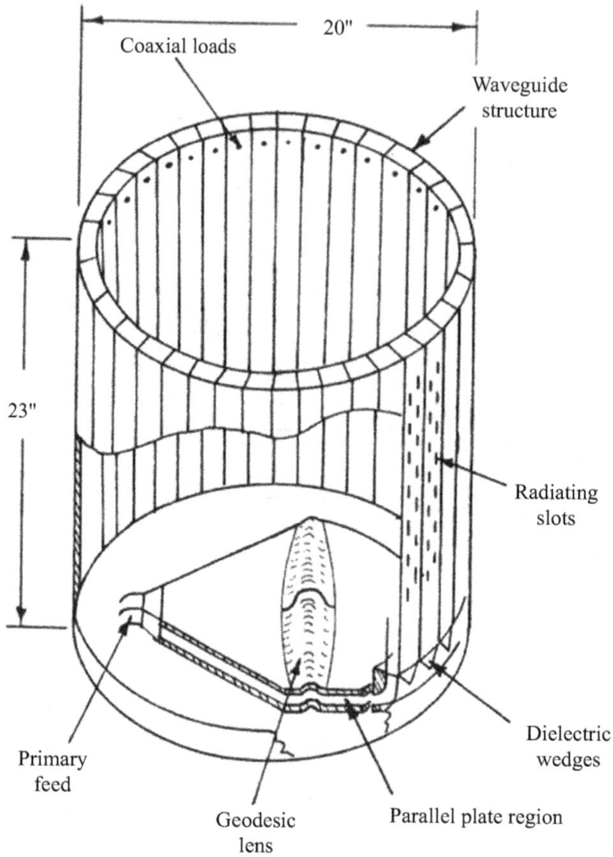

Figure 9.25 A cylindrical slotted waveguide array with a lens type rotating feed. © IEEE 1980, reprinted from [61], with permission

In linear and planar array antennas all elements point in the same direction and we usually write the antenna pattern function as a product of an element factor and an array factor. Thus, for a linear array we have

$$E(\phi) = EL(\phi)\sum_n V_n e^{jnkd \sin\phi} \tag{9.13}$$

where $EL(\phi)$ is the element pattern function, V_n is the element excitation amplitude, d is the element spacing and $k = 2\pi/\lambda$.

For a corresponding circular ring array (cf. Figure 9.27(b)), an element factor can in general not be factored out as was done in (9.13). We have instead

$$E(\phi) = \sum_n V_n EL(\phi - n\Delta\varphi) e^{jkR\cos(\phi - n\Delta\varphi)} \tag{9.14}$$

where $\Delta\varphi$ is the element angular spacing and R is the circular array radius.

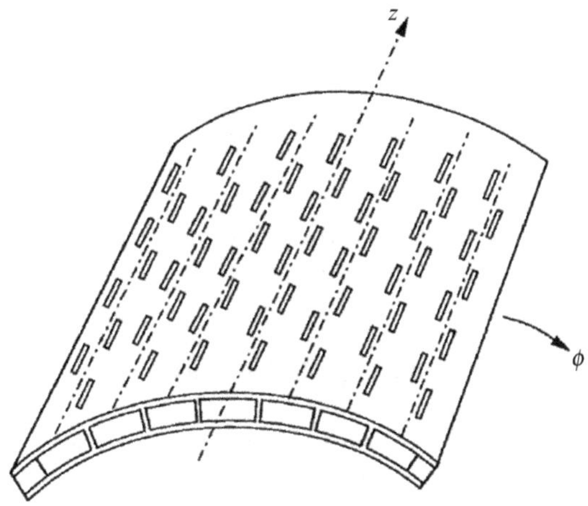

Figure 9.26 Curved slotted waveguide array antenna for radar cross section studies. © IEEE 1997, reprinted from [62], with permission

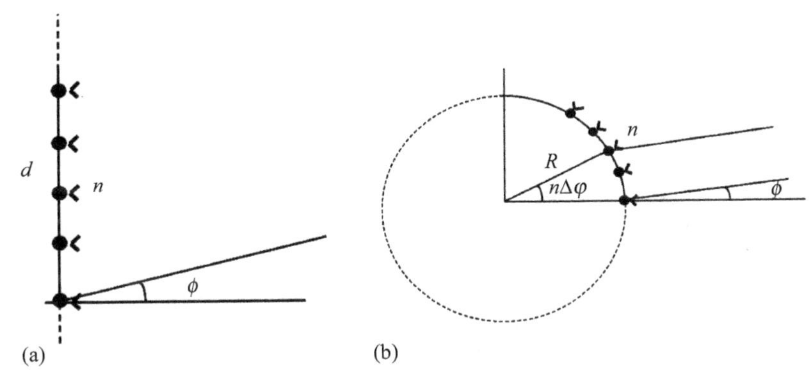

Figure 9.27 (a) Linear array; (b) part of a circular ring array

One consequence is that for conformal arrays in general the polarisation is not constant as the beam is steered to different angular positions. This may result in a need to adjust the polarisation individually for each radiating element [63]. Cf. also Section 9.3 on dual polarised waveguide arrays. However, in the special case of a circular or cylindrical array of vertical electric dipole elements parallel to the cylinder axis there is a common element factor with constant polarisation. For the array in Figure 9.25 we can see the slots as magnetic current sources, all pointing vertically. This results in a constant polarisation but there is still no common element factor that can be factored out.

Stripline-fed tapered slots [64] have been used in broadband cylindrical arrays [65]. For more complex shapes the polarisation problem gets worse. Circularly polarised elements is one possible solution. Linear polarisation from slot elements on a cone is a difficult case [66] as illustrated by Figure 9.28.

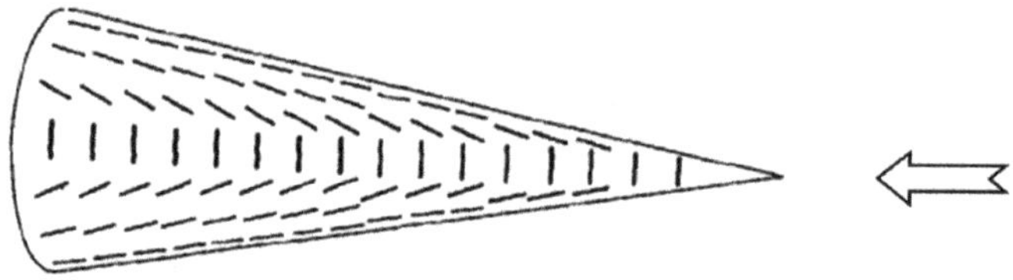

Figure 9.28 *Linear slot elements on a cone, oriented to produce vertical polarisation in the head-on direction (arrow). Adapted from [66]*

9.6 Monopulse slotted waveguide arrays

Monopulse antennas are used in tracking systems such as radar and missile seekers where the angular coordinates of detected targets shall be measured, in addition to measuring the distance. Early tracking systems used conical scan, typically a reflector antenna with a rotating feed causing the received signal to be amplitude modulated. Zero modulation resulted when the antenna was pointed correctly on the target. With *monopulse* (or *simultaneous lobing*) a single pulse provides in principle the necessary information by processing three simultaneously received signals from the antenna. This minimises the effect of amplitude noise on the measurement accuracy [54, p. 175].

Figure 9.29 illustrates the principle for a phase monopulse array antenna. The aperture is divided into four quadrants which are combined in a comparator network to provide the sum, the azimuth difference and the elevation difference signals.

The lower part of Figure 9.29 is the *comparator* consisting of four hybrids (magic Tees or equivalent) that form the sum and difference of the quadrant signals. The output signals are labelled Σ, Δ_{az}, and Δ_{el}.

The antenna patterns of the sum and difference signals depend on the illumination function. For the difference signal one aperture half is 180° out of phase with respect to the other half, that is, there is a strong discontinuity across the centre of the aperture where the amplitude is maximum. Very high sidelobes result. One could of course design an illumination function which is optimum for the difference pattern, but that would penalise the sum pattern with high sidelobes (see Figure 9.30).

A typical slotted waveguide array with its feed network is shown in Figure 9.31 [67]. The four quadrant outputs shall be connected to a waveguide comparator.

Optimum distributions for the sum function are, for example, the Taylor (sampled Taylor) and Dolph–Chebyshev distributions and for the difference function, the Bayliss distribution [68,69]. The best solution would be if the optimum distributions could be realised simultaneously and independent of each other. This is in theory possible since the distributions are orthogonal to each other. However, the resulting waveguide circuitry would be very complex.

Derneryd [70] proposed a separate central part of the array which is not included in forming the difference illumination functions (Figure 9.32). For the sum function the central part is included together with the quadrant parts. The result

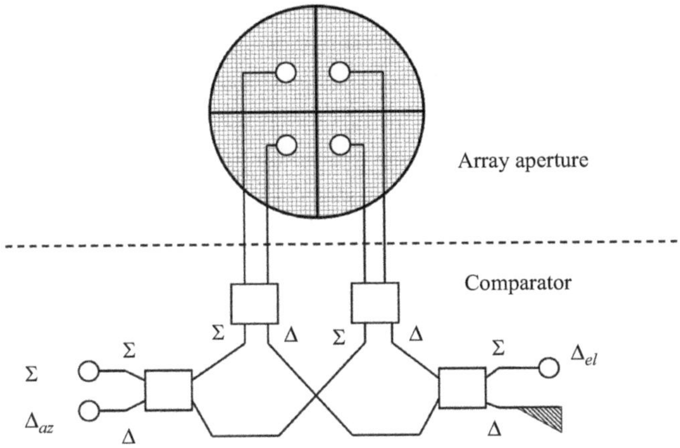

Figure 9.29 A dual plane monopulse antenna system with the aperture divided into four quadrants

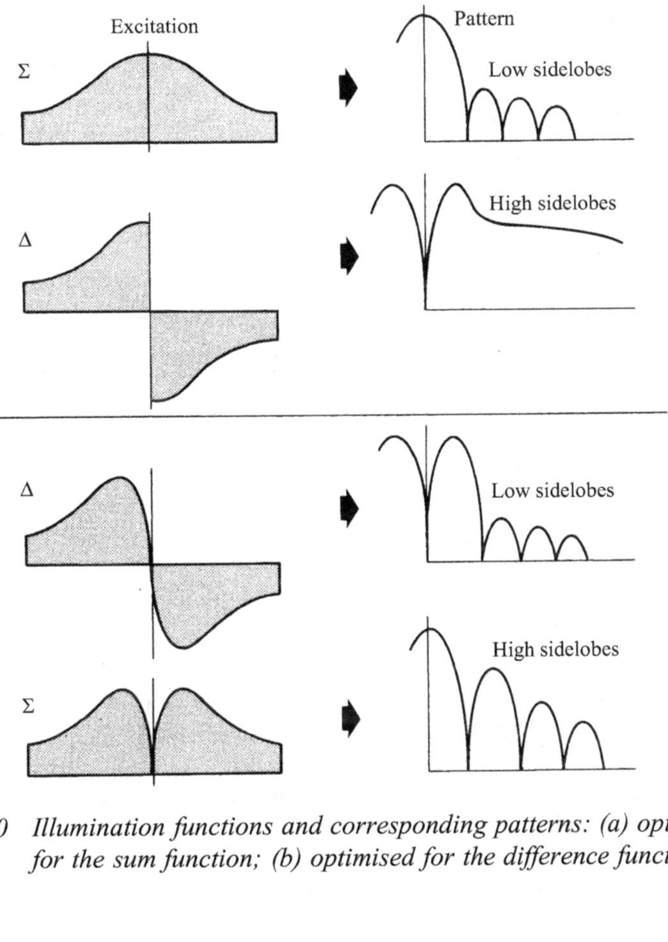

Figure 9.30 Illumination functions and corresponding patterns: (a) optimised for the sum function; (b) optimised for the difference function

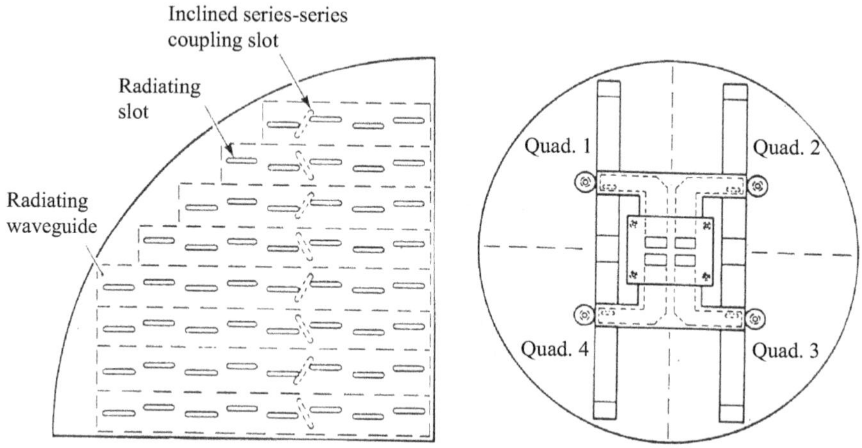

Figure 9.31 A typical slotted waveguide array antenna with one feeding point in each quadrant. One quadrant is shown to the left; the complete antenna (rear view) on the right. Courtesy of Microwave Journal, reprinted from [67], with permission

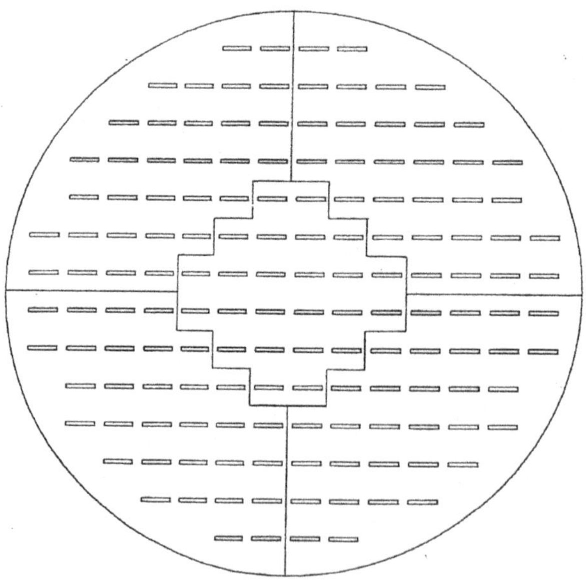

Figure 9.32 Partitioning of a monopulse array antenna with a fifth sector added. Reprinted from [70]

is a smoother (projected) difference illumination in the centre of the array (cf. Figure 9.30) and a sidelobe improvement of about 5 dB.

Quite complex feed networks are sometimes used for planar slotted waveguide arrays in order to extend the bandwidth of the antenna [71]. For a relative

Figure 9.33 Complex waveguide feed network on the rear of a large planar array. © Raytheon Company. Reproduced from [72], with permission

bandwidth of 10% the number of slots in series of a resonant slotted waveguide should not exceed about 50/10 = 5 slots, or 10 slots if centre fed (see Section 9.11). For a large array the aperture is therefore partitioned into smaller sections that are fed from a rather complex waveguide parallel feed network, sometimes in several layers (Figure 9.33).

9.7 Stripline slotted arrays

Stripline technology may seem a bit outside the slotted waveguide array subject that we have discussed so far. However, there are similarities between stripline-fed slots and waveguide-fed slots that are worth pointing out. Also, complex multilayer feed circuits can be realised in stripline in a more compact form compared to a waveguide solution. As an example a stripline flat plate monopulse antenna with partly independent sum- and difference feeds will be presented.

9.7.1 Introduction

A stripline slotted array antenna has slots in a metal ground plane fed from stripline circuitry in a layer below the slots [Figure 9.34(b)]. Just like the linear slotted waveguide array a linear stripline slot array can be envisioned with either longitudinal or transverse slots [73]. The slot positions are not bound to straight feeding waveguides, but can be chosen more freely, depending on the pattern requirements and the layout of the feeding circuit [74]. Some of the main characteristics of waveguide and stripline technologies are compared in Table 9.1.

Stripline circuits can be fabricated using printed circuit technology and are therefore less expensive than waveguide circuits. On the other hand, metal

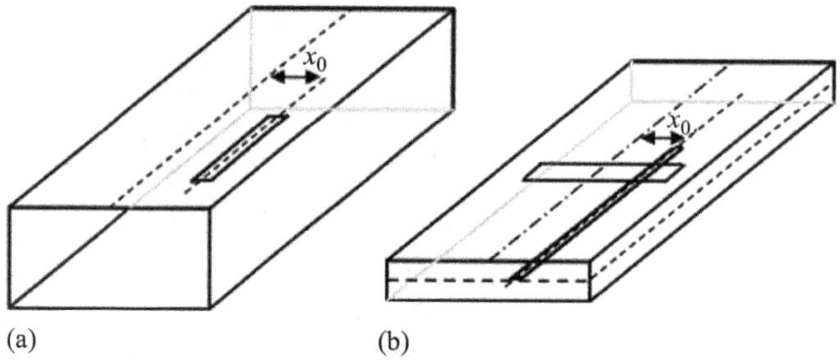

Figure 9.34 (a) Waveguide slot, offset from the waveguide centreline. (b) Stripline slot with the feed line offset from the slot centre

Table 9.1 Waveguide and stripline technologies compared

Parameter	Waveguide	Stripline
Material	Metal, aluminium	Plastic, PTFE
Fabrication	Dip brazing	Printed circuit technology
Power handling	Good	More limited
Design	Rather thick	Thin
Losses	Low	Moderate
Feed network	Complex	More flexible

Table 9.2 Selected stripline parameters

Ground plane separation	b	3.2 mm
Dielectric filling	ε_r	2.33
Strip conductor thickness	t	17 μm
Characteristic impedance	Z_0	50 Ω
Strip conductor width	w	to be determined

waveguides can withstand higher temperatures. There are also other technologies such as waveguides made of metallised CFRP, see Chapter 11, and metal-plated plastic waveguide antennas, described in Chapter 10.

In a stripline array the TEM mode is the dominant transmission line mode. In fact, the structure is similar to a coaxial line, albeit in a flat configuration. Side walls are necessary for preventing parallel plate modes to escape. The characteristic impedance of a stripline transmission line can be calculated or obtained from published graphs, see, for example, [75]. As an example we assume the following parameters (Table 9.2).

A calculation gives the result: $w = 2.56$ mm for 50 Ω impedance.

The slot can be symmetrically fed from the stripline in the centre of the slot. An offset feed position reduces the impedance [76]. The impedance depends also

(a)

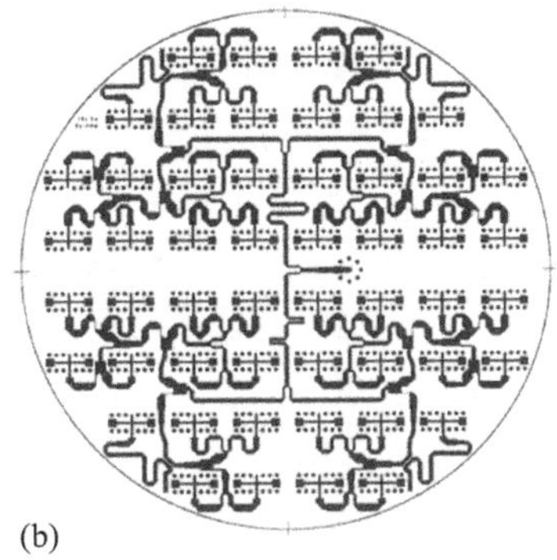

(b)

Figure 9.35 *(a) An X-band flat plate antenna in stripline technology. (b) Feed network. For clarity, both slots and stripline are here shown in black*

on the dimensions of the cavity formed under the slot, and the dielectric material [77,78]. At resonance a sinusoidal field distribution results along the slot. As we know from Chapter 4.2, the slot antenna has by itself rather high radiation impedance, typically around 500 Ω and some way of transforming the impedance to match the stripline feed is necessary. An example is shown in Figure 9.35, where the circuit layout of a stripline flat plate antenna [79] is shown. Each slot is fed from a narrow strip of length $\lambda_g/4$, which acts as an impedance transformer. The stripline continues as an open stub about $\lambda_g/4$ beyond the slot. The stub length is adjusted to tune the reactive part of the slot impedance. Alternatively, an off-centre feed position can be chosen as in Figure 9.34(b). However, this can cause problems with the circuit layout since a great deal of symmetry is lost.

9.7.2 Independent excitations

By subdividing each antenna quadrant into sectors and combining corresponding sectors in several comparator circuits a great degree of independence can be realised. The principle is illustrated in Figure 9.36. As seen in the figure each group of four sectors are combined in a comparator circuit. The outputs from the comparators are summed separately for *E*-plane difference, *H*-plane difference and Sum functions. These summation networks are optimised for their respective function. *Within* each sector the amplitude distribution remains the same for all functions.

In our design example (see the antenna in Figure 9.38) four sectors per quadrant were chosen. The amplitude weightings were decided from a detailed optimisation study where gain, sidelobe level, difference slope, number of sectors, etc., were included. A typical measured difference pattern is shown in Figure 9.37.

The slotted monopulse stripline array (Figure 9.38) is made up of three separate functional layers: the slot layer, the comparator layer and the summation layer. The total thickness is 7 mm. The slot cavities are formed by plated-through holes. The same technology is used for interconnections between layers. A corresponding antenna using ordinary metallic waveguides would be difficult to build.

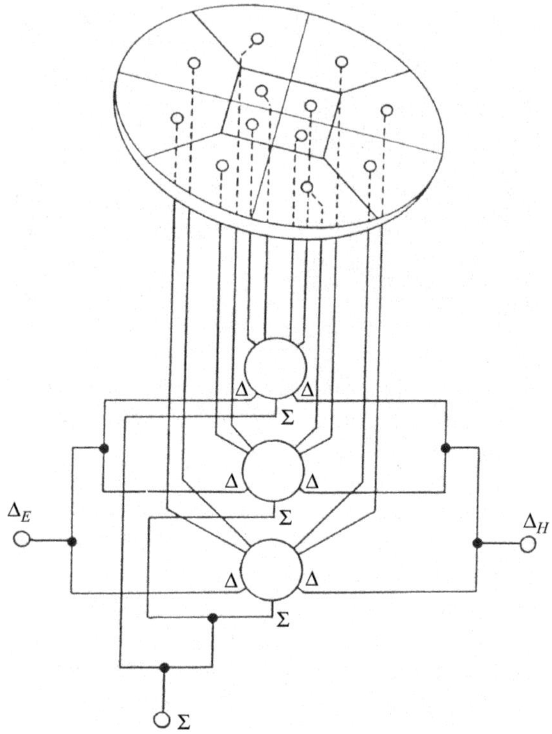

Figure 9.36 Feed principle for independent excitations with three sectors per quadrant. Reprinted from MEDE '77 [80]

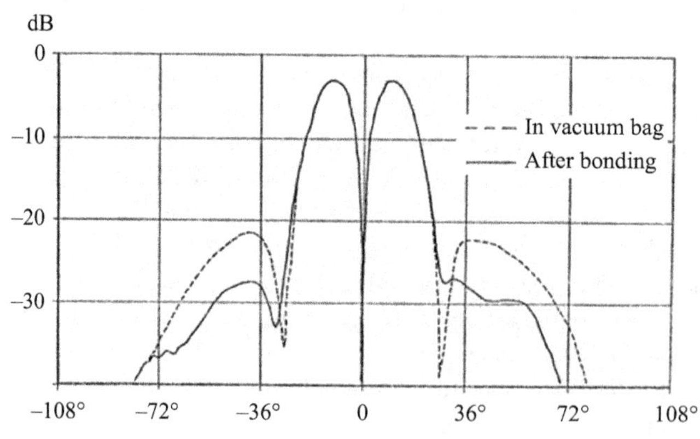

Figure 9.37 Measured difference pattern (H-plane). The dashed line was measured before bonding of the antenna layers. Reprinted from MEDE '77 [80]

Figure 9.38 Slotted monopulse stripline array antenna with three functional layers. Reprinted from MEDE '77 [80]

9.8 Scattering from slotted waveguide arrays

9.8.1 Introduction

Antennas are supposed to transmit and/or receive electromagnetic signals with adequate efficiency. Incoming signals shall be brought to the receiver(s), and not be scattered in various directions – in particular for stealth applications. It can be shown, however, that a good receiving antenna often is a good scatterer, as the following example shows. A waveguide aperture with length 39 mm and width 8 mm in a large conducting cylinder of diameter 600 mm has been analysed by matching the fields in the aperture using MoM [81]. Figure 9.39 shows the calculated scattering cross section in the E plane at a frequency of 6.8 GHz ($\lambda = 44$ mm), with and without matching of the aperture. Note that the contribution from the conducting cylinder itself is not included.

This example shows that matching of the aperture, which improves the receive function, also increases the scattering by about 4 dB. In the unmatched case the return loss was about 7 dB (transmission loss about 0.9 dB). For this analysis the single mode assumption was sufficient [81]. The matching was accomplished by means of an inductive iris a few mm below the aperture.

The *scattering cross section* σ referred to earlier is defined as

$$\sigma = \lim_{r \to \infty} 4\pi r^2 \frac{|E_{\text{scat}}|^2}{|E_{\text{inc}}|^2} \tag{9.15}$$

also called *radar cross section*. It is defined for backscattering – the monostatic case – where the illuminator and the receiver are in the same place. It is usually given in square meters, or dB > 1 m^2, dBsm. E_{inc} is the electric field strength at the target and E_{scat} is the field strength at the receiver.

There are also other cross section definitions:

- bistatic cross section – illuminator and receiver in different positions
- total scattering cross section – the sum of power scattered in all directions
- absorption cross section – total power absorbed
- extinction cross section – sum of scattered and absorbed power

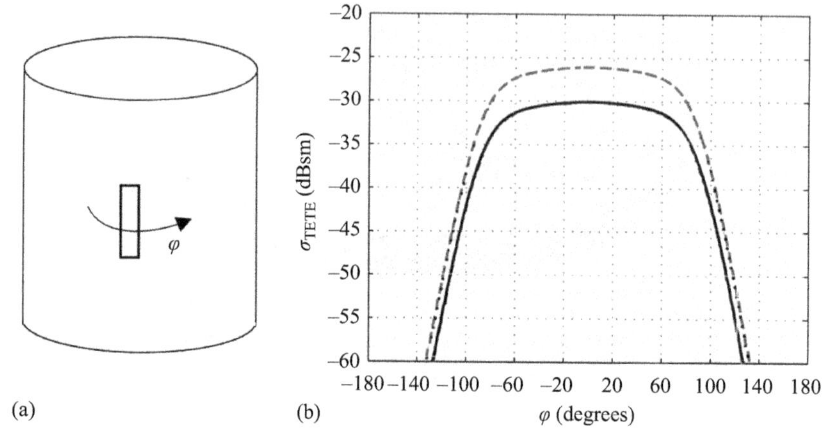

Figure 9.39 *(a) Waveguide aperture in a cylinder. (b) Calculated scattering cross sections compared. Single mode (TE_{10}) solution. Solid line: no matching; dashed line: matched aperture. © Wiley 2006. Adapted from [63], with permission*

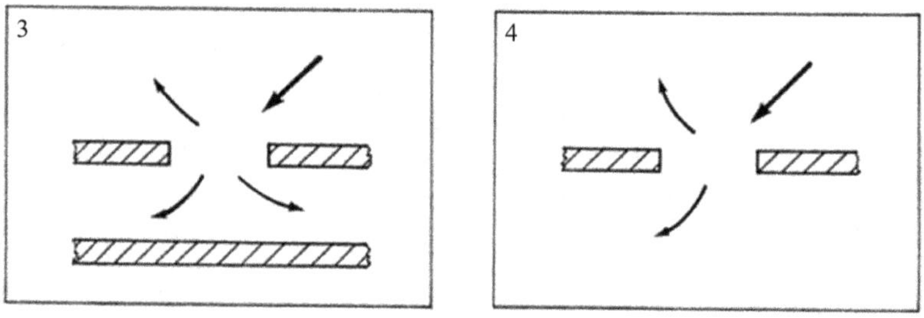

Figure 9.40 *Slot boundary problems numbers 3 and 4, waveguide slot and slot in a ground plane*

Numerous references to publications on electromagnetic scattering can be found in the *Proceedings of the IEEE*, August 1965 (special issue). See also [82] for a discussion about transmitting, receiving and scattering properties of antennas. In this section we will discuss very briefly a few aspects of antenna scattering that are related to slotted waveguide arrays.

9.8.2 Scattering by a single slot in a planar ground plane

Chapter 5 treated boundary problems for slots in rectangular waveguides. Four typical situations were outlined in Figure 5.5. We pick out no. 3 and no. 4 as relevant to scattering situations with plane wave incidence (Figure 9.40).

Let us for the time being assume that the wall thickness is very small and can be neglected. The extension to finite wall thickness is straight-forward, following the procedure in Section 5.1.4. The important field components are the longitudinal

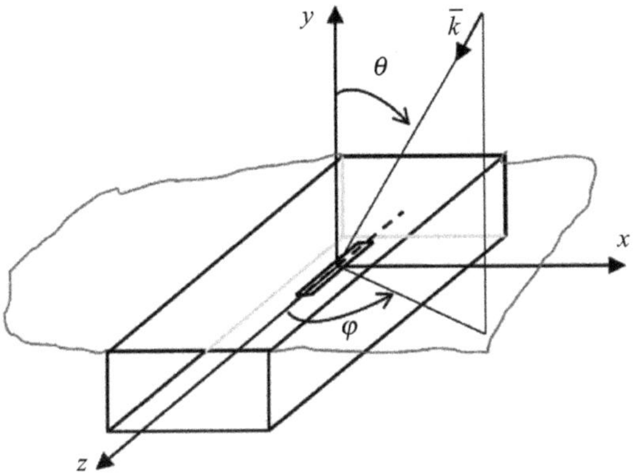

Figure 9.41 Plane wave incident on a waveguide slot

H_z and transverse E_x as discussed in Chapter 5. We denote the external field with superscript e and internal/lower field with superscript i, respectively. The boundary condition in mathematical form for problem 3 is therefore

$$H_z^{\mathrm{i}}(E_x) - H_z^{\mathrm{e}}(E_x) = H_z^{\mathrm{inc}} \qquad (9.16) = (5.2)$$

A unit amplitude plane wave incident from an angle (θ, φ) is assumed (Figure 9.41).

The E_x field is expanded as before (Chapter 5). Testing according to the moment method yields a solution for the electric field in the vector $[E] = (E_1, E_2, \ldots, E_N)$, cf. (5.8), where

$$[E] = H_0\left[[Y^i] - [Y^e]\right]^{-1} \cdot [h] \tag{9.17}$$

The excitation vector $[h]$ is here normalised; H_0 represents the amplitude of the incident field including the projection onto the slot and possible polarisation mismatch, thus

$$H_z^{\mathrm{inc}}(x, z) = H_0 \cdot e^{j(k_x x + k_z z)} \tag{9.18}$$

The excitation vector is very similar to the case with excitation from the waveguide. We obtain the elements h_s

$$h_s = w \cdot \mathrm{sinc}\left(\frac{k_x w}{2}\right) \frac{\pi s/L}{\left(\frac{\pi s}{2L}\right)^2 - k_z^2} \begin{Bmatrix} j \sin k_z L \\ \cos k_z L \end{Bmatrix} \begin{matrix} s \text{ even} \\ s \text{ odd} \end{matrix} \tag{9.19}$$

cf. (5.44).

The problem no. 4 in Figure 9.40, the slot in a ground plane (no waveguide), is easily solved in the same manner. We have $H_z^i = -H_z^e$ and thus

$$[E] = -(H_0/2) \cdot [Y^e]^{-1} \cdot [h] \tag{9.20}$$

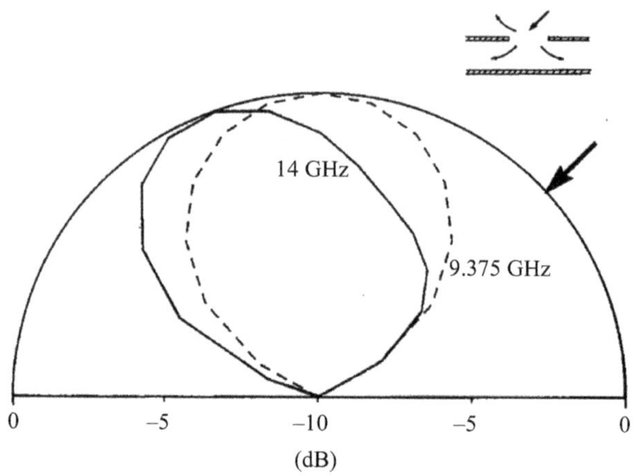

Figure 9.42 Bistatic scattering from a longitudinal slot in a waveguide. The angle of incidence is $\theta = 60°$

Now that we have the electric field distribution in the slot aperture we can derive the electric vector potential in the far field

$$F_z = \frac{\varepsilon e^{-jkr}}{2\pi r}[E] \cdot [h(k_x, k_z)] \tag{9.21}$$

where $[h(k_x, k_z)]$ is the same as the excitation vector, cf. (9.19), but with (k_x, k_z) now representing the outgoing direction. The far field is easily obtained since

$$\left(H_\theta, H_\varphi\right) = -j\omega\left(F_\theta, F_\varphi\right) \tag{9.22}$$

The bistatic scattering cross section is defined as $\sigma = \lim_{r\to\infty} 4\pi r^2 P_{\text{scat}}/P_{\text{inc}}$, where P_{inc} and P_{scat} are the incident and scattered power densities, respectively, yielding

$$\sigma = \frac{k^2 - k_z^2}{\pi Z_0}\left|[E] \cdot [h]\right|^2 \tag{9.23}$$

In this expression k_z represents the outgoing direction, which is also contained in $[h]$. The incidence direction is implicit in the solution for $[E]$.

Figure 9.42 shows a scattering diagram for a case with a waveguide slot illuminated from $(\theta, \varphi) = (60°, 0°)$. The slot length is 16 mm and thus close to resonance for the frequency 9.375 GHz. At the higher frequency an asymmetric slot field results due to the presence of higher slot modes. This is illustrated in Figure 9.43 showing the electric field distribution in the slot [83].

9.8.3 Scattering from an array of slots

The scattering cross section of a planar slotted waveguide array antenna has several contributions: the array of slots, the ground plane and the feed network behind the

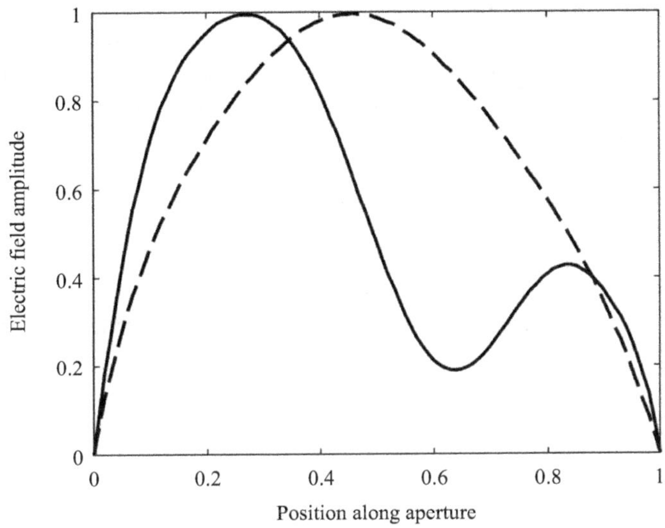

Figure 9.43 Slot electric field along the slot for the same case as in Figure 9.42. Solid line is 14 GHz, and broken line is 9.375 GHz. © IEEE 1990. Adapted from [83], with permission

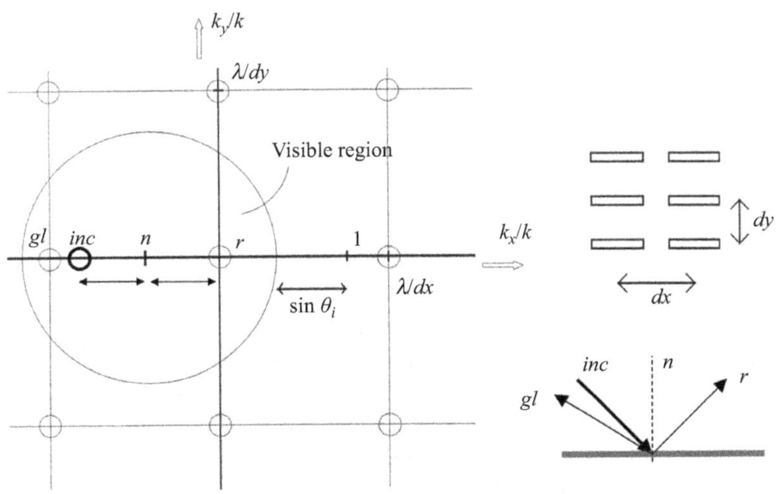

Figure 9.44 The grating lobe lattice. The incidence wave ('inc') excites the slots with a phase slope causing a reflected wave ('r'). In this example the grating lobe ('gl') is close to radiating back towards the illuminator

slots. At oblique incidence from the illuminator the slots are excited with a phase slope and radiate like a phase steered array. The grating lobe lattice, the reciprocal of the slot element lattice [84], is shown in Figure 9.44 for an array with slots in a rectangular grid.

A common spacing between waveguide slots is $\lambda_g/2$ or typically 0.7 λ_0 which prevents grating lobes to appear in the array antenna pattern. However, considering the scattering diagram the spacing must be as small as $\lambda_0/2$ to eliminate grating lobes for all directions of incidence. The general case is described by the grating equation

$$\sin\theta_{\text{inc}} + \sin\theta_{gl} = m\frac{\lambda}{d}, \tag{9.24}$$

where m is the order of diffraction. Compare Figure 9.11.

At the blazing angle the grating lobe radiates back in the direction of incidence, that is,

$$\sin\theta_{\text{inc}} = \sin\theta_{gl} = m\frac{\lambda}{2d} \tag{9.25}$$

The array ground plane scatters like a metal plate of the same size. Based on a physical optics approximation (neglecting edges) the monostatic cross section for a plate with area A is $\sigma = \frac{4\pi A^2}{\lambda^2}$ in the normal direction. The approximate angular dependence for a square plate with side L is

$$\sigma(\theta) = \sigma(0)\left[\frac{\sin(kL\sin\theta)}{kL\sin\theta}\right]^2\cos^2\theta \tag{9.26}$$

and for a circular plate with radius a

$$\sigma(\theta) = \sigma(0)\left[\frac{J_1(2ka\sin\theta)}{ka\sin\theta}\right]^2\cos^2\theta \tag{9.27}$$

see [85,86].

The simple expressions such as (9.26) for the rectangular plate work well near the normal direction. An analysis based on equivalent currents along the plate edges can predict the scattering more accurately outside the principal symmetry planes of rectangular plates, as demonstrated by Polka and Balanis [87].

A circular plate with 40 cm radius has about 3,000 m^2 radar cross section at X-band. This is quite a large cross section. The contribution from the slot array can be of the same order of magnitude [88].

It is evident that bending of the array defocuses the returns, that is, conformal arrays have an advantage here. A curved array of slotted waveguides was shown in Figure 9.26. We compare in Figure 9.45 the scattering from a planar and a cylindrical array of waveguide apertures, in both cases 40×32 apertures covering about 120° of the 600 mm diameter ($=13.6\ \lambda_0$) cylinder.

The example in Figure 9.45 is for the H-plane. In the E-plane the comparison is not so favourable for the conformal case due to the specular return for all φ values. See also [62].

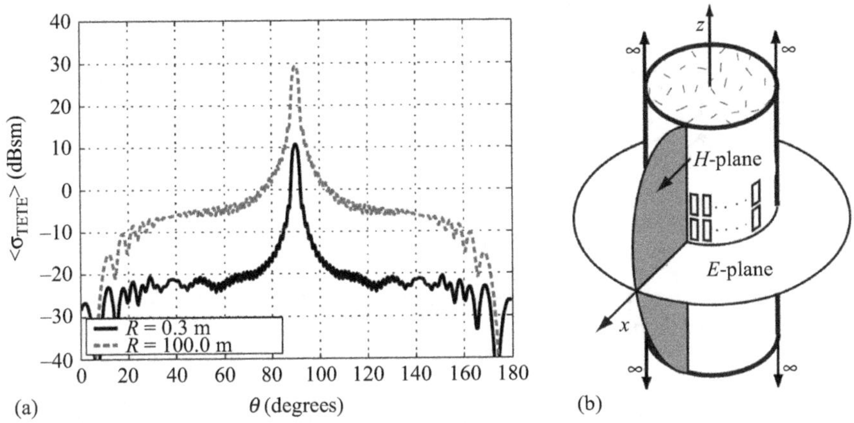

Figure 9.45 (a) Monostatic radar cross section for a waveguide array with 40 × 32 slots on a cylinder (solid line) compared with a corresponding array on a planar surface (dashed line). H-plane, (b) shows the geometry. © IEEE 2003, reprinted from [81], with permission

9.8.4 The feeding network

The elements in a slotted waveguide array antenna are fed from a network consisting of coupling slots, waveguide splitters and hybrid couplers, etc. This network is usually not well matched; in fact most of the power dividers are reactive and mismatched on the output ports. This will cause reflections towards the aperture and re-radiation as well as a perturbation of the phase and amplitude distribution. Consider, for example, a non-isolated power divider as symbolically illustrated in Figure 9.46(a). It is matched on the input port when the admittances are chosen $Y_0 = Y_2 + Y_3$. (An ideal transformer on the output ports is assumed in order to bring back all signals to the same impedance level.) A symbolic form of the divider is shown in Figure 9.46(b).

In terms of the power division ratio $q = P_2/P_3$ we can write the 3 × 3 scattering matrix:

$$S_q = \begin{pmatrix} 0 & \sqrt{\dfrac{q}{1+q}} & \dfrac{1}{\sqrt{1+q}} \\ \sqrt{\dfrac{q}{1+q}} & -\dfrac{1}{1+q} & \dfrac{\sqrt{q}}{1+q} \\ \dfrac{1}{\sqrt{1+q}} & \dfrac{\sqrt{q}}{1+q} & -\dfrac{q}{1+q} \end{pmatrix} \tag{9.28}$$

We find that an equal power splitter ($q = 1$) with no isolation has 6 dB return loss on the output ports. With a more uneven power split the mismatch increases on one of the ports, and decreases on the other. Note that transmission line lengths are not included in the above expression.

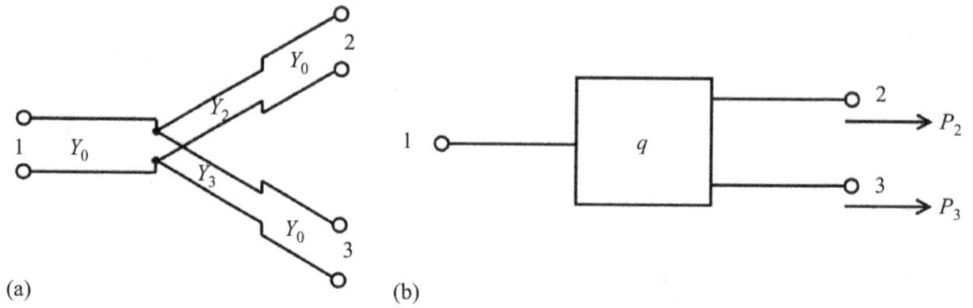

Figure 9.46 (a) A non-isolated power divider with transmission lines. (b) Symbolic version

The impact of the feeding system on the scattering from slotted waveguide array antennas is not much discussed in the literature with few exceptions; see, for example, Tittensor and Newton [88]. There are also various ways of modelling feed networks [89].

9.8.5 Numerical examples

There are not many publications dealing with the RCS performance of slotted waveguide arrays. Fan and Jin [90] and Zhang *et al.* [91] have calculated the scattering from fairly large (up to $>$1,000 slots) slotted waveguide arrays where the slotted waveguides were either shorted or matched at the waveguide ends and all slots had in most cases identical dimensions. The theoretical results have also been compared with measured results. The results verify the expected general behaviour including the grating lobe patterns.

In [92] a more realistic antenna, albeit small, was analysed. The rectangular array had six waveguides, each with 12 longitudinal slots, a feeding waveguide with coupling slots and a matched termination at the feed input. The analysis was based on solving the integral equations for the problem using MoM, in principle as described in Chapters 6 and 7. A typical computed result is shown in Figure 9.47.

In Figure 9.47 we see the specular reflection from the slot array at $-45°$ including a grating lobe at 32.5° (solid line). The specular return from the ground plane at $-45°$ is also shown. It can easily be found with the aid of the grating equation (9.24) that the slot spacing in the *H*-plane is 0.8 free space wavelengths. Very little power goes into the antenna matched load at this angle of incidence.

9.8.6 Spatial filters

Covering a planar antenna with a curved frequency selective radome will reduce the monostatic scattering peak at normal incidence, at least at out-of-band frequencies. Such frequency selective surfaces (FSS) can be designed using wire grids, dielectric layers, slot or dipole arrays, etc. Pelton and Munk [93] describe a conical metallic radome with a slotted surface with good transmission properties within a narrow frequency band. A spherical FSS surface to be mounted on a planar

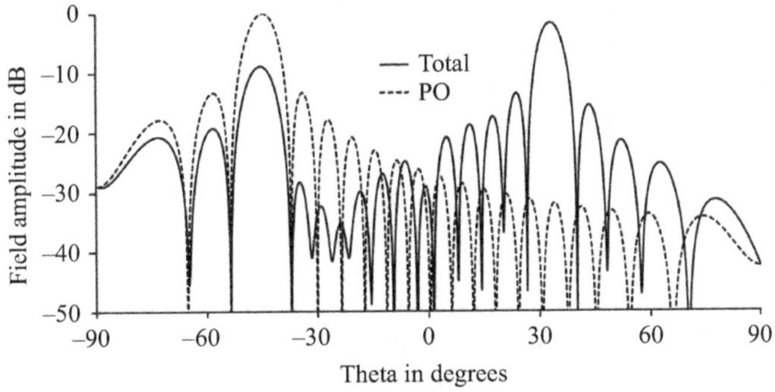

Figure 9.47 Scattered field in the H-plane for a rectangular slotted waveguide array antenna. Incidence is from $\theta_{inc} = 45°$. © IEEE 2007, reprinted from [92], with permission

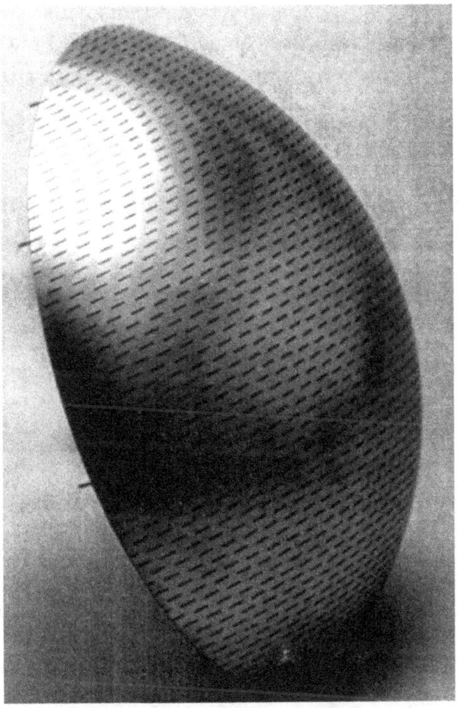

Figure 9.48 Frequency selective slotted radome. Courtesy of Saab AB

slotted waveguide array antenna has been described by Stanek and Johansson [94] (see Figure 9.48).

The FSS radome in Figure 9.48 is built up of two metal skins and a dielectric support. The 2,000 + slots in each skin are laid out in a quasi-periodic pattern due to the curvature, and designed for equal phase delay within ±5° and minimum loss within the operating frequency band. Out-of-band (and for the orthogonal polarisation) the radome scatters as a solid spherical surface. It can be viewed as a large conformal slotted array (but with no waveguides).

9.9 Second-order lobes

9.9.1 Introduction

The typical linear slotted waveguide array has longitudinal slots separated at about one-half waveguide wavelength (0.5 λ_g along the waveguide). The slot excitation is governed by the amount of slot offset from the waveguide centreline, with equal phase radiation obtained by alternating the offset directions (Figure 9.49).

Since the radiator spacing of 0.5 λ_g corresponds to about 0.7 λ_0 in free space the first grating lobe appears at an angle corresponding to $\sin\theta_g = 1/0.7 \approx 1.4$, that is, outside visible space (cf. Figure 9.11). However, considering the alternating slot arrangement, the array period is actually twice that of the single slot, that is, the array element is strictly speaking *the slot pair*. Thus, a grating lobe will appear at $\sin\theta = 0.7$ or circa 45° from broadside. However, in the vertical plane containing the array symmetry line such a grating lobe will not appear. But outside this plane the phase differences due to the slot offsets will result in increased sidelobes, or 'butterfly lobes' [95,96] (see Figure 9.50). This phenomenon, also named 'Gruenberg lobes', has been studied by several authors [97–101].

9.9.2 Analysis

For the simple case shown in Figure 9.49 the radiation patterns with and without slot offsets are easily calculated. Assuming the slots to be radiating isotropically

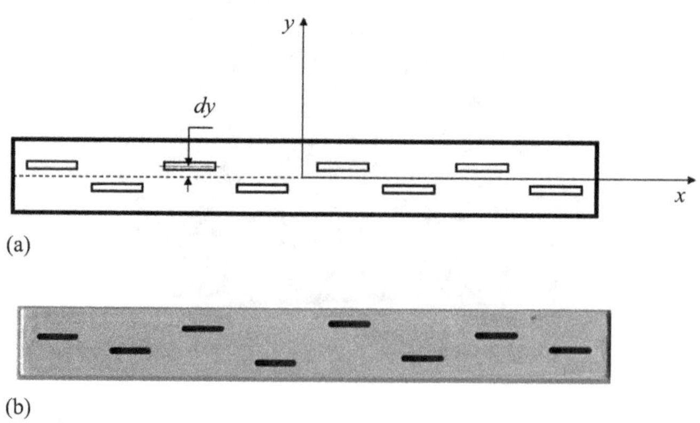

Figure 9.49 (a) A linear slot array, with the slot offset from the centreline indicated (dy). (b) A corresponding real antenna (with tapered excitation)

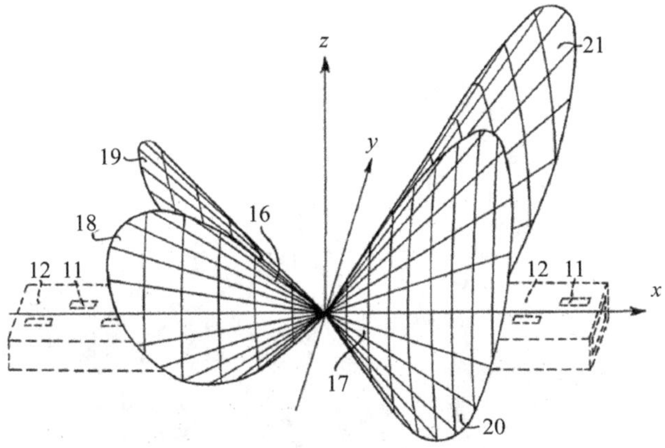

Figure 9.50 Butterfly lobes (numbers 18, 19, 20 and 21 in the figure) due to slot offsets showing up outside the cardinal planes. Reprinted from [96]

with identical polarisation we only need to examine the array factor. Two examples are shown in Figure 9.51.

As predicted, the secondary lobe appears at the grating lobe location $k_x/k = \frac{\lambda}{2dx}$ where dx is the slot spacing along the waveguide. The magnitude of the disturbance is proportional to the slot offset from the centreline; it follows the cone $k_x = \text{constant}$ and increases linearly with the transverse coordinate k_y.

It is possible to reduce the secondary lobe amplitude somewhat by breaking up the symmetry of the array; see Section 7.6. A possible layout from Ahlbom *et al.* [18] is shown in Figure 9.52. This type of layout applied to our eight-slot example (Figure 9.49) given earlier gives the result shown in Figure 9.53. See also [97,101].

A phase steered array with slotted waveguides as radiating elements (as in Figure 9.1) will also suffer from secondary lobes due to the slot offsets. As the main beam is steered the secondary lobe will be steered as well. It may then be moved into a region where the sidelobes otherwise are quite low (see Figure 9.54).

In Figure 9.55 the radiation over the full k_x/k_y-plane is shown for the same array, now with a –26 dB Dolph–Chebyshev taper in the *E*-plane and a slot offset of ± 0.1 wavelengths. The waveguide width is 0.77 λ and the main beam is steered 11.5° in the *E*-plane (about the maximum possible for normal waveguides; cf. Figure 9.2).

9.9.3 *Reducing the secondary lobes*

The larger the offset the larger is the butterfly lobe. The slot offset can be reduced by designing each slotted waveguide for a higher impedance (less offset) and use an impedance transformer in the feed system. Also long waveguides with many slots have small offsets and thus secondary beams are not so problematic.

With slots on the centreline of the waveguide no secondary lobes are generated. For exciting the slots an internal structure in the waveguide can then be used. One possibility is 'meandering' the waveguide [96,102] while keeping the slots on

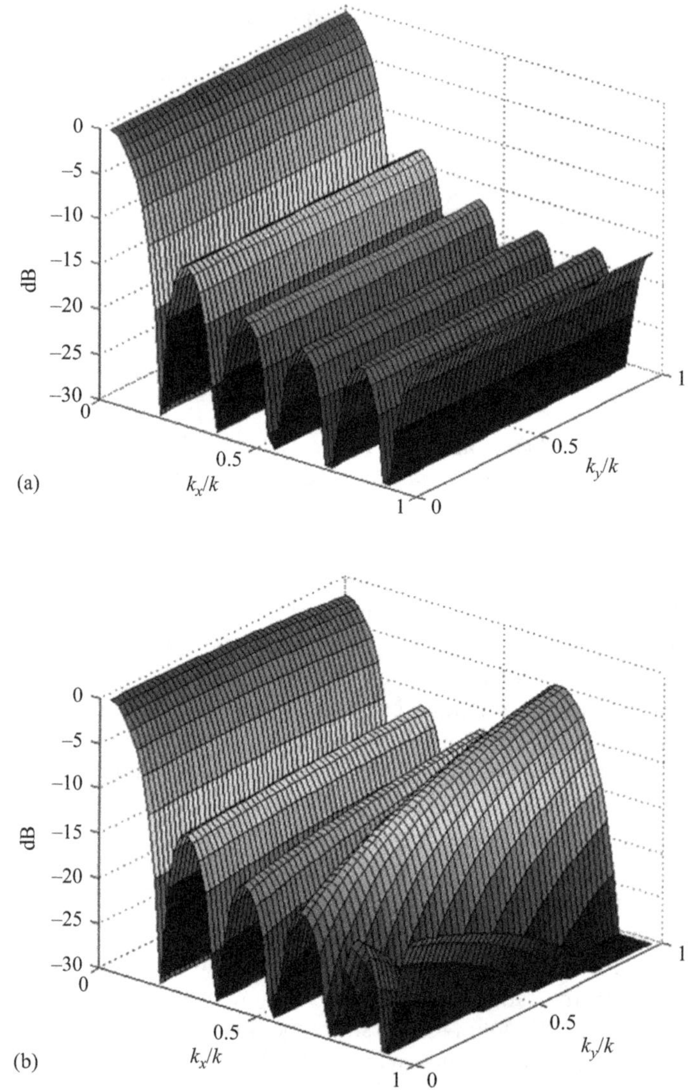

Figure 9.51 Radiation patterns in one quadrant with (a) no slot offset, (b) slot offset dy = ±0.1 wavelengths. Eight slots, spacing along waveguide dx = 0.7 wavelengths

a straight line (Figure 9.56). Another possibility is to use post or iris excited slots (Figures 9.57 and 9.58).

The iris can be inductive or inductive-capacitive (compound iris) as in Figure 9.58 (see Section 8.4). The iris can be tuned together with the slot to improve the overall bandwidth of the radiator [103,104]. The ridge waveguide with alternating depths discussed in connection with Figure 9.4 represents still another solution.

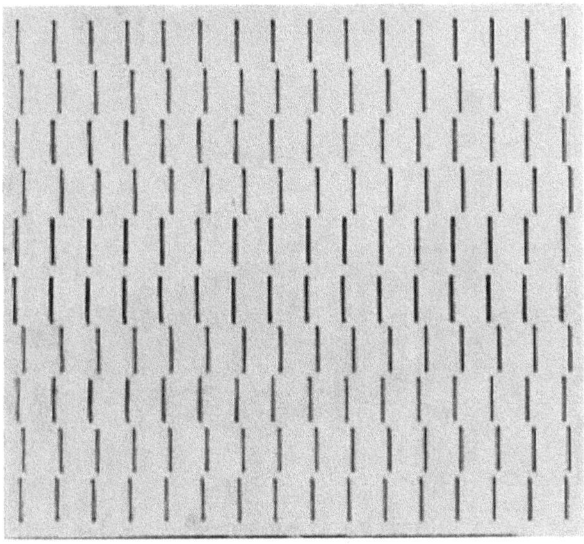

Figure 9.52 Slot panel. The upper half is a mirror image of the lower half. Courtesy of Saab AB

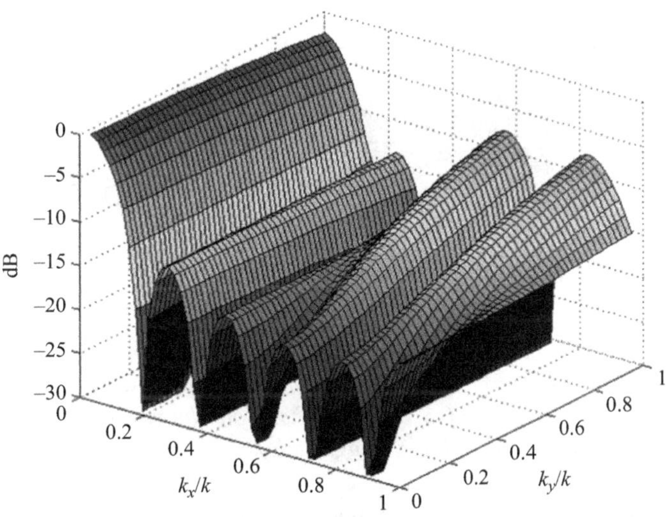

Figure 9.53 Computed pattern for a single waveguide with mirrored slot layout. Cf. Figure 9.51(b)

All these examples refer to shunt slots in the broad wall of the waveguide. The secondary beam problem occurs also in shunt slots *in the narrow wall* of the waveguide. In those, instead of controlling the excitation with slot offsets, the excitation is controlled by rotating the narrow wall slots. Second-order beams

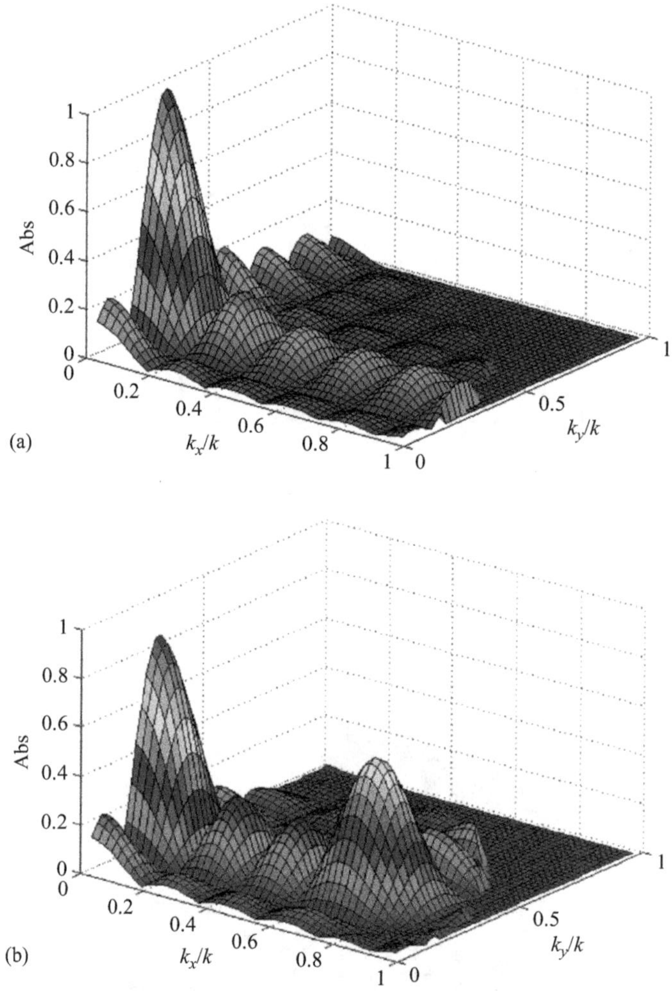

Figure 9.54 Array factor for a planar array of eight slotted waveguides. The beam is steered 17.5° in the E-plane. (a) No slot offset. (b) With a slot offset = ±0.3 wavelengths. Note the vertical axis scale is here in linear voltage (magnitude)

are therefore cross-polarised. Non-rotated slots can be excited by probes, wires, posts or irises inside the waveguide, similar to what we have described for the longitudinal slot array [47] (cf. Figures 9.20 and 9.59). More about the many different slot types and their characteristics is discussed in Section 8.1.

In an array using slots that are offset or rotated, the secondary lobes can also be eliminated by using external chokes or baffles. These will filter out the higher modes that are the source of the secondary radiation (cf. Section 9.3 on Dual

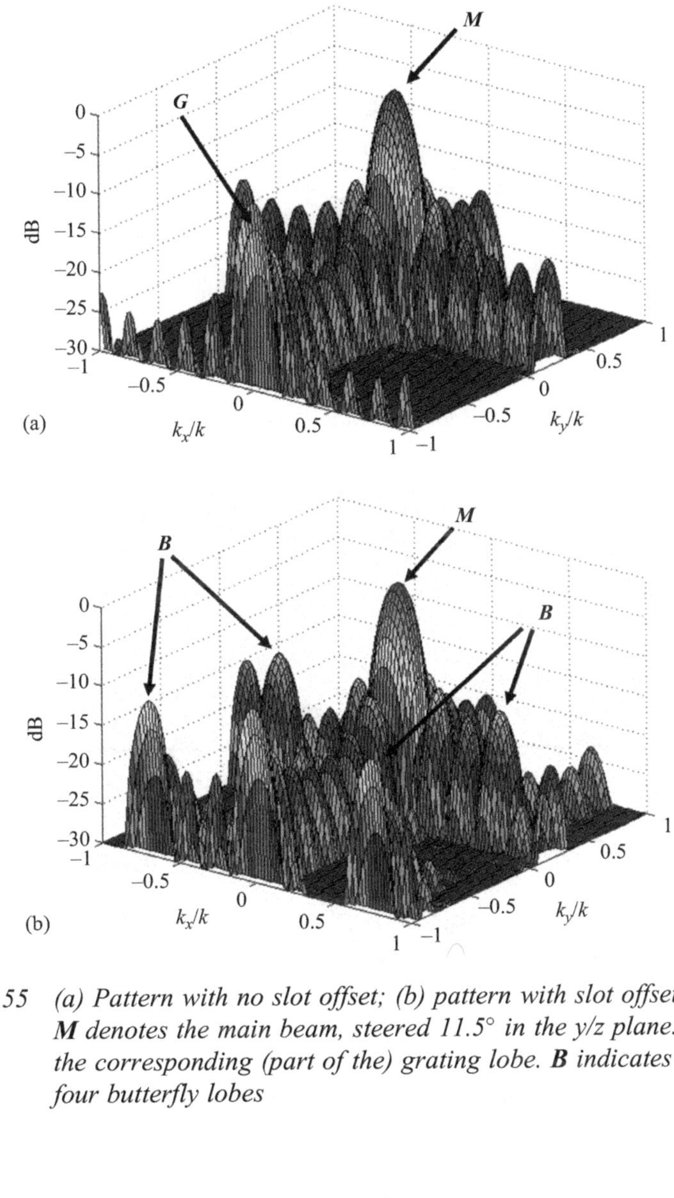

Figure 9.55 *(a) Pattern with no slot offset; (b) pattern with slot offset.* ***M*** *denotes the main beam, steered 11.5° in the y/z plane.* ***G*** *is the corresponding (part of the) grating lobe.* ***B*** *indicates the four butterfly lobes*

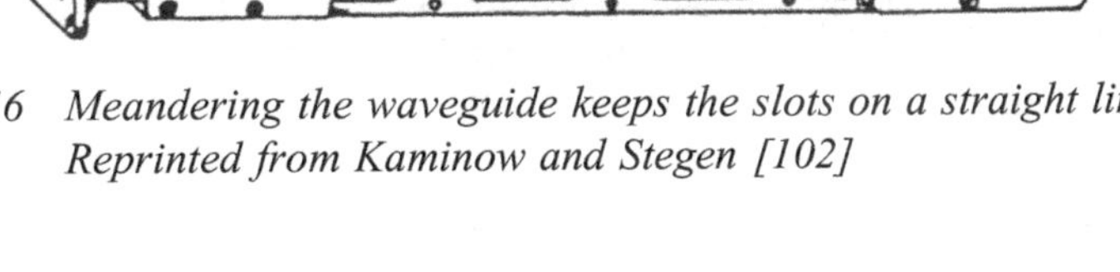

Figure 9.56 *Meandering the waveguide keeps the slots on a straight line. Reprinted from Kaminow and Stegen [102]*

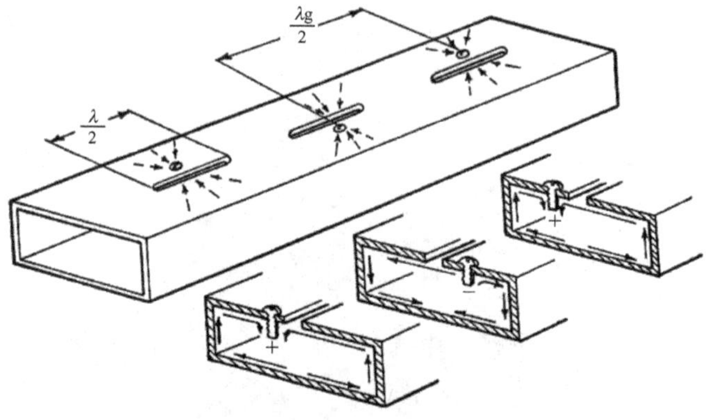

Figure 9.57 Post-excited longitudinal slots. Reprinted from [34, p. 301]

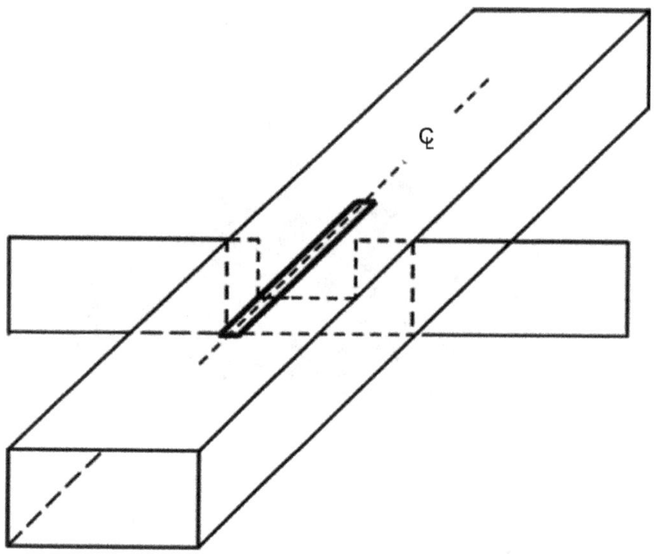

Figure 9.58 Iris-excited slot in a rectangular waveguide

polarised waveguide arrays). Baffles for a longitudinal slot array were described in a patent by Gruenberg [106,107] (Figure 9.60).

For planar arrays with rotated edge wall slots the external space between adjacent waveguides can act as a choke which reduces the currents associated with cross-polarisation radiation. An optimum choke depth is about a quarter of a wavelength [10].

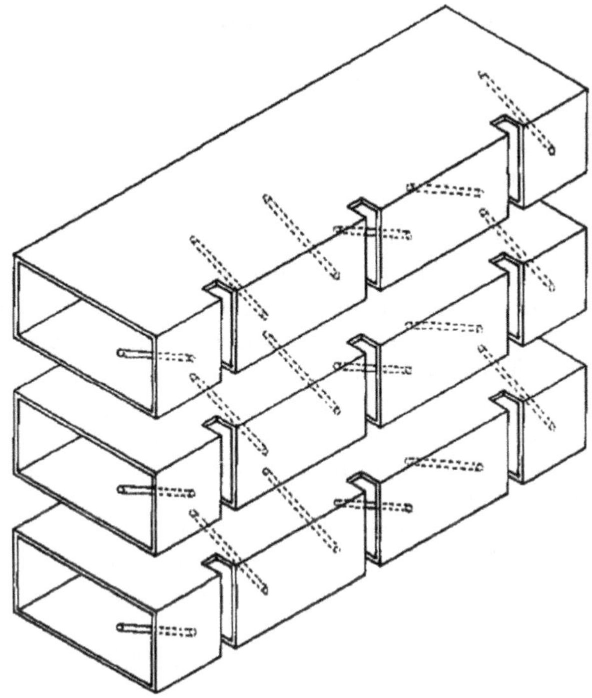

Figure 9.59 Non-tilted edge slots with wire excitation. © IEEE 1990, reprinted from [105], with permission

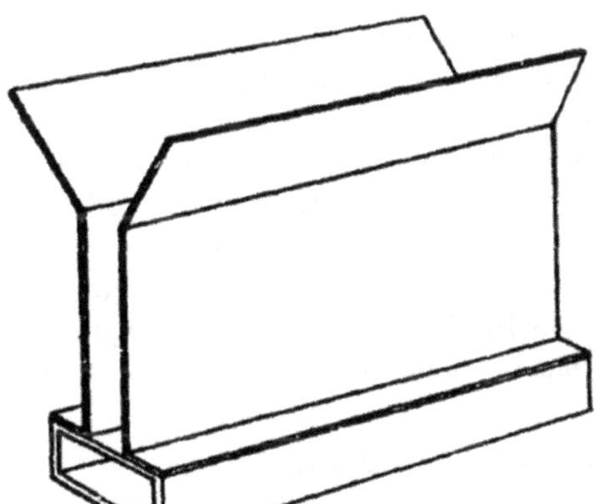

Figure 9.60 External baffles for suppression of secondary lobes. Reprinted from [102]

9.10 Omnidirectional slotted waveguide array antennas

In previous sections we have considered antenna arrays made up of a number of slots in a linear, planar or curved configuration with the purpose of forming a narrow directive beam and sometimes to steer the beam(s) over several angles. There are, however, also applications that require a fixed omnidirectional beam, for example radio and television broadcasting systems, local area networks (LAN) and radar beacon systems [108]. An omnidirectional antenna is sometimes also used together with a directive antenna for sidelobe cancelling functions [54]. With *omnidirectivity* we here mean coverage of 360° in one plane, typically the horizontal or azimuth plane. True omnidirectivity over 4π sr (strictly isotropic) is not realisable for a single antenna even though it is used as a reference for 0 dB antenna gain [109–111].

In this section we will deal with the problem of creating omnidirectivity using slotted waveguide arrays. First we will treat slots in rectangular and circular waveguides and then discuss the general problem of realising omnidirective patterns from slots in conducting cylinders. This has bearing upon slotted circular waveguides as well as cylindrical and circular array antennas in general. The interested reader is referred to the literature on conformal antennas [63]. The external conductance for resonant slots in conducting cylinders of varying curvature has been analysed by Wait [112,113]. With increasing curvature the conductance increases for circumferential slots and decreases for axial slots. The case with a flat infinite ground plane was discussed in Section 4.2.

9.10.1 Slotted rectangular waveguide array

A single longitudinal slot in a rectangular waveguide has a fairly omnidirectional pattern in the *E*-plane. Several slots along the length of the waveguide will increase gain and sharpen the *H*-plane pattern. The region on the rear side of the waveguide will be poorly covered, however. Adding slots on the rear side of the waveguide just like the front face can solve this problem [114]. Further improvements are obtained if a reduced height waveguide is chosen [115,116] and in particular if all slots are aligned along the centreline of the waveguide. This last alternative will require probe excitation of the slots or a similar technology (cf. Figures 9.57 and 9.58 in Section 9.9). A satellite communication antenna using this approach is shown in Figure 9.61.

Many more solutions using special slotted waveguides have been proposed. One example is the longitudinally slotted rectangular waveguide bent to a circular form ('Ring Waveguide'), thus providing omnidirectional coverage in the *H*-plane [117].

9.10.2 Slotted circular waveguide array

An omnidirectional antenna based on a slotted *circular* waveguide should perform even better than the slotted rectangular waveguide due to the rotational symmetry of the former. To preserve symmetry the chosen waveguide mode should be rotationally symmetric, for example, the TM_{01} mode (Figure 9.62(a)). The fundamental coaxial mode can be used as well [118]. Unfortunately, the (non-symmetric) TE_{11} mode is also supported by the cylindrical waveguide since it has a lower cut-off

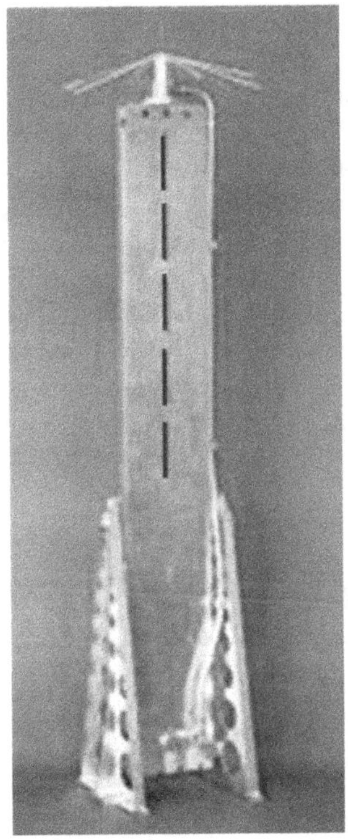

Figure 9.61 The omnidirectional rectangular waveguide antenna (breadboard) for the ISEE-B satellite. Courtesy of RUAG Space AB

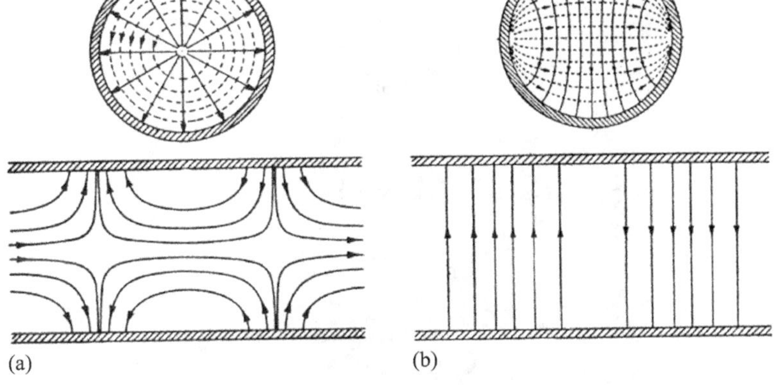

Figure 9.62 (a) The lowest order rotationally symmetric mode in a circular waveguide is TM_{01}. (b) The lowest order propagating mode in a circular waveguide is TE_{11}. Solid lines indicate electric field; broken lines are magnetic field lines [120]

frequency than the TM_{01} [119,120]. A proper symmetric feeding system is therefore required to prevent excitation of this mode.

With TM_{01} (or a coaxial waveguide) the currents flow along the length of the guide. For exciting axial slots probes can be used as shown in the example in Figure 9.63. Here additional screws/probes are also seen in between each set of radiating slots, claimed to improve impedance matching. One can also use inclined or crossed slots as was shown by Grabherr and Huder [121] and others.

A beacon antenna [122] for navigational purposes is shown in Figure 9.64. It has an upper circular slotted array for horizontal polarisation (X-band). The lower circular array for S-band has just one row of inclined slots covering both horizontal and vertical polarisations. The biconical horn arrangement provides beam sharpening in elevation; the vertical coverage for both antennas is about 20°.

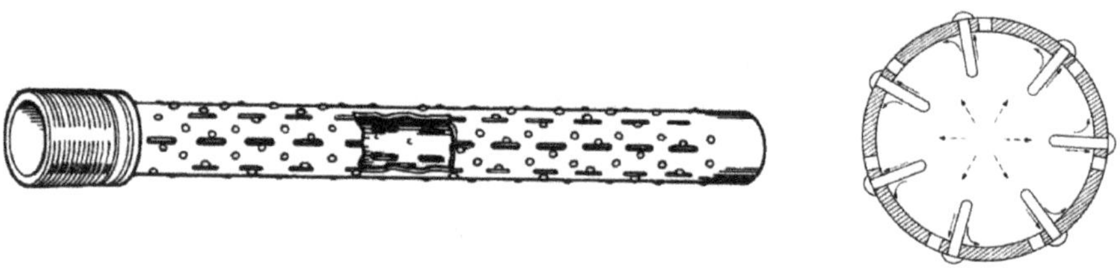

(a) (b)

Figure 9.63 (a) Slotted cylindrical waveguide array; (b) cross section showing probes for exciting slots [34]

Figure 9.64 A radar beacon antenna system using circular waveguide arrays and an electronics package and radome for X- and S-band operation. Courtesy of Telefonaktiebolaget L M Ericsson, reprinted from [122]

9.10.3 The pattern ripple problem

An interesting question is: how many slots are needed around a given cylinder size, assuming a specified maximum pattern ripple in azimuth? It turns out that 'the more the better' is not always true as we shall see. The problem has been studied by many authors, mainly for circular dipole arrays [123]. The use of directive elements instead of omnidirectional dipole elements complicates the analysis somewhat but improves the bandwidth and pattern stability [124,125].

Let us assume that we have a slot array in a cylindrical body as shown in Figure 9.65(a).

There are N slots around the cylinder with radius R. Each slot is horizontally polarised with a TE_{10}-type electric field. The spacing between consecutive slots is $d = 2\pi R/N$, that is, $Nd/\lambda = kR$ where k is the propagation constant, $k = 2\pi/\lambda$.

For the circular/cylindrical geometry the excitation function $V(\varphi)$ is rotationally symmetric, that is, it repeats every 360° in azimuth (φ-direction). It can therefore be represented by a Fourier series:

$$V(\varphi) = \sum_{-\infty}^{+\infty} C_m e^{jm\varphi} \tag{9.29}$$

The coefficients C_m are phase mode amplitudes; the mth mode has m times 2π phase variation along the circumference. The radiated far field can similarly be written as a Fourier series:

$$E(\phi) = \sum_{-\infty}^{\infty} A_m e^{jm\phi} \tag{9.30}$$

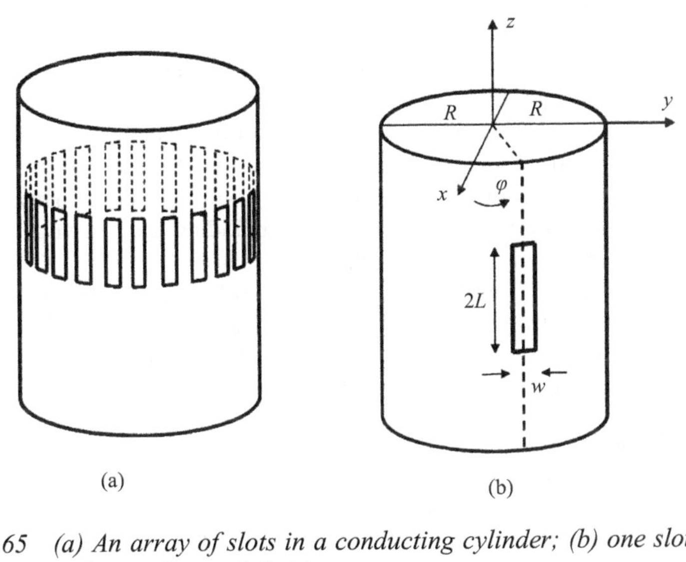

Figure 9.65 (a) An array of slots in a conducting cylinder; (b) one slot with coordinate definitions

The radiated pattern from an N-element circular array can also be written as a summation of the element contributions:

$$E(\phi) = \sum_{n} V_n EL(\phi - n\Delta\varphi)e^{jkR\cos(\phi - n\Delta\varphi)} \qquad (9.31) = (9.14)$$

where $EL(\phi)$ is the element pattern and $\Delta\varphi = 2\pi/N$ is the element angular spacing around the circle. The element pattern is the same for all elements although the elements point in different directions. Since it also is periodic, period 2π, the element pattern can be written in Fourier series form as

$$EL(\alpha) = \sum_{p=-\infty}^{\infty} D_p e^{jp\alpha} \qquad (9.32)$$

Inserting this into (9.31) yields

$$E(\phi) = \frac{1}{2\pi}\int_{-\pi}^{\pi} V(\varphi) \sum_{p=-\infty}^{\infty} D_p e^{jp(\phi-\varphi)} e^{jkR\cos(\phi-\varphi)} d\varphi \qquad (9.33)$$

Inserting also the excitation $V(\varphi)$ according to (9.29) and taking out just the contribution from phase mode number m yields the expression

$$E_m(\phi) = C_m e^{jm\phi} \sum_{p=-\infty}^{\infty} D_p \frac{1}{2\pi}\int_{-\pi}^{\pi} e^{j(m-p)(\varphi-\phi)} e^{kR\cos(\varphi-\phi)} d\varphi \qquad (9.34)$$

or

$$E_m(\phi) = C_m \left[\sum_{p=-\infty}^{\infty} D_p j^{m-p} J_{m-p}(kR)\right] e^{jm\phi} = A_m e^{jm\phi} \qquad (9.35)$$

where J_m is the Bessel function of the first kind and order m.

This gives us an important relation between the slot excitation, that is, the excitation modes (C_m, near field) and the radiation modes (A_m, far field). Ideally, for an omnidirectional pattern, we only want one radiation mode, preferably A_0. With discrete elements, however, we most likely will also have many excitation modes C_m, and our hope is that most of them, including their harmonics, will be attenuated by the factor in brackets in (9.35).

The coefficients D_p represent a mode spectrum for the radiating element. Different elements: isotropic, electric dipole, rectangular slot, etc., have different spectra. For the slot in Figure 9.62, assuming a type TE_{10} excitation, the far-field azimuth pattern function can be written [113,126]:

$$E_\phi = \frac{2LV_0 e^{-jkr}}{\pi^3 Rr} \sum_{-\infty}^{\infty} \frac{j^n e^{n\phi}}{H_n^{(2)'}(kR)} \qquad (9.36)$$

where V_0 is the peak of the sinusoidal voltage across the slot with length $2L$, and $H_n^{(2)'}$ is the derivative of the Hankel function of the second kind and order n.

The element azimuth pattern according to (9.36) is shown in Figure 9.66(a) compared with the approximation ($1 + \cos \phi$). The corresponding mode spectra are shown in Figure 9.66(b) and (c).

As we see, the radiation patterns (and the corresponding mode spectra) are quite similar, from which we can conclude that the ($1 + \cos \phi$) function is a satisfactory pattern approximation for the axial slot case. Using this approximation it is a simple matter to calculate the full circular array pattern for different numbers of slots and cylinder sizes. The resulting ripple amplitude for several cases is shown in Figure 9.67. A contour plot is presented in Figure 9.68.

The results show that increasing the number of slots for a given cylinder size in general makes the ripple smaller. There are, however regions where, for example, six slots is better than seven slots. Similar results have been published by Chu [127] and Croswell and Knop [128]. With omnidirectional elements (i.e. vertical dipoles) it has been shown that quite often an odd number of elements is better than an even number. This is not so often the case with directional elements.

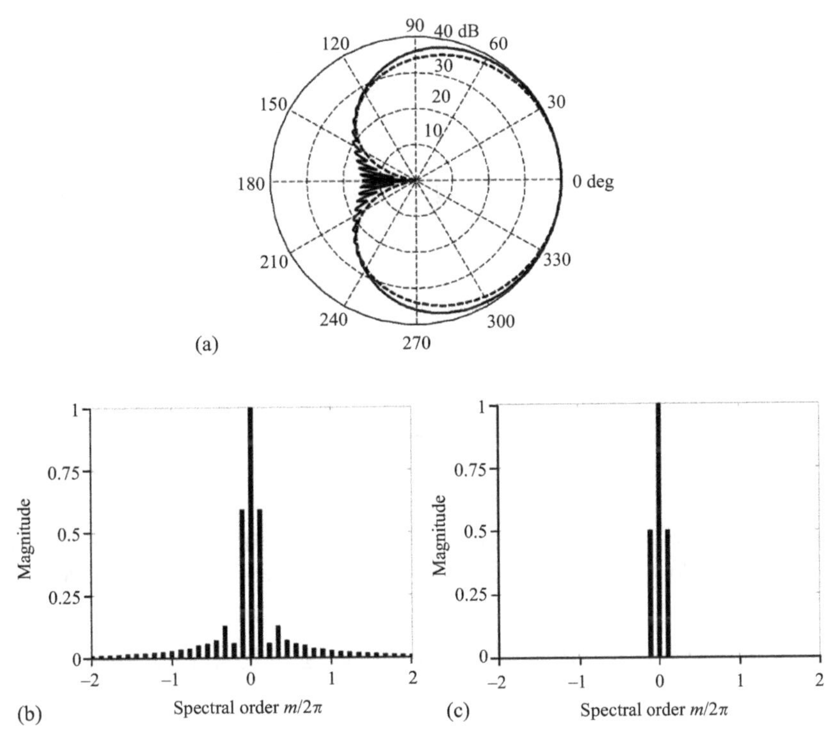

Figure 9.66 (a) Axial slot pattern (solid line) and function ($1 + \cos \phi$) (dashed line). Mode spectra for slot (b) and ($1 + \cos \phi$) function (c).

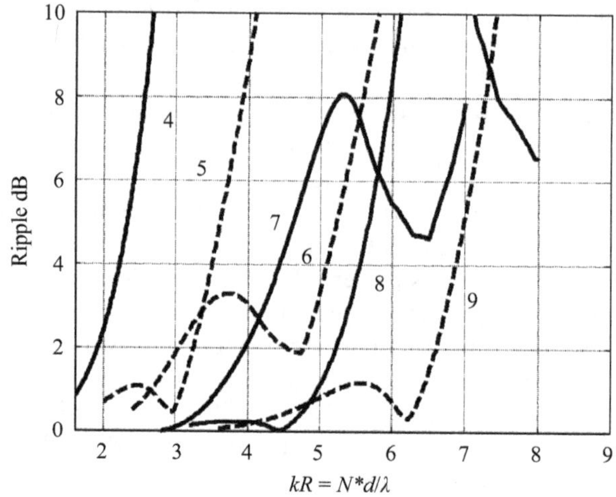

Figure 9.67 Maximum ripple versus cylinder size kR for several numbers N of slots with assumed $(1 + \cos \varphi)$ element patterns

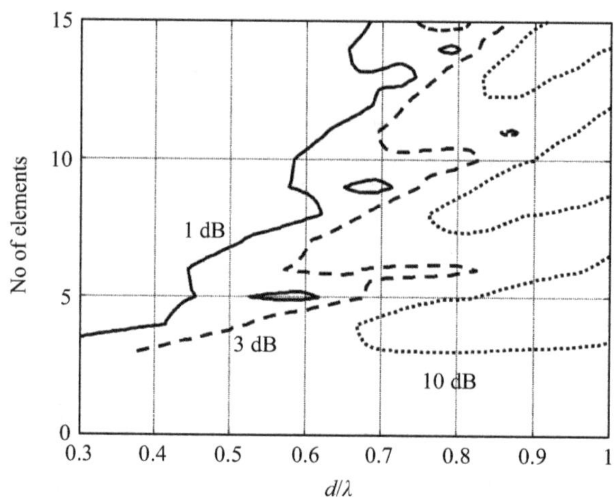

Figure 9.68 Ripple levels indicated by contour lines: solid curve 1 dB, dashed 3 dB and dotted 10 dB. $(1 + \cos \varphi)$ element patterns assumed

9.11 On the bandwidth of slotted waveguide array antennas

9.11.1 Introduction

Slotted waveguide array antennas are not intrinsically broadband since in the most common designs the radiating slots are fed by a resonant waveguide structure. Besides, the slot itself is a resonant $\lambda/2$ long radiator. These limitations may not always be a problem for applications where bandwidths of the order of 10% or less are required. In any case, the bandwidth 'problem' deserves our attention and many of the chapters in this book have already addressed this issue.

In this last section of the chapter we discuss bandwidth limiting factors from an antenna system point of view, without going into the details of specific designs. Design rules are addressed in a broad sense, so that the reader can get a feeling for what is important and what phenomena are involved. Cross-references to relevant parts of the book and references to the literature are given. The index at the end of the book also provides aid to find more information.

9.11.2 *The slot itself*

A slot antenna is the dual of the dipole antenna, and they are both about half wavelength long when working in their resonant mode; cf. Chapter 4. The bandwidth of the slot (dipole) alone is typically about 10%. In special cases up to 20% or 30% has been reported. Some kind of feeding structure or cavity is usually required which may limit the bandwidth.

The bandwidth is defined from the maximum and minimum frequencies within which acceptable performance is realised, often in terms of impedance matching: for example, VSWR less than 2 or return loss better than 10 dB. Other limitations can be pattern performance (gain, sidelobe level, beam direction, etc.) or polarisation performance (cross-polarisation level), etc.

Slots fed by a rectangular waveguide, respectively by a stripline waveguide were discussed and compared in Section 9.7 (see Figure 9.35). For both types the cavity formed by the feeding system has a significant impact on the slot impedance and bandwidth [77,128]. In the stripline case the feeding strip ends at about a quarter of a wavelength from the slot, transforming the strip open line to an effective short at the slot location. Hence, this stub imposes a frequency limitation just as the slot itself. Some increase in bandwidth is possible by adding one or more (shorter) slots beyond the first slot [129]. Each of the shorter slots sees a shorter stub, as they should (see Figure 9.69). Altogether a bandwidth of, for example, 9–12.5 GHz in the X-band, that is, about 30%, has been noted for a VSWR less than 2 [78]. Multiple slots have also been used in waveguides [130]. Another broadband shape is offered by the dumbbell slot.

Many different shapes of slots have been investigated. The dumbbell slot is both short and wide (Figure 9.69(b)). The circular ends make manufacturing easier if the slot is to be machined. The general rule is that the resonant length of a wide slot is shorter than that of a narrow slot. A very narrow slot resonates at one-half wavelength, while a thicker slot is somewhat shorter at resonance. There are, however, exceptions to this rule when the slot cavity or feeding waveguide is taken into account.

9.11.3 *The waveguide-fed longitudinal slot*

We expect a wide slot to have a larger bandwidth than a narrow slot, also when excited from a waveguide. This is exemplified in Figure 9.70 for longitudinal slots

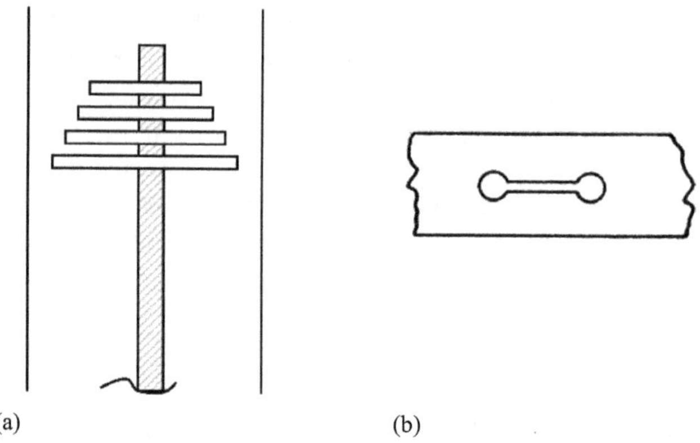

Figure 9.69 (a) Stripline feed (hatched) with multiple slots for increased bandwidth; (b) dumbbell slot

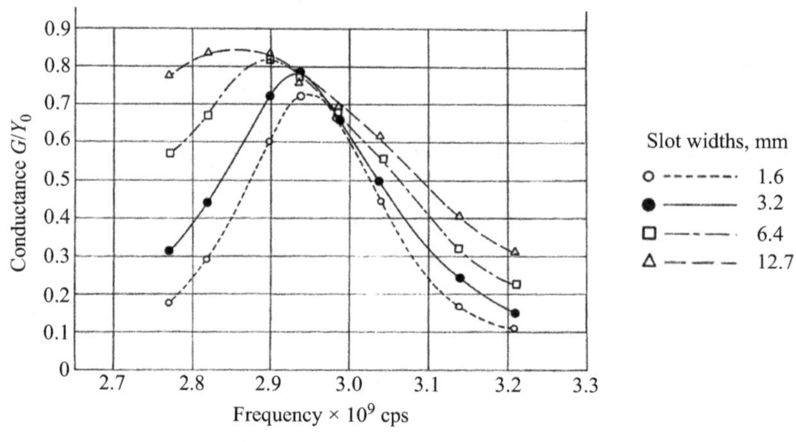

Figure 9.70 Conductance curves for longitudinal slots in a rectangular waveguide with different widths. From Dodds and Watson, National Research Council of Canada. Adapted from [34]

in a rectangular waveguide, showing the conductance of slots with widths ranging from 1/16 in to ½ in (1.6–12.7 mm). As expected, the conductance curve is flatter the wider the slot. The slot length was kept constant, about 50 mm, and the offset was 19.8 mm. The susceptance follows a similar behaviour [34].

For longitudinal slots the resonant length dependence on the width also depends on the slot offset. General rules are hard to establish; an example is shown in Figure 9.71.

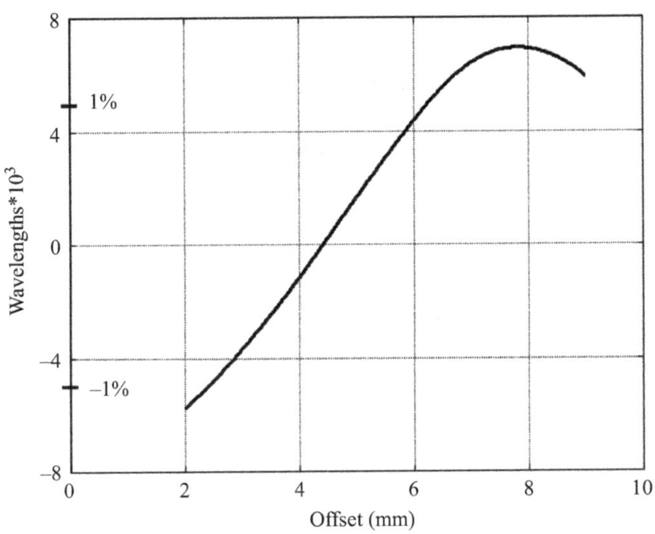

Figure 9.71 *Calculated change in resonant length when the slot width is changed from 3 mm to 1.6 mm, for different slot offsets. Standard X-band waveguide with wall thickness 1.27 mm*

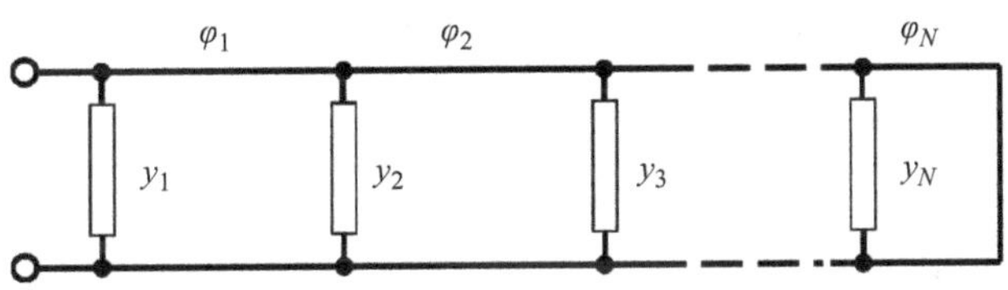

Figure 9.72 *Equivalent network for an end-fed linear shunt slot array antenna*

9.11.4 The linear resonant array antenna

We assume that the slotted waveguide is fed from one end. An equivalent network for this case is shown in Figure 9.72. The nominal slot spacing is one-half guide wavelength corresponding to the electrical length φ_i and the last slot is a quarter of a guide wavelength from the short. At the centre frequency all slot susceptances y_i are real and sum up to the waveguide conductance G_0 – we have a matched case.

A similar network applies to a series slot array antenna where the waveguide is loaded with series impedance elements. In both cases the analysis of the network can be simplified by a suitable matrix representation [131]; here we have chosen wave matrices [132,133]. A wave matrix relates the wave amplitudes on the input and output sides of a two-port network as shown in Figure 9.73.

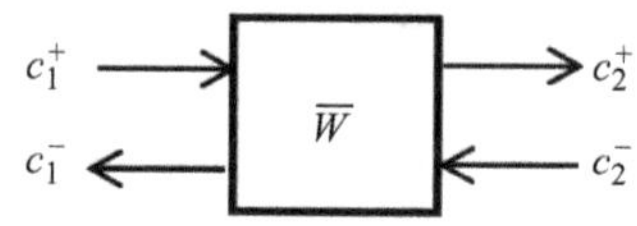

Figure 9.73 The input and output ports of a wave matrix

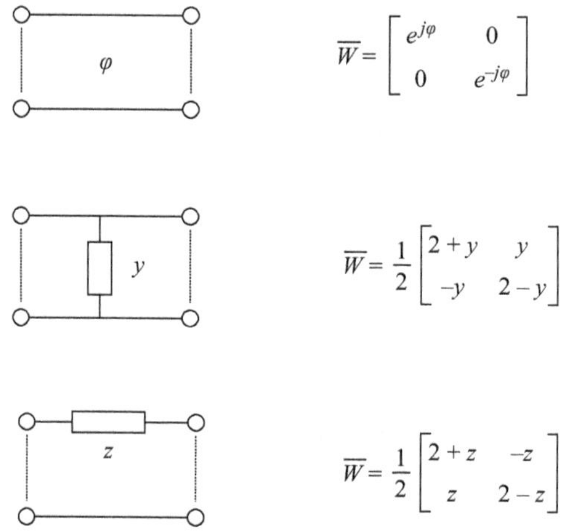

Figure 9.74 Wave matrices for some simple two ports

We can write

$$\begin{pmatrix} c_1^+ \\ c_1^- \end{pmatrix} = \bar{W} \begin{pmatrix} c_2^+ \\ c_2^- \end{pmatrix} \tag{9.37}$$

or in short form

$$\bar{c}_1 = \bar{W} \cdot \bar{c}_2 \tag{9.38}$$

The total wave matrix for the complete ladder network such as shown in Figure 9.72 is found by multiplying the matrices of the individual blocks (9.39).

$$\bar{W}_{\text{tot}} = \prod_{i=1}^{N} \bar{W}_i \tag{9.39}$$

From the total wave matrix the total transmission and reflection are obtained according to

$$\begin{cases} T = 1/W_{11} \\ \Gamma = W_{21}/W_{11} \end{cases} \tag{9.40}$$

The wave matrices for some elementary networks are shown in Figure 9.74.

A short as in Figure 9.72 can be represented by letting $c_2^- = -c_2^+$; cf. Figure 9.73.

The matrix multiplications are conveniently executed with modern computers. It can be noted that in the case when all sections of the ladder network are identical we have simply the matrix of one section raised to the power N corresponding to the number of sections. This case can be formulated in a compact form using the theory of iterated networks [134] involving Dolph–Chebyshev polynomials of the second kind. However, quite often the sections are not identical, for instance when the array is tapered, and the matrix multiplications (9.39) will have to be executed.

Using the wave matrix approach we have calculated the input return loss of a linear array as represented in Figure 9.72 for different numbers of slots and different frequencies off resonance. For a given maximum return loss we can then find the array bandwidth. Note that the slot conductances were identical and assumed to be constant, that is, not depending on the frequency. This may be a valid approach if the variation in frequency is small. The mutual coupling was neglected.

The results were compared with calculations based on an approximate formula according to Takeshima and Isogai [135]. See also Hamadallah [136]. Define df as (half) the relative bandwidth in per cent and introduce the variable

$$a = \frac{2(3+q^2)}{q(2+q^2)} \quad \text{where } q = \pi N df/100 \tag{9.41}$$

The return loss (reflection factor) according to [135] is then obtained as

$$\Gamma = \frac{1+\sqrt{1+a^2}}{1+a^2+\sqrt{1+a^2}} \tag{9.42}$$

The two approaches agree quite well; Figure 9.75, cf. also Figure 6.2.

It must be emphasised that these calculations are approximate. However, they indicate the general behaviour in terms of frequency sensitivity and number of slots fed in series. The following effects are not accounted for

- slot admittance is frequency sensitive
- the dispersion of the waveguide itself
- mutual coupling among slots
- impact on radiation patterns

A commonly used expression for the relative bandwidth in % is 50/N, where N is the number of slots fed in series. Examination of Figure 9.75 reveals that 60/N is a good measure for the (total) bandwidth if −10 dB return loss is postulated. If −15 dB is required 30/N gives a good match.

9.11.5 Overloading

The term ‘overloading’ refers to choosing the slot conductance values higher than G_0/N where G_0 is the waveguide admittance, for instance $R \cdot G_0/N$, with an overload factor of R [34]. The input admittance of the array will then be increased and

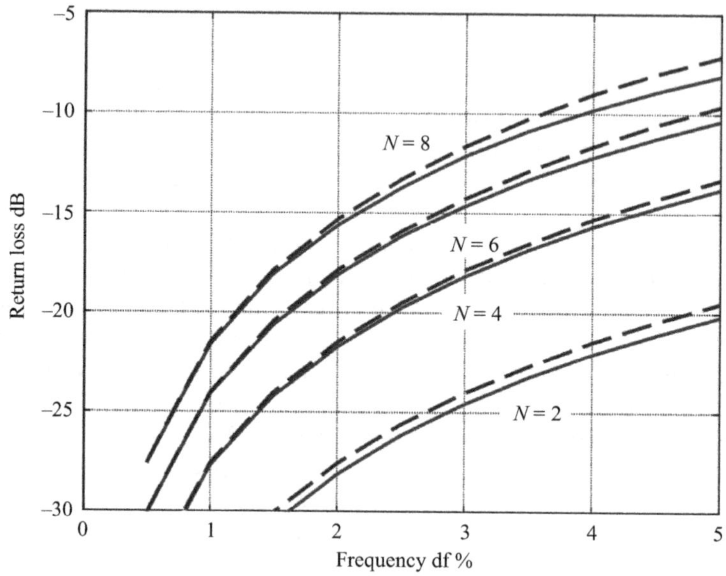

Figure 9.75 The nominal (half) bandwidth in per cent is obtained from the return loss versus frequency for arrays with two, four, six and eight slots. The solid line is according to the wave matrix analysis; the broken line according to the formula (9.42). No overloading

an impedance transformer is required on the input in order to match the array. The transformer could be a waveguide section with a quarter wave length and admittance $G_t = \sqrt{R}G_0$ [131]. A similar overloading scheme can be implemented for arrays with series slots where the slots are represented by series elements.

Results obtained by overloading by a factor of $R = 2$ are shown in Figure 9.76 and compared with the previous case with no overloading.

Overloading improves the impedance match also for an amplitude tapered array. The amplitude tapering is usually introduced in order to improve the pattern characteristics and reduce the sidelobe level. However, in a study by Derneryd and Petersson [137] regarding a typical planar array antenna it was concluded that overloading should be avoided when maximum bandwidth from a sidelobe point of view is desired. In the practical case a careful trade off will have to be made.

9.11.6 Mutual coupling

As mentioned earlier, mutual coupling had been neglected in the bandwidth calculations. This is justified for linear arrays with longitudinal slots since the coupling is in the H-plane where the coupling is low; cf. Section 4.5. For a planar array there will be strong coupling in the E-plane, which will have to be accounted for in the design. The coupling reduces the bandwidth of the radiating slot compared to the isolated slot as demonstrated by the example shown in Figure 9.77.

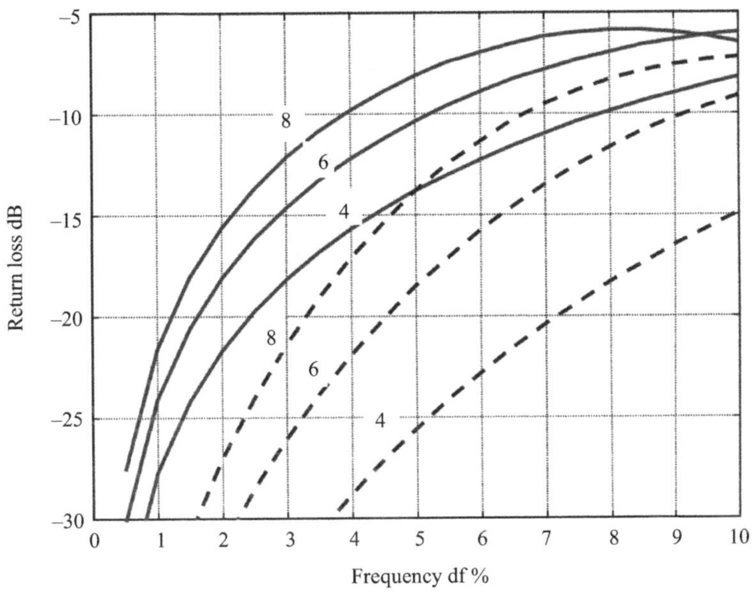

Figure 9.76 The return loss versus frequency for arrays with four, six and eight longitudinal slots. The wave matrix analysis was used; the solid line is without overloading, the dashed line with an overloading factor of 2

It has been found that breaking up the periodicity of a planar slot array can reduce the bandwidth limitation caused by the mutual coupling. Instead of having identical radiating waveguides with alternating slot offsets, adjacent waveguides can be designed to be mirror images of each other. In one example [139,140] the improvement was about a doubling of the reflection bandwidth, from 1.2% to 2.5% at −15 dB return loss (at the expense of some grating lobes in the diagonal planes).

9.11.7 The planar array and feeding arrangements

A planar array must have some sort of network for feeding the radiating waveguides. This network is usually a series feed with coupling slots as exemplified in Figure 3.3 and other places. For large arrays the aperture will have to be broken down into sub-arrays resulting in a relatively complex feed network. As already indicated, a rough bandwidth estimate can be obtained from the formula

$$B(\%) = C1/N \qquad (9.43)$$

where B is the bandwidth in per cent, N is the average number of slots in series and $C1$ is a constant, say $C1 = 30$. For a planar array N shall include both the coupling slots in the feeding network and the radiating slots. Since the slot spacing is one-half guide wavelength or about 0.7 λ_0 where λ_0 is the free space wavelength we can estimate the approximate number of slots in series as $N = D/1.4\lambda_0$, where D is

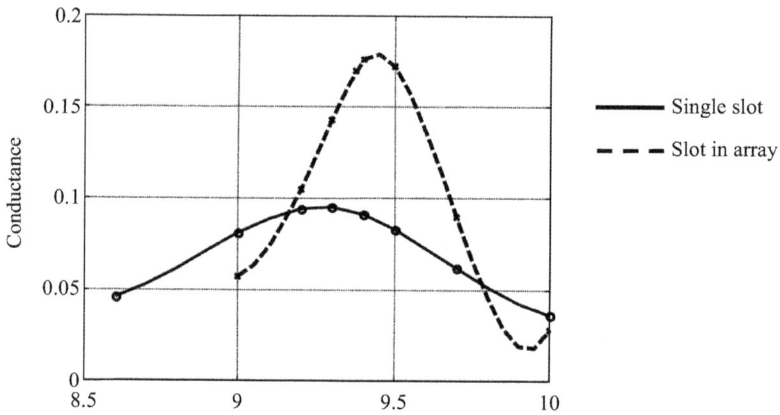

(a)

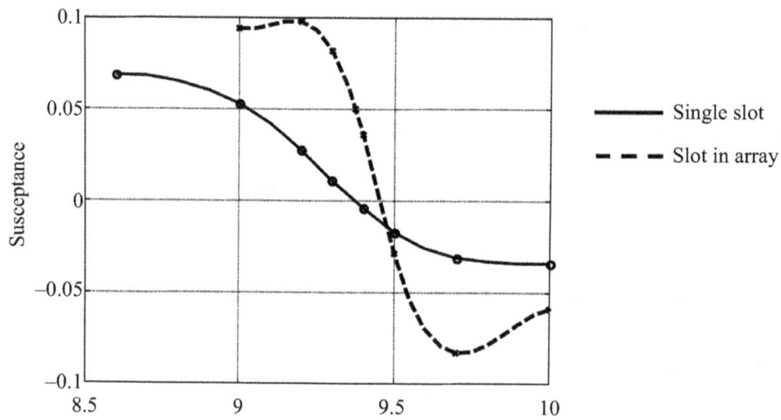

(b)

Figure 9.77 (a) Normalised conductance and (b) normalised susceptance versus frequency for an isolated longitudinal slot compared to a slot in an infinite array. Standard X-band (R 100) waveguide, slot width 1.59 mm, total slot length 14.76 mm, offset 2 mm from centreline. Slot spacing in the array 0.7 λ_0 along waveguides, 0.75 λ_0 across [138]

the array diameter, provided that there is only one feeding point for the complete circular array. If there are more feeding points we can write the bandwidth in GHz as

$$B(\text{GHz}) = 250\sqrt{N_p}/D \tag{9.44}$$

where N_p is the number of feeding points per quadrant and D is the diameter in millimetre (see Figure 9.78).

In this analysis we have not included the radiation pattern performance, tapering effects, overloading or the precise shape of the array. In a detailed

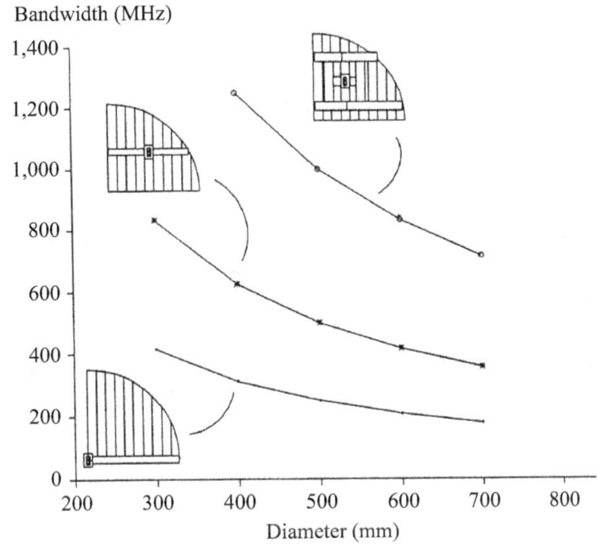

Figure 9.78 *The approximate bandwidth in MHz as a function of the array diameter for different number of feeding points per quadrant: 1/4, 1 and 4. See also Figure 9.31*

investigation the influence from coupling slots, mutual coupling, waveguide short design, impedance levels and transformers, etc., will have to be included [141].

The ideal solution could be to use a corporate feed broadband power divider to feed all radiating waveguides in parallel, or even to feed all individual slots by such a power divider. In one solution by Zhou *et al.* [142] a 1:16 waveguide power divider feeds 2×2 sub-arrays with radiating wide slot apertures. Better than 30% bandwidth (return loss < -10 dB) was obtained at 12 GHz. Another interesting feed network technology is based on barline transmission lines which essentially is a coaxial line technology. It is more broadband and offers better design flexibility than the traditional waveguide feed using rotated coupling slots. Dual barline power dividers 1:8 were used to feed the dual polarised slotted waveguides that make up the panels of a large C-band scatterometer instrument [47].

9.11.8 Travelling wave arrays

Travelling waveguide arrays differ very much from resonant waveguide arrays. In Section 9.2 on frequency scanned arrays the performance of linear slotted arrays with a matched load instead of a short at the waveguide end was discussed. In this case there is only one forward travelling wave in the waveguide and a good match is obtained at the input port, except when the slot spacing approaches one-half waveguide wavelength. Design of both travelling wave arrays and standing wave arrays with several examples is treated in Chapter 6.

References

[1] Amitay N., Galindo V., Wu C. P. *Theory and Analysis of Phased Array Antennas*. Wiley Interscience, New York, 1972.

[2] Josefsson L., Derneryd A., Karlsson E. R. 'A phase steered slotted waveguide array with non-squinting beam'. *Proceedings Military Microwaves*. London, July, 1990, pp. 251–256.

[3] Sen B., Ranga Rao K. S. 'Impedance properties of narrow radiating slots in broadface of dielectric loaded rectangular waveguide'. *Proceedings of the IEEE Antennas and Propagation Symposium*, Vancouver, Canada, 1985, pp. 191–194.

[4] Rengarajan S. R., Steinbeck M. 'Longitudinal slots in dielectric filled rectangular waveguides'. *Proceedings of the IEEE Antennas and Propagation Symposium*, London, Ontario, Canada, 1991, pp. 1276–1279.

[5] Joubert J., Coetzee J. C. 'The use of inhomogeneously loaded rectangular waveguide to overcome grating lobes and distribution taper limitations in transverse slot arrays'. *Proceedings European Microwave Conference*. 1992, pp. 870–875.

[6] Cohn S. B. 'Properties of ridge wave guide'. *Proceedings of the IRE*. 1947; **35**(8):783–788.

[7] Hopfer S. 'The design of ridged waveguides'. *IRE Transactions on Microwave Theory and Techniques*. 1955;**3**(5):20–29.

[8] Hu A. Y., Lunden C. D. 'Rectangular-ridge waveguide slot array'. *IRE Transactions on Antennas and Propagation*. 1961;**9**(1):102–105.

[9] Shnitkin H. 'Unique joint stars phased-array antenna'. *Proceedings of the IEEE Antennas and Propagation Symposium*, Dallas, USA, 1990, pp. 678–681.

[10] Hilburn J. L., Kinney R. A., Emmett R. W., Prestwood F. H. 'Frequency-scanned X-band waveguide array'. *IEEE Transactions on Antennas and Propagation*. 1972;**20**(4):506–509.

[11] Bostrom D. E., Castaneda J. A., Hodges R. E., Fujii P. 'Large X-band electronically scanned array'. *Proceedings of the IEEE Antennas and Propagation Symposium*, Los Angeles, CA, USA, 1981, pp. 213–216.

[12] Elsallal M. W., Herting B. J., West B. J. 'Planar edge slot waveguide antenna array design using COTS EM tools'. *Proceedings of the 2007 Antenna Applications Symposium*, Allerton Park, IL, USA, 2007.

[13] Conciauro G., Bressan M., Zuffada C. 'Waveguide modes via an integral equation leading to a linear matrix eigenvalue problem'. *IEEE Transactions on Microwave Theory and Techniques*. 1984;**32**(11):1495–1504.

[14] Falk K. 'Admittance of a longitudinal slot in a ridge waveguide'. *IEE Proceedings H – Microwaves, Antennas and Propagation*. 1988;**135**(4):263–268.

[15] Garb K., Kastner R., Meyerova R. 'Analysis of planar arrays of shunt slots in ridged waveguides'. *Electronics Letters*. 1994;**30**(7):533–534.

[16] Spradley J. L. 'A volumetric electrically scanned two-dimensional microwave antenna array'. *IRE International Convention Record*. 1958;**6**:204–212.

[17] Derneryd A., Lorentzon T. 'Design of a phase/frequency scanned array antenna with non-resonant slotted ridge waveguide elements'. *Proceedings of the IEEE Antennas and Propagation Symposium*, London, Ontario, Canada, 1991, pp. 1728–1731.

[18] Ahlbom S., Andersson P., Lagerlöf R. 'A Swedish airborne early warning system based on the Ericsson ERTEYE radar'. *Ericsson Review*. 1995;**72**(2): 54–63.

[19] Josefsson L. 'Radar antenna R&D in Sweden'. *7th European Conference on Antennas and Propagation (EUCAP 2013)*, Gothenburg, Sweden, April 2013, pp. 3166–3170.

[20] Cullens E. D., Ranzani L., Vanhille K. J., Grossman E. N., Ehsan N., Popović Z. 'Micro-fabricated 130–180 GHz frequency scanning waveguide arrays'. *IEEE Transactions on Antennas and Propagation*. 2012;**60**(8): 3647–3653.

[21] Dion A. (1958), 'Nonresonant slotted arrays'. *IRE Transactions on Antennas and Propagation*. 1958;**6**(4):360–365.

[22] Ishimaru A., Tuan H.-S. 'Theory of frequency scanning of antennas'. *IRE Transactions on Antennas and Propagation*. 1962;**10**(2):144–150.

[23] Ramsay J. F., Popovich B. V. 'Series-slotted waveguide array antennas'. *IEEE International Convention Record*. March 1963, pp. 30–55.

[24] Elliott R. S. *Antenna Theory and Design*. Prentice-Hall, Englewood Cliffs, NJ, 1981.

[25] midwaysailor.com/midwaymyphotos4/restormisc.html. Accessed November 2016. Courtesy of Troy Prince.

[26] Hossain M. G. S., Tsunemitsu Y., Hirokawa J., Ando M. 'Grating lobe suppression by a dipole layer over a waveguide transverse slot array with reflection-canceling posts'. *Proceedings of the IEEE Antennas and Propagation Symposium*, Washington, DC, USA, 2005, pp. 720–723.

[27] Solbach K. 'Below-resonant-length slot radiators for traveling-wave-array antennas'. *IEEE Antennas and Propagation Magazine*. 1996;**38**(1):7–14.

[28] Ando M., Zhang M., Lee J., Hirokawa J. 'Design and fabrication of millimeter wave slotted waveguide arrays'. *Proceedings of the fourth European Conference on Antennas and Propagation (EuCAP 2010)*, Barcelona, Spain, April 2010, pp. 1–6.

[29] Simmons A. J. 'Circularly polarized slot radiators'. *IRE Transactions on Antennas and Propagation*. 1957;**5**(1):31–36.

[30] Getsinger W. J. 'Elliptically polarized leaky-wave array'. *IRE Transactions on Antennas and Propagation*. 1962;**10**(2):165–171.

[31] Josefsson L., Derneryd A., Lagerlöf R. 'Electronic scanning with slotted waveguide arrays'. *Proceedings of the Radarcon 90 Conference*, Adelaide, Australia, April 1990.

[32] O'Loughlin W. R., Kim Y. U., Elliott R. S. 'Pattern synthesis for a forward-fire/backward-fire linear array'. *IEEE Transactions on Antennas and Propagation*. 1989;**37**(6):721–727.

[33] Sparks R. A. 'Recent microwave applications to radar in the U.S.' *Proceedings of the 19th European Microwave Conference (EuMA)*, 1989, pp. 96–106.

[34] Silver S. ed. *Microwave Antenna Theory and Design*. MIT. Rad. Lab Series, Vol. 12, 1949.

[35] Hougardy H. H., Shanks H. E. 'Arbitrarily polarized slot array'. *IRE Wescon Convention Record*. Los Angeles, USA, 1958, pp. 157–160.

[36] Sangster A. J. 'Circularly polarized linear waveguide array'. *IEEE Transactions on Antennas and Propagation*. 1973;**21**(5):704–705.

[37] Ajioka J. S., Joe D. M., Tang R., Wong N. S. 'Arbitrarily polarized slot radiators in bifurcated waveguide arrays'. *IEEE Transactions on Antennas and Propagation*. 1974;**22**(2):196–200.

[38] Ajioka J. S., Joe D. M., McFarland J. L. 'Slot radiators in septated waveguide'. *IEEE Transactions on Antennas and Propagation*. 1984;**32**(3):247–251.

[39] Sangster A. J., Smith P., Jacobs R., Wang H. 'Moment method solutions for polarization agile slots in bifurcated waveguide'. *Electromagnetics*. 1999; **19**(1):91–107.

[40] Josefsson L. 'A dual polarised slotted waveguide array antenna'. *Proceedings URSI Electromagnetic Theory Symposium*, Sydney, Australia, August 1992.

[41] Josefsson L. 'A grating lobe filter for transverse slot arrays'. *Proceedings of the IEEE Antennas and Propagation Symposium*, London, Ontario, Canada, June 1991, pp. 1156–1159.

[42] Josefsson L. 'A waveguide transverse slot for array applications'. *IEEE Transactions on Antennas and Propagation*. 1993;**41**(7):845–850.

[43] Pedersen J. F., Hannan P. W. 'A metal-grid 5 × 5 foot angular filter'. *Proceedings of the IEEE Antennas and Propagation Symposium*, Albuquerque, USA, 1982, pp. 471–474.

[44] Josefsson L. *Waveguide Antenna which Includes a Slotted Hollow Waveguide*. US Pat. 5 541 612, 1996.

[45] Wood P. J., Sultan N., Seguin G. 'A dual-polarized reconfigurable-beam antenna for the DSAR synthetic aperture radar'. *Proceedings of the IEEE Antennas and Propagation Symposium*, Baltimore, USA, 1996, pp. 1716–1719.

[46] Derneryd A., Lagerstedt A. 'Novel slotted waveguide antenna with polarimetric capabilities'. *Proceedings of the Geoscience and Remote Sensing Symposium (IGARSS)*, Florence, Italy, 1995, pp. 2054–2056.

[47] Magnusson P., Di Salvo M., Scarchilli C. 'A dual polarised slotted waveguide antenna for satellite based wind scatterometer instruments'. *7th European Conference on Antennas and Propagation (EuCAP)*, Gothenburg, Sweden, April 2013, pp. 3449–3452.

[48] Zahn R., Stangl M., Rostan F. 'The X-band SAR instrument demonstrator: intermediate test results'. *Proceedings Millennium Conference on Antennas & Propagation (AP 2000)*, Davos, Switzerland, April 2000.

[49] Roederer A., Mosig J., Balling P., Vandenbosch G., Keizer W., Kabacik P. *COST Action 260. Summary and Final Report*, May 2001.

[50] Alexander D. K., Sichelstiel B. A. 'A unique waveguide phased array with independently steered beams'. *1970 Phased-Array Antenna Symposium (digest)*, Polytechnic Institute of Brooklyn, NY, June 1970.

[51] Butler J., Lowe R. 'Beamforming matrix simplifies design of electronically scanned antennas'. *Electronic Design*. 1961;**9**(7):170–173.

[52] Blass J. 'Multidirectional Antenna – a new approach to stacked beams'. *IRE International Convention Record*. 1960, pp. 48–50.

[53] White W. D. 'Pattern limitations in multiple-beam antennas'. *IRE Transactions on Antennas and Propagation*. 1962;**10**(4):430–436.

[54] Skolnik M. I. *Introduction to Radar Systems*, McGraw-Hill, New York, 1962.

[55] Torrero E. A. 'The bypass of 3-D radar'. *IEEE Spectrum*. 1976;**13**(10): 49–52.

[56] van't Klooster C. G. M., Karlsson I., Bonnedal M., Albertsen N. C. 'Multiple beam slotted waveguide antenna for spaceborne synthetic aperture radar'. *Proceedings of the IEEE Antennas and Propagation Symposium*, London, Ontario, Canada, 1991, pp. 531–534.

[57] Ingvarson P., Zachrisson J. 'Swedish space antenna projects'. *7th European Conference on Antennas and Propagation (EUCAP 2013)*, Gothenburg, Sweden, April 2013, pp. 3049–3053.

[58] Simmons A. J., Giddings O. M. 'A multiple-beam two-dimensional waveguide slot array'. *IRE International Convention Record*. 1963, pp. 56–69.

[59] Saltzman H., Stavis G. 'A dual beam planar antenna for Janus type Doppler navigation systems'. *IRE National Convention Record*. 1958, pp. 240–247.

[60] Miller J. R., Forman R. J. 'A planar slot array with four independent beams'. *IEEE Transactions on Antennas and Propagation*. 1966;**14**(5):560–566.

[61] Bodnar D. G., Barnes W. J., Corey L. E. 'A lightweight HWL radar antenna'. *Proceedings IEEE International Radar Conference*, Arlington, VA, USA, 1980, pp. 259–262.

[62] Fan G., Jin J-M. 'Scattering from a cylindrically conformal slotted waveguide array antenna'. *IEEE Transactions on Antennas and Propagation*. 1997;**45**(7):1150–1159.

[63] Josefsson L., Persson P. *Conformal Array Antenna Theory & Design*. IEEE Press/Wiley, 2006.

[64] Lewis L. R., Fassett M., Hunt J. 'A broadband stripline array element'. *Proceedings of the IEEE Antennas and Propagation Symposium*, Atlanta, GA, USA, June 1974, pp. 335–337.

[65] Rehnmark S. 'Instantaneous bearing discriminators with omnidirectional coverage and high accuracy'. *IEEE Microwave Theory and Techniques International Symposium*, Washington, DC, USA, 1980, pp. 120–122.

[66] Kummer W. H. Special issue on conformal arrays (Preface). *IEEE Transactions of Antennas and Propagation*. 1974;**22**(2):1–3.

[67] Muhs H. P. 'mm-wave antenna'. *Microwave Journal*. 1985;**28**(7):191–194.

[68] Bayliss E. T. 'Design of monopulse antenna difference patterns with low sidelobes'. *Bell Systems Technical Journal*. 1968;**47**(5):287–289.

[69] Hansen R. C. *Phased Array Antennas*. Wiley, New York, 1998.
[70] Derneryd A. 'A technique for low sidelobes in monopulse planar array antennas'. *Proceedings of the 1986 Antenna Applications Symposium*, Allerton Park, IL, USA, 1986.
[71] Watson C. K., Ringer K. 'Feed network design for airborne monopulse slot-array antennas'. *Microwave Journal*. 1988;**31**(6):129–145.
[72] https://commons.wikimedia.org/wiki/File:An_apg65_h.jpg. Retrieved 9 June 2017.
[73] Robertson R. S., Elliott R. S. 'The design of transverse slot arrays fed by the meandering strip of a boxed stripline'. *Proceedings of the IEEE Antennas and Propagation Symposium*, Vancouver, Canada, 1985, pp. 111–114.
[74] Barnett R. I., Elliott R. S. 'A feasibility study of stripline-fed slots arranged as a planar array with circular grid and circular boundary'. *IEEE Transactions on Antennas and Propagation*. 1989;**37**(12):1510–1515.
[75] Cohn S. B. 'Characteristic impedance of the shielded-strip transmission line'. *IRE Transactions on Microwave Theory and Techniques*. 1954;**2**(2): 52–57.
[76] Breithaupt R. W. 'Conductance data for offset series slots in stripline'. *IEEE Transactions on Microwave Theory and Techniques*. 1968;**16**(11): 969–970.
[77] Lagerlöf R. 'Stripline fed slots'. *Proceedings of the second European Microwave Conference*, Stockholm, August, 1971.
[78] Lagerlöf R. 'Optimization of cavities for slot antennas'. *Microwave Journal*. 1973;**16**(10):12C–12F.
[79] Josefsson L., Moeschlin L., Svensson T. 'A stripline flat plate antenna with low sidelobes'. *Proceedings of the IEEE Antennas and Propagation Symposium*, Atlanta, GA, USA, 1974, pp. 282–285.
[80] Josefsson L., Moeschlin L., Sohtell V. 'A monopulse flat plate antenna for missile seeker'. *Proceedings of the Military Electronic Defence Expo (MEDE)*, Wiesbaden, Germany, September 1977.
[81] Thors B., Josefsson L. 'Radiation and scattering trade-off design for conformal arrays'. *IEEE Transactions on Antennas and Propagation*. 2003;**51**(5): 1069–1076.
[82] Bach Andersen J., Vaughan R. G. 'Transmitting, receiving, and scattering properties of antennas'. *IEEE Antennas and Propagation Magazine*. 2003; **25**(4):93–98.
[83] Josefsson L. 'Slot coupling and scattering'. *Proceedings of the IEEE Antennas and Propagation Symposium*, Dallas, USA, 1990, pp. 942–945.
[84] Diamond B. L. 'A generalized approach to the analysis of infinite planar array antennas'. *Proceedings of the IEEE*. 1968;**56**(11):1837–1851.
[85] Kerr D. E. (ed.) *Propagation of short radio waves*, MIT Rad. Lab. Series, Vol. 13, McGraw-Hill, New York, 1947.
[86] Ross R. A. 'Radar cross section of rectangular flat plates as a function of aspect angle'. *IEEE Transactions on Antennas and Propagation*. 1966;**14**(3): 329–335.

[87] Polka L. A., Balanis C. A. 'Non-principal plane scattering from perfectly conducting, flat, rectangular plates'. *Proceedings of the IEEE Antennas and Propagation Symposium*, San José, CA, USA, 1989, pp. 1278–1281.

[88] Tittensor P. J., Newton M. J. 'Prediction of the radar cross-section of an array antenna'. *Proceedings of the Sixth International Conference on Antennas and Propagation (ICAP)*, 1989, pp. 258–262.

[89] Levy R. 'Determination of simple equivalent circuits of interacting discontinuities in waveguides or transmission lines'. *IEEE Transactions on Microwaves, Theory and Techniques*. 2000;**48**(10):1712–1716.

[90] Fan G., Jin J-M. 'Scattering from a large planar slotted waveguide array antenna'. *Electromagnetics*. 1999;**19**(1):109–130.

[91] Zhang L., Yuan N., Zhang M., Li L-W., Gan Y-B. 'RCS computation for a large array of waveguide slots with finite wall thickness using the MoM accelerated by P-FFT algorithm'. *IEEE Transactions on Antennas and Propagation*. 2005;**53**(9):3101–3105.

[92] Rengarajan S. R. 'Scattering and receiving characteristics of waveguide slot array antennas'. *Proceedings of the IEEE Radar Conference*, Waltham MA, USA, 2007, pp. 532–534.

[93] Pelton E. L., Munk B. A. 'A streamlined metallic radome'. *IEEE Transactions on Antennas and Propagation*. 1974;**22**(6):799–803.

[94] Stanek T., Johansson F. S. 'Analysis and design of a hemispherical metallic radome'. *Proceedings of the Workshop on EM Structures*, Nottingham, UK, 1995.

[95] Gruenberg H. 'Second-order beams of slotted waveguide arrays'. *Canadian Journal of Physics*. 1953;**31**(1):55–69.

[96] Gruenberg H. '*Microwave Antenna*'. U.S. patent 2 639 383, May 19, 1953.

[97] Kurtz L. A., Yee J. S. 'Second-order beams of two-dimensional slot arrays'. *IRE Transactions on Antennas and Propagation*. 1957;**5**(4):356–362.

[98] McCormick G. C. 'A two-dimensional slotted array'. *IRE Transactions on Antennas and Propagation*. 1958;**6**(1):26–35.

[99] McCormick G. C. 'The effect of the size of a two-dimensional array on second-order beams'. *IRE Transactions on Antennas and Propagation*. 1958;**6**(3):297–298.

[100] Derneryd A. 'Butterfly lobes in slotted waveguide antennas'. *Proceedings of the IEEE Antennas and Propagation Symposium*, Blacksburg, VA, USA. 1987, pp. 360–363.

[101] Herting B. J., Elsallal M. W., West J. B. 'Sub-array mirroring for the mediation of second-order beams in planar edge slot waveguide arrays'. *Proceedings of the 2007 Antenna Applications Symposium*, Allerton Park, IL, USA, 2007.

[102] Kaminow I. P., Stegen R. J. *Waveguide Slot Array Design*. Hughes Aircraft Company, Techn. Memo. No. 348, July 1954 (available as NTIS Report AD 63600).

[103] Tang R. 'A slot with variable coupling and its application to a linear array'. *IRE Transactions on Antennas and Propagation*. 1960;**8**(1):97–101.

[104] Dudley D. G. 'An iris-excited slot radiator in the narrow wall of rectangular waveguide'. *IRE Transactions on Antennas and Propagation*. 1961;**9**(4): 361–364.

[105] Hashemi-Yeganeh S., Elliott R. S. 'Analysis of untilted edge slots excited by tilted wires'. *IEEE Transactions on Antennas and Propagation*. 1990; **38**(11):1737–1745.

[106] Gruenberg H. 'Theory of waveguide-fed slots radiating into parallel-plate regions'. *IRE Transactions on Antennas and Propagation*. 1952;**3**(1):63–66.

[107] Gruenberg H. *Microwave Antenna*. U.S. patent 2 659 005, Patented Nov. 10, 1953.

[108] Roberts A. (ed.) *Radar Beacons*. MIT. Rad. Lab Series, Vol. 3, 1947.

[109] Galindo V., Green K. 'A near-isotropic circularly polarized antenna for space vehicles'. *IEEE Transactions on Antennas and Propagation*. 1965; **13**(6):872–877.

[110] Scott W. G., Soo Hoo K. M. 'A theorem on the polarization of null-free antennas'. *IEEE Transactions on Antennas and Propagation*. 1966;**14**(3): 587–590.

[111] Kraus J. D. *Antennas*, 2nd edition, McGraw-Hill, New York, 1988, p. 89.

[112] Wait J. R. 'On the conductance of slots'. *IRE Transactions on Antennas and Propagation*. 1956;**4**(2):124–127.

[113] Wait J. R. *Electromagnetic Radiation from Cylindrical Structures*. Pergamon Press, London, 1959.

[114] Gayen R. K., Das S. D., Das A. C. 'A high gain-bandwidth, nearly omnidirectional waveguide slot array antenna'. *Proceedings of the IEEE Antennas and Propagation Symposium*, Vancouver, BC, Canada. 2015, pp. 222–223.

[115] Li J-Y., Li L-W. 'Analysis of omnidirectional waveguide slots array antennas'. *Proceedings of the 6th International Symposium on Antennas, Propagation and EM Theory*, ISAPE'03, Beijing, October 28–November 1, 2003, pp. 38–41.

[116] Solbach K. 'Simulation of slot array antennas using standard network analysis program'. *IEEE Antennas and Propagation Magazine*. 1991;**33**(6):45–47.

[117] Croswell W. F., Knop C. M., Hatcher D. M. 'A dielectric-coated circumferential slot array for omnidirectional coverage at microwave frequencies'. *IEEE Transactions on Antennas and Propagation*. 1967;**15**(6): 722–727.

[118] Fiet O. O., Scudder R. M. *Ultrahigh-Frequency Broadcast Antenna System*. US patent 2 658 143, November 1953.

[119] Lee C. S., Lee S. W., Chuang S. L. 'Plot of the modal field distribution in rectangular and circular waveguides'. *IEEE Transactions on Microwave Theory and Techniques*. 1985;**33**(3):271–274.

[120] Marcuvitz N. *Waveguide Handbook*. MIT. Rad. Lab Series, Vol. 10, 1951.

[121] Grabherr W., Huder B. 'Omnidirectional slotted-waveguide array antenna'. *Proceedings of the 29th European Microwave Conference*, Munich, Germany. 1999, pp. 134–137.

[122] Morwing B. 'AGA-ERICON – a marine radar beacon'. *Ericsson Review.* 1981;**58**(4):180–187.
[123] Knudsen H. L. 'Radiation from ring quasi-arrays'. *IRE Transactions on Antennas and Propagation.* 1956;**4**(3):452–472.
[124] Davies D: E. N. 'Circular arrays: their properties and potential applications'. *IEE Proceedings of the Second International Conference on Antennas and Propagation*, April 1981, pp. 1–10.
[125] Rahim T., Davies D. E. N. 'Effect of directional elements on the directional response of circular antenna arrays'. *IEE Proceedings H – Microwaves, Antennas and Propagation.* 1982;**129**(1):18–22.
[126] Harrington R. F. *Time-Harmonic Electromagnetic Fields*, IEEE Press, New York, 2001.
[127] Chu T-S. 'On the use of uniform circular arrays to obtain omnidirectional patterns'. *IRE Transactions on Antennas and Propagation.* 1959;**7**(**4**): 436–438.
[128] Croswell W. F., Knop C. M. 'On the use of an array of circumferential slots on a large cylinder as an omnidirectional antenna'. *IEEE Transactions on Antennas and Propagation.* 1966;**14**(3):394–396.
[129] Lagerlöf R., Sjöholm L., Svensson T. *Antenna for Strip Transmission Lines*, US patent 3 713 165, filed Oct. 5, 1970.
[130] Ohira M., Miura A., Ueba M. '60-GHz wideband substrate-integrated-waveguide slot array using closely spaced elements for planar multisector antenna'. *IEEE Transactions on Antennas and Propagation.* 2010;**58**(3): 993–998.
[131] Montgomery C., Dicke R., Purcell E. *Principles of Microwave Circuits.* MIT Rad. Lab. Series, Vol. 8, McGraw-Hill, New York, 1948.
[132] Collin R. E. *Foundations for Microwave Engineering.* McGraw-Hill, Inc., New York, 1966.
[133] Collin R. E. *Field Theory of Guided Waves.* IEEE Press, New York, 1991.
[134] Pease M. 'The iterated network and its application to differentiators'. *Proceedings of the IRE.* 1952;**40**(6):709–711.
[135] Takeshima T., Isogai Y. 'Frequency bandwidth of slotted array aerial system'. *Electron. Eng.* February 1969, pp. 201–204.
[136] Hamadallah M. 'Frequency limitations on broad-band performance of shunt slot arrays'. *IEEE Transactions on Antennas and Propagation.* 1989;**37**(7):817–823.
[137] Derneryd A., Petersson R. 'Bandwidth characteristics of monopulse slotted waveguide antennas'. *Fourth International Conference on Antennas and Propagation* (ICAP 85), 16–19 April, 1985, pp. 27–30.
[138] Josefsson L. *Slots in Waveguide Arrays Analyzed by Expanding the Fields in Waveguide Modes and Floquet Modes*. Rpt No. AP-202, 1983, EE Dept, UCLA, USA.
[139] Svensson B., Snygg G., Holmberg P. 'A low cost, high performance point-to-point slotted waveguide array'. *Microwave Journal*, 1999;**42**(11):1–5.

[140] Rengarajan S. R., Zawadzki M. S., Hodges R. E. 'Bandwidth enhancement of large planar slot arrays'. *Proceedings of the IEEE Antennas and Propagation Symposium*, Honolulu, HI, USA, 2007, pp. 4405–4408.

[141] Rengarajan S., Chatterjee S. 'Bandwidth characteristics of waveguide-fed planar slot arrays'. *Proceedings URSI General Assembly*, Chicago, IL, USA, 2008.

[142] Zhou S-G., Huang G-L., Peng Z-H., Li-Jian Y. 'A wideband full-corporate-feed slot planar array'. *IEEE Transactions on Antennas and Propagation*. 2016;**64**(5):1974–1978.

Chapter 10

Slot arrays in special waveguide technologies

In this chapter, we discuss slot arrays in two types of parallel plate waveguides propagating the TEM mode. In the first an electromagnetic wave propagates in the radial direction inward or outward in a circular parallel plate waveguide, while slots cut in one of the parallel plates radiate. Such an array is convenient for large apertures requiring high gain and mass production. The second type has a rectangular shape and propagates the TEM mode. Slots cut in one of the plates radiate. We then present slot arrays in substrate integrated waveguides (SIW) which have a great potential for easy integration with planar devices. Finally slot arrays in gap waveguides are discussed.

10.1 Slot arrays in circular parallel plate waveguides

The concept of a radial line slot array, first introduced by Kelly in 1957 [1] and published in 1961 by Goebels and Kelly [2], is shown in Figure 10.1. A standing wave TM_{10} mode is excited between a pair of circular flat plates by a circular waveguide feed at the centre. Annular slots cut at certain radial values produce broadside radiation with the desired polarisation, including dual polarisation and circular polarisation by exciting the appropriate waveguide modes. In a subsequent paper [3] they presented a monopulse antenna divided into four 90° sectoral waveguides as shown in Figure 10.2. Much of the later developments of the radial line slot antennas were made at the Tokyo Institute of Technology, by Goto, Ando, Hirokawa and others. We will discuss designs of radial line slot antennas for circular and linear polarisations. The design of a radial line slot antenna for monopulse applications will also be presented.

10.1.1 Circularly polarised radial line slot antenna

The radial line slot antennas are easy to manufacture, for applications such as direct broadcast satellite (DBS) receive antennas where inexpensive mass production is desired. They have been investigated by numerous researchers and have been published extensively. They have found applications in radar, satellite communications, wireless local area networks (WLAN), local multiport distribution systems (LMDS) and plasma etching, among others, at microwave and millimetre-wave frequencies. The two-layer radial line antenna is well suited to the design of

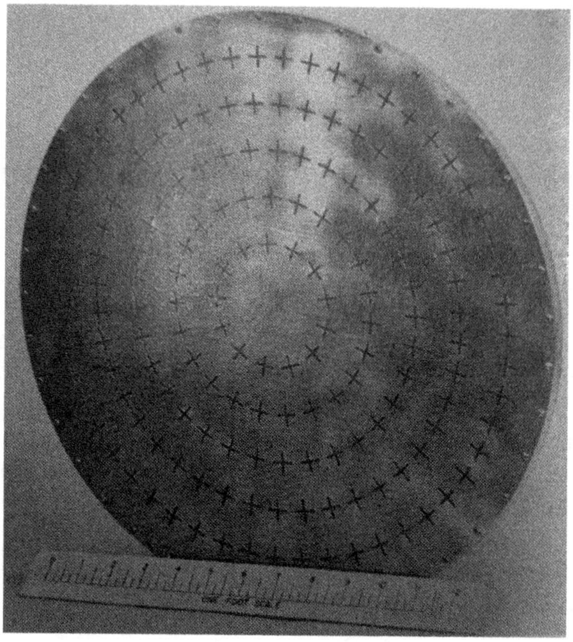

Figure 10.1 A typical radial line slot antenna. © IEEE 1961, reprinted from [2], with permission

Figure 10.2 Monopulse radial line slot array. © IEEE 1964, reprinted from [3], with permission

nearly uniform aperture distributions required for high gain applications. In this structure the increase in power density of the wave as it propagates inward is offset by the power radiated by the slots. In a single-layer radial waveguide antenna power flows radially outward. It is better suited to the design of a tapered amplitude distribution. Since a systematic design of the two-layer radial line slot antenna was presented first, we start with a discussion of that structure. Figure 10.3 shows a radial line slot antenna consisting of a two-layer circular parallel plate waveguide [4]. The lower parallel plate waveguide is excited by an outward travelling radial TEM mode. In the upper parallel plate waveguide consisting of radiating slots on the upper conductor, the TEM mode propagates inward. It is then terminated in an absorber at the centre.

Figure 10.3(a) shows the two-layer antenna structure. Figure 10.3(b) illustrates the power flow direction in both parallel plate waveguides. The antenna has a coaxial cable to radial waveguide adaptor and a 180° *E*-bend between lower and upper radial lines. In the upper waveguide orthogonal slot pairs couple power from the radial waveguide and radiate a circularly polarised broadside beam. An absorbing material placed near the centre at the radial value ρ_m of the upper

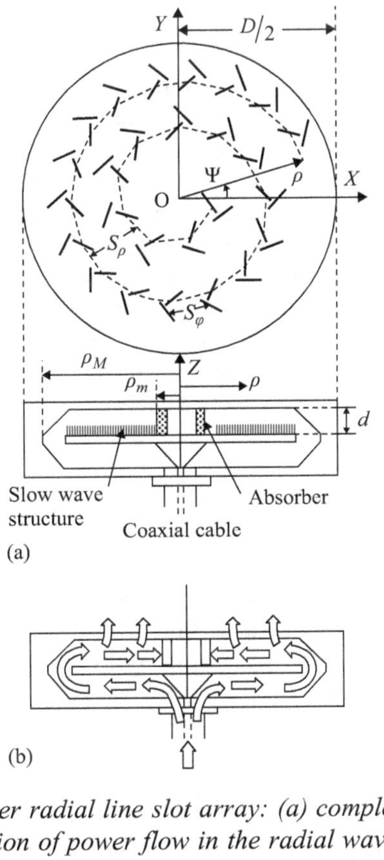

Figure 10.3 A two-layer radial line slot array: (a) complete antenna structure; (b) direction of power flow in the radial waveguides. © IEEE 1985, reprinted from [4], with permission

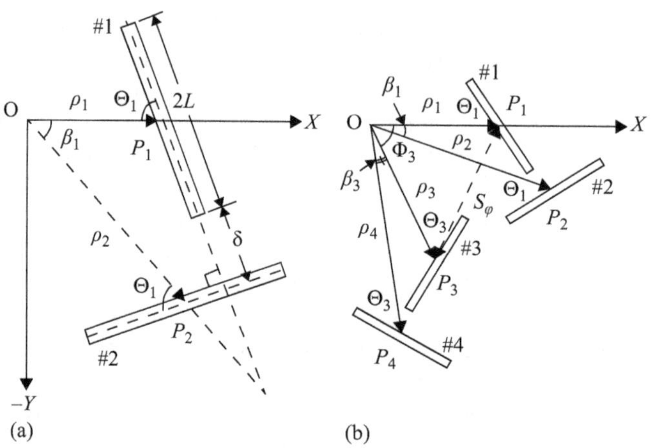

Figure 10.4 Slot arrangements in the upper wall: (a) first slot pair; (b) first two slot pairs.

waveguide dissipates the remaining power in the travelling wave. In order to produce a good match at the coaxial cable input, the coaxial cable to radial waveguide adaptor and the 180° E-bend have to be designed to have very little reflection. In addition, reflection from the slot pairs shall be minimised as well.

The arrangement of slot pairs on the top plate of the upper parallel plate waveguide is illustrated in Figure 10.4. The two slots of each slot pair are orthogonal to each other with nearly the same amplitude of excitation and 90° phase difference so as to produce a circular polarisation. The current distribution in the plate for an inward travelling wave will be in the form of a Hankel function of the first kind and order zero, whose phase term may be approximated by a cylindrical phase function. Therefore the radial coordinates of the centres of the slot pair are spaced a quarter guide wavelength so that their excitation phases are 90° apart as shown by (10.1) and (10.2) [4].

$$k_g(\rho_2 - \rho_1) = \pi/2 \tag{10.1}$$

$$k_g = 2\pi/\lambda_g \tag{10.2}$$

where λ_g is the guide wavelength and k_g is the radial wavenumber in the upper parallel plate waveguide. Excitations of slots 1 and 3 lag 90° with respect to those of 2 and 4, respectively. Similarly all other slot pairs have a quarter wavelength spacing between the radial coordinates of the centres of the slots. The two slot pairs shown in Figure 10.4 as well as all others with similar orientation will radiate right circular polarisation.

In order to eliminate the grating lobes, one may employ a slow-wave structure in the upper parallel plate waveguide by loading it with a dielectric or by using a corrugated structure in the lower plate of the parallel plate waveguide. The azimuthal spacing between successive slot pairs, S_φ is approximately equal to the radial spacing S_ρ. The first slot of each slot pair makes the same angle with respect to the radial line from the origin to the slot centre (e.g. $\Theta_1 = \Theta_3$). In order to produce a broadside beam, the two slot pairs in Figure 10.4(b) should radiate in phase in the direction normal to the aperture. This requires that the tilt angle of the second slot pair with respect to the first compensate the excitation phase difference as specified by (10.3).

$$k_g(\rho_3 - \rho_1) = \Phi_3 + 2n\pi \tag{10.3}$$

where n is an integer. Slot pairs are arranged over a spiral with one guide wavelength radial spacing between successive turns of the spiral. The reflections from different slot pairs cancel out since they originate from different radial values just like slots in a linear travelling wave array. The slot pairs are non-resonant so that they couple a small amount of power from the propagating TEM wave.

We will now describe a specific design of the coaxial cable to the radial waveguide adaptor and the *E*-bend presented by Ando *et al.* [5]. Figures 10.5 and 10.6 show a section of the adaptor and the *E*-bend, respectively, which are rotationally

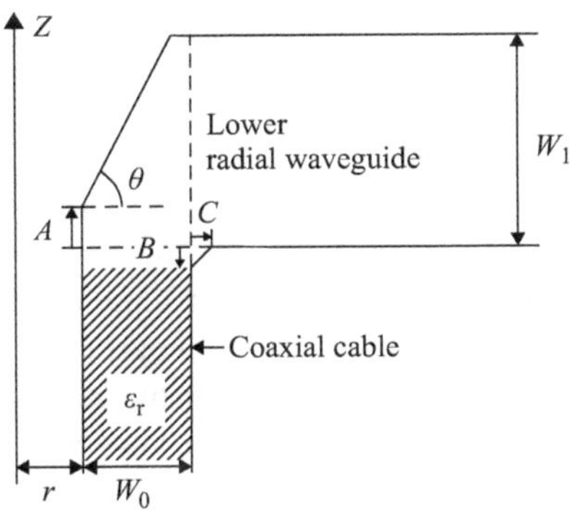

Figure 10.5 Coaxial cable to radial waveguide adaptor ($A = 0$, $B = C = 1.5$ mm, $\theta = 51.3°$, $W_0 = 7.5$ mm, $W_1 = 12.5$ mm, $r = 3$ mm). © IEEE 1987, reprinted from [5], with permission

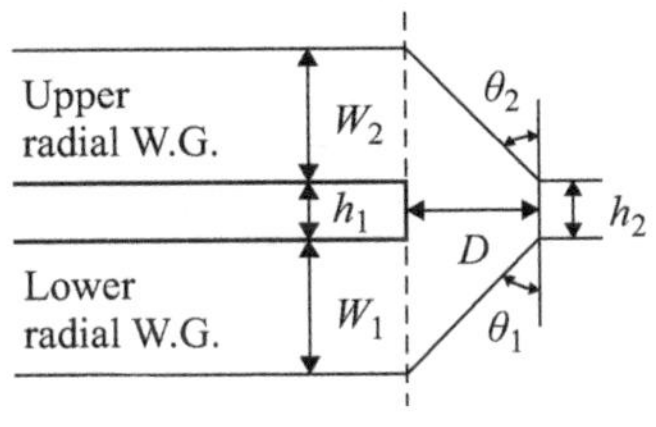

Figure 10.6 180° E-bend between lower and upper radial lines ($D = 12$ mm, $\theta_1 = 43.8°$, $\theta_2 = 38.7°$, $W_1 = 12.5$ mm, $W_2 = 15$ mm, $h_1 = h_2 = 4$ mm). © IEEE 1987, reprinted from [5], with permission

symmetric with respect to a vertical axis. The radii of the inner and outer conductors of the coaxial cable are r and $r + W_0$, respectively.

Ando *et al.* [5] used the commercial finite element analysis code HFSS to design the parameters A, B, C, θ, W_0, W_1 and r of the adaptor and D, θ_1, θ_2, W_1, W_2, h_1 and h_2 of the bend at 6 GHz. The reflection coefficients of these devices were found to be less than 0.1 in the frequency range of 5.5–6.5 GHz. One could use any computational electromagnetic code to design these devices with a starting value obtained by scaling these parameters for the frequency range of interest.

10.1.2 Slot coupling and the attenuation coefficient of the propagating wave

The power coupled by each slot pair from the inward travelling wave for radiation is designed to provide the required aperture distribution. In many applications, it is desired to have a uniform amplitude distribution to maximise the gain of the antenna. Assuming weak slot coupling and neglecting reflections, the incident wave at each slot pair may be expressed in the form

$$f(\rho) = \exp(jk_g\rho + \alpha\rho)/\sqrt{\rho} \tag{10.4}$$

where α is the attenuation coefficient of the propagating TEM wave due to radiation from slot pairs.

Figure 10.7 shows the amplitude distribution $f(\rho)$ as a function of ρ for different values of α from 0.03 to 0.15 for a nine wavelength radius aperture. As seen a value of $\alpha = 0.15$ produces an approximately uniform distribution while a value of $\alpha = 0.03$ provides an 11 dB taper. The slot lengths are designed to have the desired coupling by using an approximate infinite array model [6], with a unit cell slot pair as shown in Figure 10.8. The waveguide model has conducting walls at the top and bottom and periodic walls on the sides. This is an approximate model for the slot pairs on a spiral. One can use a computational electromagnetics code such as HFSS to analyse such a structure and obtain the slot lengths to realise the desired

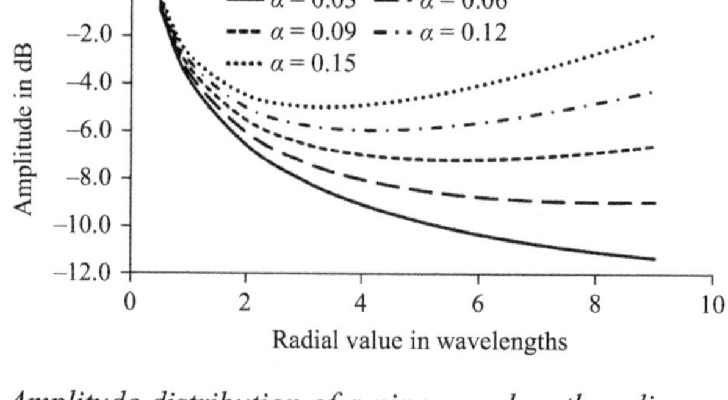

Figure 10.7 Amplitude distribution of a nine wavelength radius antenna for different values of the attenuation coefficient α

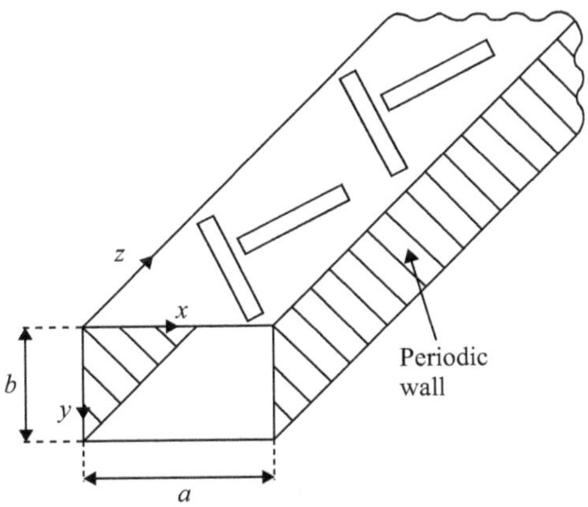

Figure 10.8 Waveguide model for designing the coupling from a slot pair. © IEE 1990, reprinted from [6], with permission

coupling factor and the aperture taper. The factor $1/\sqrt{\rho}$ and the attenuation term in (10.4) will not appear in the analysis model as they are not relevant in the rectangular waveguide. Let the coupling coefficient of the nth slot pair be c_n and the field incident at the nth slot pair be E_n. The power radiated by the nth slot is $(E_n c_n)^2$. It is easy to show that the coupling coefficient c_n is related to the attenuation coefficient by $(1 - c_n^2)^{1/2} = e^{\alpha(\rho_n - \rho_{n-1})}$. If the radial spacing between adjacent slot pairs is the same, the coupling coefficients are equal.

10.1.3 Linearly polarised radial line slot antenna

The two-layer structure with a coaxial cable feed in the previous section is also applicable to produce a linearly polarised antenna. In this case slot pairs are arranged on concentric circles spaced a half-guide wavelength. Figure 10.9 shows an x-directed linearly polarised radial line slot antenna in the top conductor of the upper parallel plate waveguide, and Figure 10.10 shows the first two slot pairs. The electric current in the slot plate induced by the travelling TEM mode is radially directed. The slot 4 does not radiate since it is along the current direction of the TEM mode whereas slot 3 radiates x-polarised field. The relationship between angles θ_1 and θ_2 that slots 1 and 2 make with the radial line is derived below for x-polarised radiation from the slots [7].

The co-polarised and cross-polarised radiations from the slot pairs 1 and 2 are given by $\beta_1 \sin(\theta_1 + \phi) + \beta_2 \sin(\theta_2 + \phi)$ and $-\beta_1 \cos(\theta_1 + \phi) - \beta_2 \cos(\theta_2 + \phi)$, respectively, where the coupling factors β_1 and β_2 of slots 1 and 2 are proportional to $\sin \theta_1$ and $-\sin \theta_2$, respectively, because of the half-guide wavelength radial spacing between them. This does not take into account the radial dependence of the amplitude of the travelling wave field and the associated current. The radial dependence of the amplitude in the previous discussion of Figure 10.7 and the waveguide model used for calculating the coupling are valid for this linearly

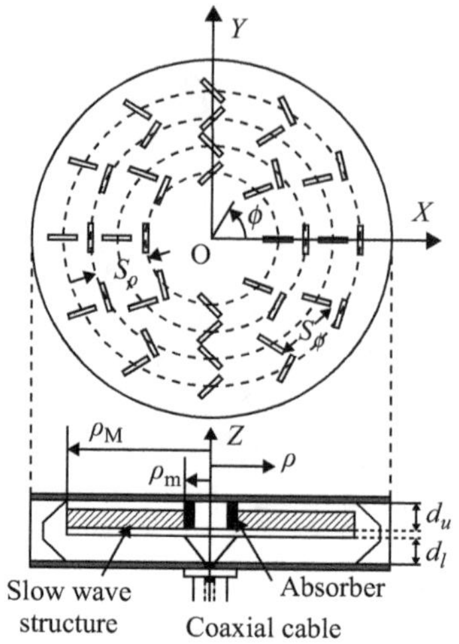

Figure 10.9 The slot pairs cut on concentric circles. © IEEE 1988, reprinted from [7], with permission

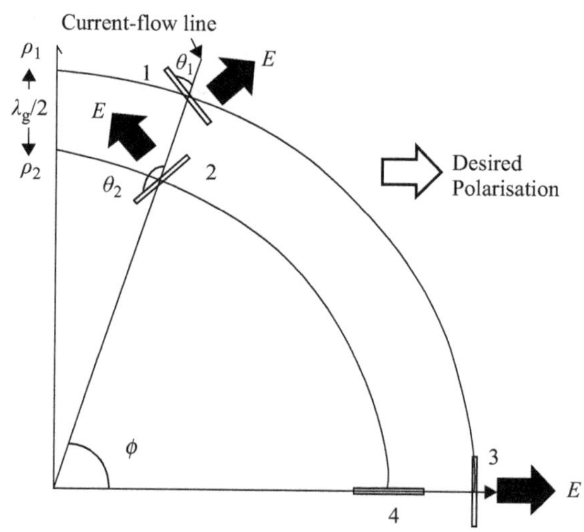

Figure 10.10 First two slot pairs. © IEEE 1988, reprinted from [7], with permission

polarised slot array also. The co-polarised and cross-polarised radiations for the slot pair are given by (10.5) and (10.6).

$$\sin\theta_1 \sin(\theta_1 + \phi) - \sin\theta_2 \sin(\theta_2 + \phi) = 1 \tag{10.5}$$

$$-\sin\theta_1 \cos(\theta_1 + \phi) + \sin\theta_2 \cos(\theta_2 + \phi) = 0 \tag{10.6}$$

These two equations are satisfied when

$$\theta_1 = \pi/2 - \phi/2 \tag{10.7}$$

and

$$\theta_2 = \pi - \phi/2 \tag{10.8}$$

The radial distance between adjacent slot pairs is kept one guide wavelength. If the guide wavelength is kept less than the free space wavelength by dielectric loading, grating lobes are suppressed. The azimuthal distance S_ϕ between adjacent slots is kept one guide wavelength, just like that of the radial line slot antenna for circular polarisation (see Figure 10.3).

10.1.4 Design of slot coupling coefficients

Let us assume that there are N rings of slot pairs with a guide wavelength spacing between successive rings. Since the parallel plate waveguide is loaded with a dielectric to make the guide wavelength smaller than the free space wavelength, grating lobes are avoided. The spacing between adjacent slot pairs in the azimuthal direction is also less than the free space wavelength for broadside beams, and the number of slots in each ring is proportional to the value of the radius. Let the field of the incident wave arriving at the outermost (Nth) ring of radius ρ_N be

$$A_N = \exp\left(jk_\rho \rho_N\right)/\sqrt{\rho_N} \tag{10.9}$$

If there are M_N slot pairs in this ring and if the coupling coefficient, defined in Section 10.1.2, of each slot is b_N, the incident field in the $(N-1)$st ring is

$$A_{N-1} = \exp\left(jk_\rho \rho_{N-1}\right)\left[1 - M_N b_N^2\right]^{1/2}/\sqrt{\rho_{N-1}} \tag{10.10}$$

The number of slots in a ring is proportional to the radius of the ring. The reflections from the slots are ignored in this analysis. The coupling coefficient b_{N-1} of each slot in this ring is related to C_N and C_{N-1}, the aperture excitation coefficients of Nth and $(N-1)$st rings, respectively, as given in (10.11).

$$b_N/\sqrt{\rho_N} = \left(b_{N-1}/\sqrt{\rho_{N-1}}\right)\left[1 - M_N b_N^2\right]^{1/2}(C_N/C_{N-1}) \tag{10.11}$$

Recurrence relations like this are written for all rings up to the first or the innermost ring. The wave going past the innermost ring of slots will be dissipated in the absorber.

The value of b_N is initially estimated and subsequently updated in an iterative fashion until the power dissipated in the absorber becomes a small value such as 5% of the power incident in the Nth ring of slots.

Figure 10.11 shows the values of coupling coefficients, b_i, where $i = 1-20$ denote the ring number. The phase of the aperture distribution is assumed to be

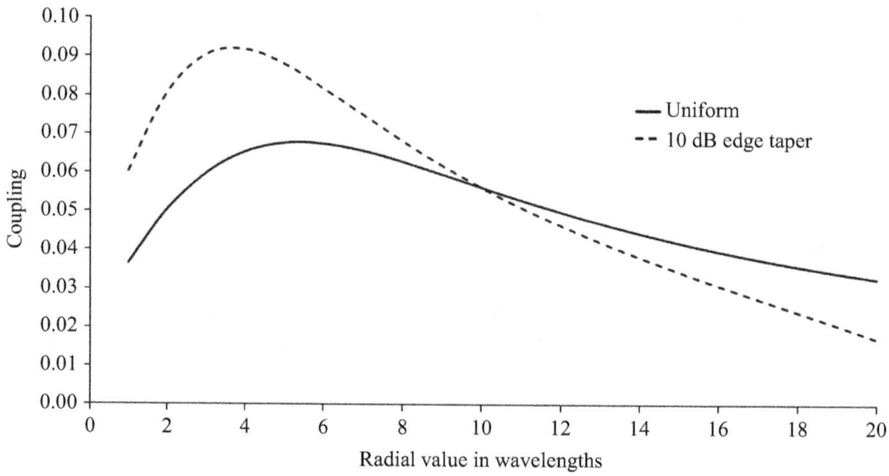

Figure 10.11 Slot coupling coefficient as a function of the radial value for an inward travelling wave in a 20 wavelength radius double-layer radial waveguide

uniform whereas two amplitude distributions, uniform and 10 dB taper, have been considered for the circular aperture [8]. For the outermost slots, coupling increases as the radius decreases for both uniform and tapered apertures due to decreasing power in the travelling wave as it propagates inward. For tapered apertures, the coupling coefficient starts with a smaller value for the outermost ring but the rate of increase is greater than that of the uniform distribution since the aperture excitation increases as the radius decreases. The coupling coefficient reaches a peak and starts decreasing for innermost rings since the ratio ρ_{n-1}/ρ_n decreases rapidly for small values of n.

The incident travelling wave amplitude in dB in each ring is shown in Figure 10.12. For both cases the incident power in the outermost ring is the same. For uniform distribution, the incident wave power decreases more rapidly than for the tapered distribution as greater power is radiated by the outermost rings in the former. Both cases were designed to dissipate -14 dB power at the centre relative to the incident wave power at the 20th ring. Therefore both plots coalesce at the centre.

Since the slot pairs are spaced one wavelength radially, the reflections from them add up. Therefore the reflection coefficient of this antenna type is high. It is possible to reduce the reflection coefficient at the input port by introducing additional pairs of slots called reflection cancelling slots on the radiating wall or on the opposite wall in the case of a single-layer parallel plate waveguide or by changing the slot spacing in the radial direction as a function of the azimuthal angle, thereby producing a squinted beam off broadside. In one case the squinted beam design produced 22 dB return loss compared to 4 dB of the corresponding broadside beam design [9]. The improved return loss caused 1 dB increase in gain for the squinted beam design. It is convenient to produce the squinted beam in the polarisation

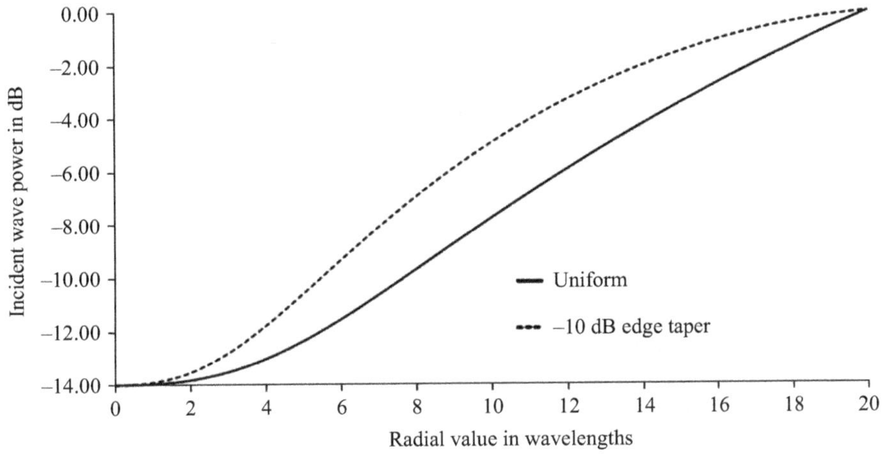

Figure 10.12 Incident wave power in dB as a function of the radial value for an inward travelling wave in a 20 wavelength radius double-layer radial waveguide

plane, for example, for an x-polarised beam pointing at an angle (θ_t, $\phi_t = 0$). The spacing, S_ρ, between two adjacent slot pairs in the radial direction is given next.

$$-k_\rho S_\rho + k_0 S_\rho \sin\theta_t \cos\phi = -2\pi \tag{10.12}$$

where k_0 is the free space wavenumber.

$$S_\rho = \frac{\lambda_g}{1 - \frac{1}{\sqrt{\varepsilon_r}} \sin\theta_t \cos\phi} \tag{10.13}$$

S_ρ reduces to the previous value of λ_g for a broadside beam or $\theta_t = 0$. It is recommended that the beam tilt θ_t be about a couple of beamwidths of the antenna, if possible.

10.1.5 Single-layer radial line slot arrays

Single-layer radial line slot antennas have been designed to produce circular polarisation or linear polarisation using the same principles as those for the double-layer radial line slot antennas discussed previously. In the single-layer array the travelling wave propagates outward while slots couple radiated power. The slots near the centre have very small amounts of coupling whereas the outer slots have large amounts of coupling. It is not possible to design this array with an attenuation coefficient such as that shown in Figure 10.7. Takahashi *et al.* showed that the coupling increases very rapidly as the radial value increases [10]. Equations similar to (10.9)–(10.11) are used for outward radially propagating TEM mode. We then obtain results for coupling coefficients and incident wave power for linearly polarised slot pairs in concentric rings over a 20 wavelength radius aperture.

The coupling coefficient shown in Figure 10.13 increases rapidly as the wave propagates outward. Since the incident wave amplitude decreases as the radius

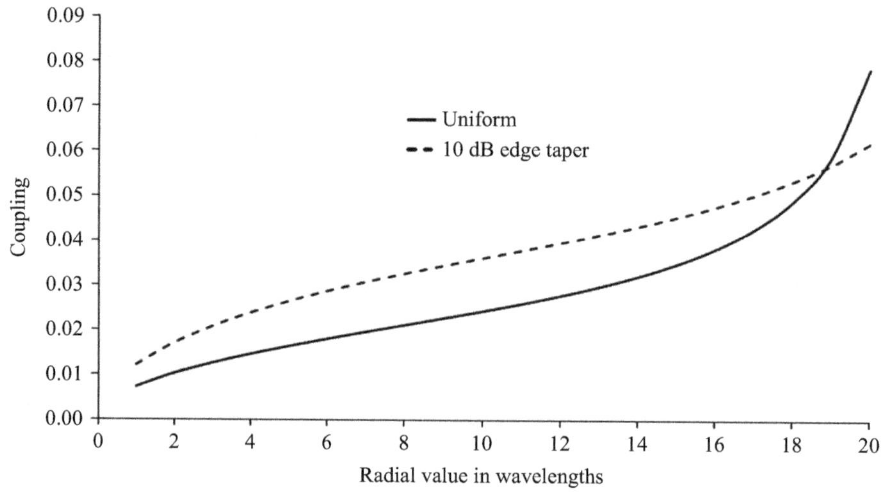

Figure 10.13 Coupling coefficient as a function of the radial value for an outward travelling wave in a 20 wavelength radius single-layer radial waveguide

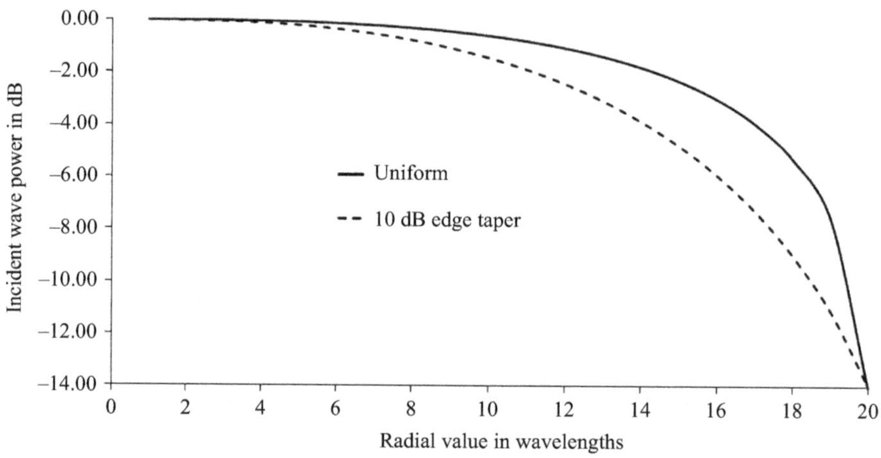

Figure 10.14 Incident wave power in dB as a function of the radial value for an outward travelling wave in a 20 wavelength radius single-layer radial waveguide

increases, larger amounts of coupling coefficient are needed to produce the desired aperture distribution even though there are greater number of slots per ring as the radius increases. For the tapered distributions, the coupling coefficient is larger than that of the uniform distribution for small radial values while it is less for large values of the radius.

Figure 10.14 illustrates the incident wave power as a function of the radial value. For small radial values the slot coupling coefficient and the radiated power are small and hence the incident wave power is nearly constant whereas for large radial values the incident wave power drops rapidly. For tapered distributions, the power level is lower since greater power is radiated from the centre of the aperture. Both plots coalesce at $N = 20$ since they are designed to have the same amount of power absorbed at the outer end of the parallel plate waveguide.

10.1.6 Radial line antennas for monopulse applications

A single-layer radial waveguide with outward travelling wave is considered here. A similar analysis can be made for a two-layer radial waveguide slot array. For monopulse applications we need four feeds, A_1, A_2, A_3, and A_4 all having the same radial value and azimuthal locations of 0°, 90°, 180° and 270°, respectively, as shown in Figure 10.15 [11].

The self and mutual impedances of the feed probes have been derived in [11] using the induced emf method. The field distribution of the outward travelling wave derived by Miyashita and Katagi [11] is used here. The coupling between feeds i and j is given by

$$S_{ij} = C_0 Z_{ij} \tag{10.14}$$

where

$$C_0 = \frac{2/Z_0}{(1 + Z_{ii}/Z_0)(1 + Z_{jj}/Z_0)} \tag{10.15}$$

Z_0 is the characteristic impedance of each feed line, and Z_{ii} and Z_{jj} are the self-impedances of the input ports i and j, respectively.

For the sum pattern with all four feeds excited equally ($A_1 = A_2 = A_3 = A_4 = A$), the incident travelling wave is equivalent to that produced by a single centred probe

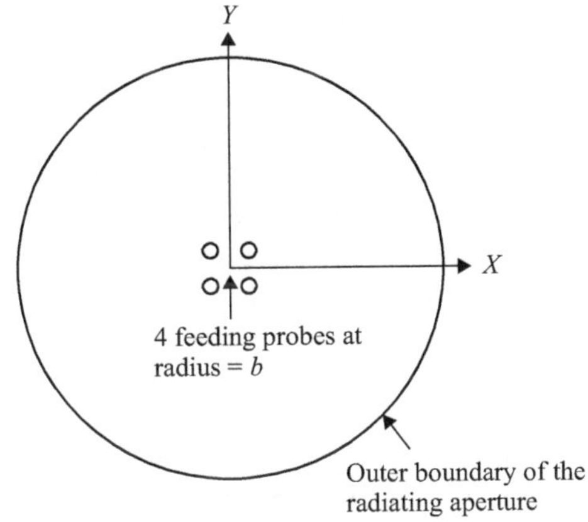

Figure 10.15 Feed probes for monopulse application. © IEEE 1996, reprinted from [11], with permission

feed. For the azimuth difference pattern in the x–z plane ($A_1 = -A_2 = -A_3 = A_4 = A$) the incident travelling wave field is equivalent to that produced by a centred probe feed weighted by a cos ϕ term. The pattern in the x–z plane produces a null in the broadside direction with a major lobe on each side. Similarly for the elevation difference pattern in the y–z plane ($A_1 = A_2 = -A_3 = -A_4 = A$) the incident travelling wave field is equivalent to that produced by a centred probe feed weighted by a sin ϕ term. The pattern in the y–z plane produces a null in the broadside direction with a major lobe on each side. The design of the radiating slots may use any of the methods discussed in the previous sections.

10.1.7 Accurate analysis and design techniques

An efficient full-wave method of moments analysis of radial line slot antennas was presented by Albani *et al.* [12]. They split the Green's function of a large circular cavity into a sum of parallel plate waveguide contributions and a rim effect contribution expanded in terms of radial modes. In a subsequent paper [13] an approximate asymptotic formula was presented for the mutual admittance of slots so that the numerical integration is reduced by more than 90% in computing the mutual coupling between slot pairs. An automatic design of circularly polarised radial line slot antennas for a specific aperture distribution was demonstrated using a full wave analysis and optimisation [14].

10.2 Slot arrays in rectangular parallel plate waveguides

Slots cut in a parallel plate waveguide propagating a planar travelling wavefront is less expensive to manufacture than a conventional planar slot array made up of several rectangular waveguides, although the latter is preferred for high-performance designs. Figure 10.16 shows a planar array of slots cut in a parallel plate waveguide [15]. The TEM mode in the parallel plate waveguide is excited by an array of coupling slot pairs cut in the broad wall of a rectangular waveguide. The rectangular waveguide is placed underneath the parallel plate waveguide. The spacing between slots in a pair is kept a quarter guide wavelength so that the reflection from each slot pair in the feed waveguide at the design frequency is reduced substantially. The adjacent slots pairs are spaced one guide wavelength so that excitation phases are equal. The feed waveguide and the parallel plate waveguide are both loaded by dielectrics such that the guide wavelength of the feed waveguide is lower than that of the parallel plate waveguide which is in turn less than that of free space wavelength. The feed waveguide may be excited by a coaxial cable at its centre or it may be excited by a shunt series type coupling slot from another waveguide.

Figure 10.16 shows the feed waveguide excited by a coaxial cable. The TE_{10} mode wave propagating from the centre of the feed waveguide towards both ends is equal in amplitude and phase but the propagation directions are opposite. Therefore, the coupling slot pairs are tilted in opposite directions on the two sides, while all the slot pairs on one side are tilted in the same direction. Both ends of the feed waveguide have matched terminations and so have both ends of the parallel plate

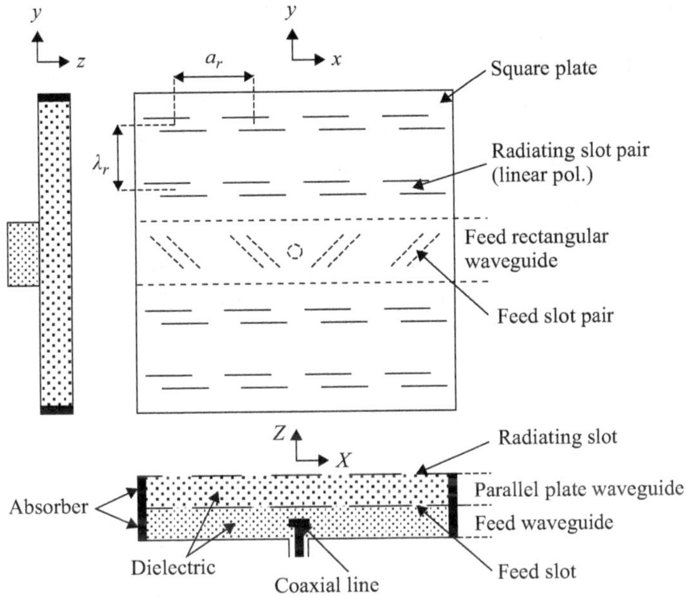

Figure 10.16 Slot array fed by a parallel plate waveguide. © IEEE 1992, reprinted from [15], with permission

waveguide. Although it is possible to have reflection cancelling slots, absorbers at the ends of the travelling wave arrays provide a better gain bandwidth and pattern bandwidth than those employing reflection cancelling slots or shorts at the end. The a dimension of the feed waveguide, a_f, and the dielectric constant ε_f are chosen to satisfy the inequality (10.16) so that only a single mode propagates in the feed waveguide.

$$\frac{\lambda_0}{2\sqrt{\varepsilon_f}} < a_f < \frac{\lambda_0}{\sqrt{\varepsilon_f}} \tag{10.16}$$

In order to have a nearly uniform TEM mode to propagate in the parallel plate waveguide, the spacing between the coupling slot pairs should be less than the TEM mode wavelength in the parallel plate waveguide, that is, $\lambda_{gf} < \frac{\lambda_0}{\sqrt{\varepsilon_r}}$. Typically

$$\lambda_{gf} = \frac{\lambda_0}{\sqrt{\varepsilon_f}}\sqrt{2} \tag{10.17}$$

Therefore,

$$\varepsilon_r < \frac{\varepsilon_f}{2} \tag{10.18}$$

The design of coupling slot pairs in the feed waveguide is carried out using the procedure described in Chapter 6. Generally, a uniform aperture distribution is desired in many applications. The scattering parameters of a given slot pair may be obtained from a computational electromagnetic code such as HFSS as a function of slot lengths and spacings. The tilt angle is kept at 45°. The spacing between slots in a pair is nominally a quarter guide wavelength while the slots are parallel. The lengths (non-resonant) and spacings are adjusted so that the reflection from the slot pair is zero in the feed waveguide. For uniform aperture distribution, the coupling into the parallel plate waveguide is the same from every slot pair and the power going into the absorber in the feed waveguide should be small.

Figures 10.17 and 10.18 display a four-port coupler and the analysis model of a feed slot pair, in which the parallel plate waveguide is replaced by a rectangular waveguide with two periodic side walls. Such a model can also be used for the case

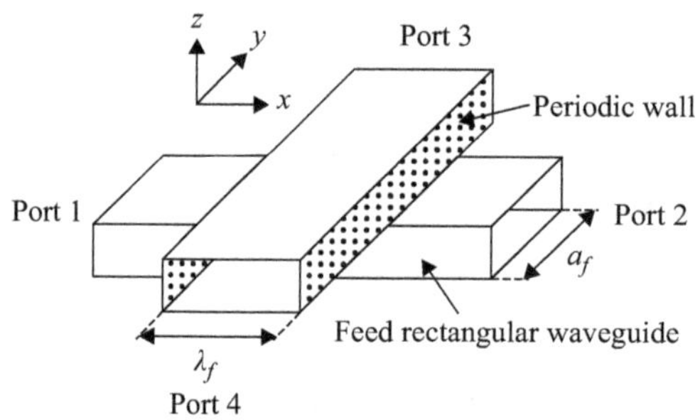

Figure 10.17 Analysis model for the feed slot pair. © IEEE 1992, reprinted from [15], with permission

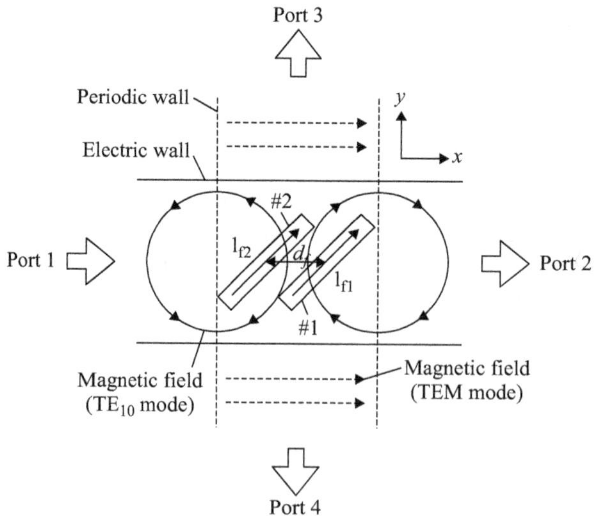

Figure 10.18 Feed slot pair in the common wall of a four-port coupler. © IEEE 1992, reprinted from [15], with permission

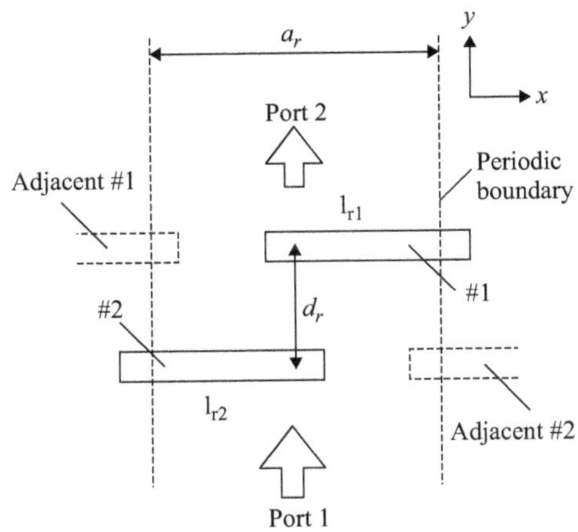

Figure 10.19 Analysis model for a radiating slot pair. © IEEE 1992, reprinted from [15], with permission

of a non-uniform aperture distribution as long as the aperture distribution is slowly varying in both directions. Scattering parameters of the four-port coupler are computed for a range of values of slot lengths and spacing for use in the design.

Radiating slot pairs in the parallel plate waveguide also produce no significant reflection at the design frequency when the spacing between the slots in a pair is a quarter guide wavelength just like the coupling slot pairs. The distance between the adjacent slot pairs is kept at one wavelength. For a relatively large aperture, an infinite array mutual coupling model may be employed to design the radiating slot pairs, as shown in Figure 10.19. The amount of coupling from each radiating slot pair is designed to produce the desired aperture distribution and to have a small amount of power going into the absorber at the ends. This design procedure is similar to that of travelling wave linear arrays discussed in Chapter 6.

The bandwidth of the parallel plate slot array shown in Figure 10.16 for 1 dB gain drop is about 3%, and it is possible to achieve a return loss greater than 10 dB in that frequency range. In the centre-fed parallel plate slot arrays such as the ones discussed here, there are no radiating slots in the centre and hence the sidelobe level in the E-plane is rather high, about 10−11 dB. The end-fed parallel plate waveguide array has a major drawback since beam squint with frequency will limit the gain bandwidth substantially. The sidelobe level may be improved by choosing a tapered aperture distribution in the centre-fed configuration [16].

10.3 Slot arrays in substrate integrated waveguides

Dielectric-filled rectangular waveguides with sidewalls replaced by an array of metallised posts are called SIW [17]. They are convenient for easy integration of planar devices and rectangular waveguide components. The dielectric-filled rectangular waveguides can be designed with a very small b dimension and are easily transitioned to microstrip transmission lines. If the number of metallised vias is greater than 10 per wavelength, such waveguides essentially behave like dielectric-filled rectangular waveguides with metallic side walls. Based on a study of dispersion characteristics of the substrate integrated rectangular waveguide, the cut-off frequencies of the first two modes were obtained as a function of the structural parameters, as shown in (10.19) and (10.20) [18].

$$f_{c10} = \frac{c}{2\sqrt{\varepsilon_r}}\left(W - \frac{D^2}{0.95b}\right)^{-1} \tag{10.19}$$

$$f_{c20} = \frac{c}{\sqrt{\varepsilon_r}}\left(W - \frac{D^2}{1.1b} - \frac{D^3}{6.6b^2}\right)^{-1} \tag{10.20}$$

where c is the velocity of light in free space, D is the diameter and b is the spacing of adjacent vias. W is the distance between the two arrays of vias, that is, nominally the a dimension of the corresponding dielectric-filled waveguide. If the spacing between vias is sufficiently small, that is, greater than 10 vias per wavelength, the effective value of the width of the waveguide is given by

$$W_{\text{eff}} = W - \frac{D^2}{0.95b} \tag{10.21}$$

Thus an SIW is equivalent to a dielectric-filled metallic rectangular waveguide with width W_{eff}. Then the values of offsets and lengths of the radiating slots in a dielectric waveguide can be determined using the design procedures discussed in Chapters 6 and 7.

An example of a planar array of slots in a two-layer SIW configuration is shown in Figure 10.20 [19]. They used the conventional design of a slot array in equivalent dielectric-filled rectangular waveguides. Transitions from microstrip line to SIW using tapered lines were found to work well. In the lower layer a feed waveguide is formed using the SIW technology with centred-inclined coupling slots. Coupling coefficients for such slots are found using (7.2) in Chapter 7.

A single-layer SIW slot array for 60 GHz is shown in Figure 10.21 [20]. The feed waveguide couples into radiating waveguides through coupling windows at one end. Even though such an end-fed arrangement has bandwidth limitations, it has the advantage of being a low-cost device because of the ease of fabrication of the single-layer device. A conductor backed coplanar waveguide (CBCPW) is used to feed the antenna, transitioning into a feed SIW.

SIW technology, that is, arrays of metallised vias in place of conducting walls, has also been employed in parallel plate waveguide slot arrays [21]. The low-temperature co-fired ceramics (LTCC) manufacturing technique is suited to make SIW slot arrays, especially at millimetre-wave frequencies and even higher frequencies in the range of hundreds of GHz, with the availability of low-loss dielectric materials [22,23].

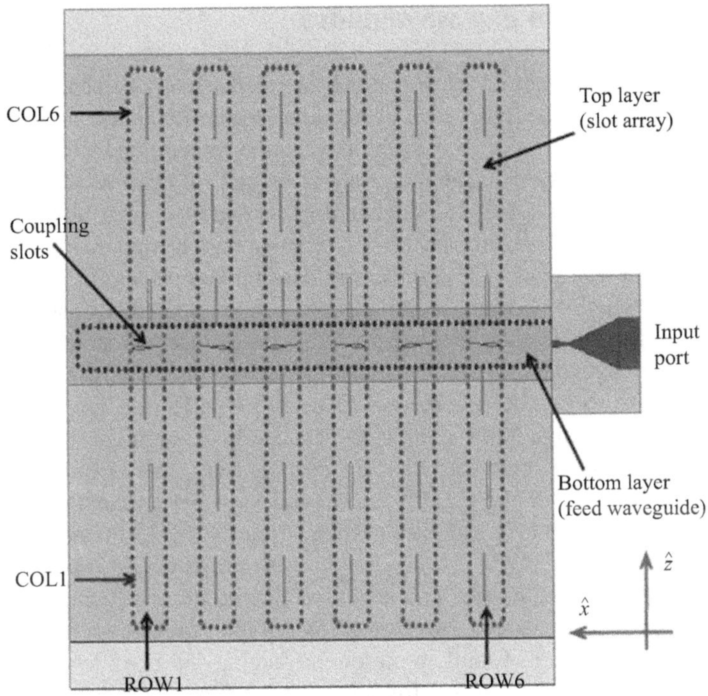

Figure 10.20 A two-layer SIW slot array at 60 GHz. Reproduced from [19], courtesy of The Electromagnetics Academy

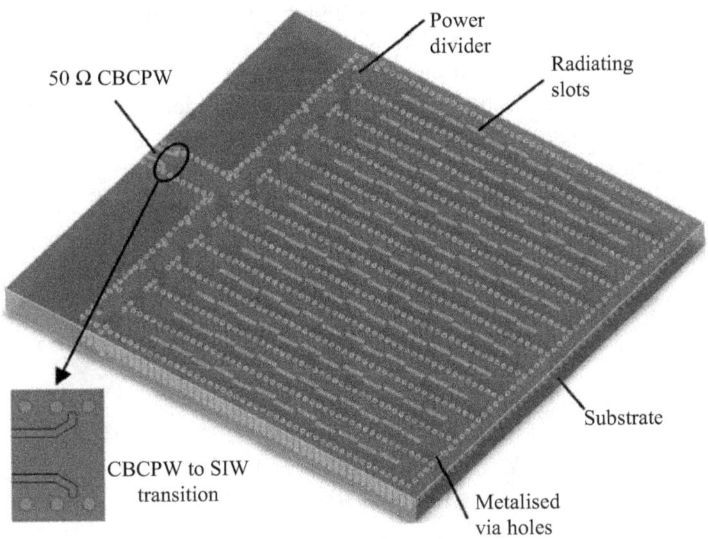

Figure 10.21 A single-layer SIW slot array at 60 GHz. © IEEE 2010, reprinted from [20], with permission

10.4 Slot arrays in gap waveguides

Slotted rectangular waveguide array antennas are difficult to fabricate at millimetre-wave frequencies and above because of manufacturing tolerances. In addition, it is difficult to ensure good electrical contact between the waveguide wall containing the slots and the side walls when the slot plate is brazed into the waveguide channel (see Section 11.2). Although the SIW technology is widely used at high frequencies, such waveguides exhibit significant losses. Gap waveguides propagate quasi-TEM mode over a wide bandwidth [24,25]. Power loss in a gap waveguide is comparable to that in a rectangular waveguide [26], and it is much less than that for an SIW. Therefore, the gap waveguide technology has a lot of promise at millimetre-wave frequencies and beyond.

Figure 10.22 shows the geometry of the gap waveguide consisting of two parallel plates and a ridge in the middle. On each side of the ridge there is a bed of nails. If the spacing 'a' between adjacent nails is very small compared to their height 'd', that is, $a \ll d$, and if there are several nails per wavelength, the bed of nails may be considered as a high impedance surface [27]. In that case the region above the bed of nails behaves as a stop band for electromagnetic wave propagation for a broad range of frequencies and a quasi-TEM mode propagates between the ridge and the conductor above. Therefore, it is possible to have an airgap between the top plate and the bed of nails, thereby eliminating the need for any electrical contact in the structure. The cut-off frequency of the first mode is given by $d = \lambda/4$ while that of the next mode is given by $d + h = \lambda/2$. By choosing a small value of h, one can realise nearly an octave of bandwidth for single-mode propagation.

Zaman and Kildal [28] describe a single-layer low-loss corporate feed network for a slot array. Figure 10.23 shows the corporate power divider and a four-element linear array of slots in the top plate shown on the right. The gap waveguide without the ridge is excited by a coaxial cable probe and the waveguide is transitioned to ridge gap waveguide by stepped ridges. The power divider is similar to microstrip power dividers with quarter wave impedance matching sections and mitred corners. The ends of the ridge waveguides are terminated in Ts before they excite the slots. Corrugated sections at the end of the E-planes reduce the edge diffracted fields in the back region.

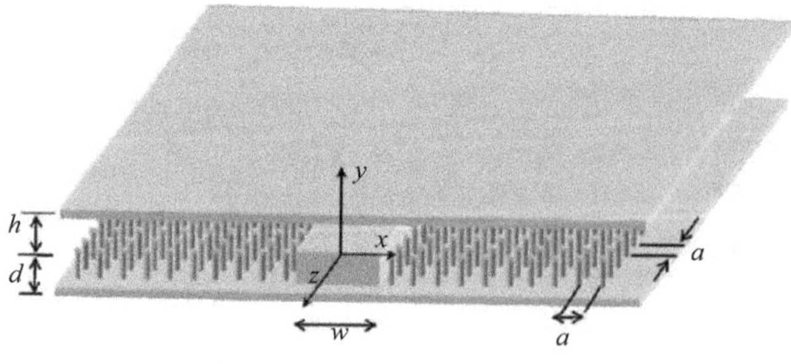

Figure 10.22 The geometry of the gap waveguide. © IEEE 2011, reprinted from [25], with permission

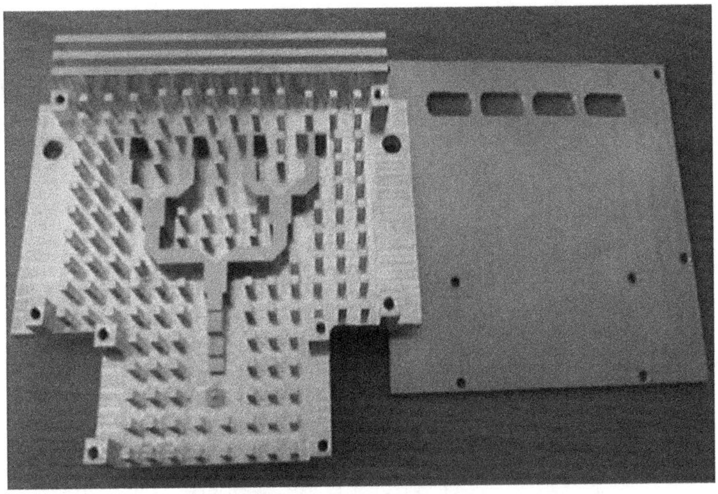

Figure 10.23 *A linear array of four slots excited by a ridge gap waveguide with a corporate feed. © IEEE 2014, reprinted from [28], with permission*

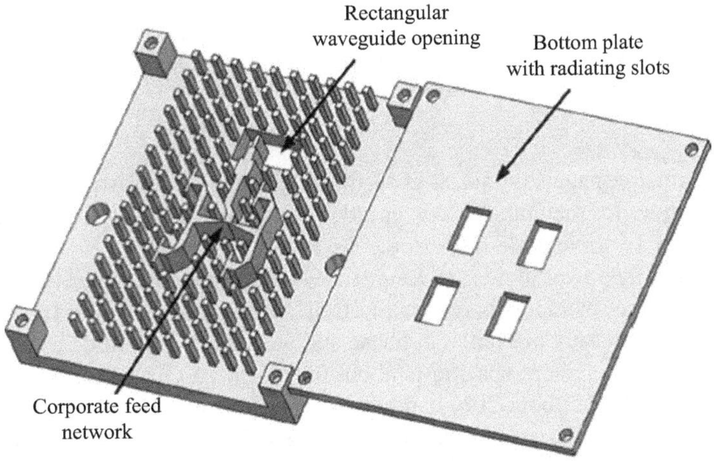

Figure 10.24 *A planar array of 2 × 2 slots excited by a ridge gap waveguide with a corporate feed. © IEEE 2014, reprinted from [28], with permission*

In the design of the bends of the *T*-junctions, a few pins were removed or relocated locally. This does not affect the performance of the gap waveguide structures. For designing the antennas without grating lobes, one row of pins has been used at some locations. Even with one row of pins the leakage of energy to neighbouring elements is 20 dB down.

Figure 10.24 shows a 2 × 2 planar array in the ridge gap waveguides. It has a 90° transition from ridge gap waveguide to rectangular waveguide. Since the ridge

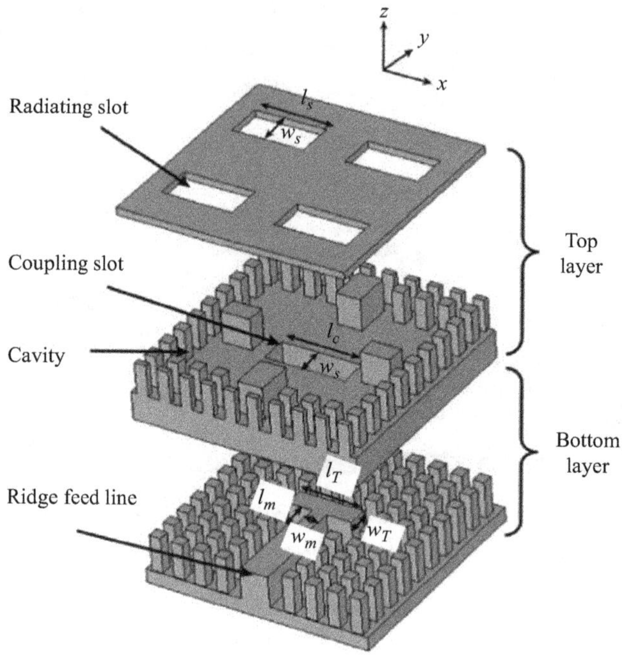

Figure 10.25 The 2 × 2 array unit cell excited by a cavity. © IEEE 2014, reprinted from [29], with permission

gap waveguide propagates a quasi-TEM mode just like a microstrip line, the ridge gap waveguide to rectangular waveguide transition is designed similar to a microstrip line to waveguide transition.

The four-way power divider is designed by tapering the ridge sections gradually for the impedance match. This results in 20 dB return loss over 12−15 GHz band. The 2 × 2 element array is excited in phase and with equal amplitude by a corporate feed network. The element spacing is about 0.875λ at 15 GHz. Both the 1 × 4 and 2 × 2 arrays produced about 20% bandwidth for better than 10 dB return loss.

Figure 10.25 shows a unit cell comprising 2 × 2 slots excited by a cavity [29]. The bottom layer of Figure 10.25 has a ridge waveguide terminated in a *T* section. A coupling slot in the middle excites the cavity which in turn excites the four radiating slots in the top plate. Sixteen such unit cells make up an 8 × 8 array of slots in ridge gap waveguide technology with a corporate power divider as illustrated in Figure 10.26 [29].

The 8 × 8 array discussed earlier has uniform aperture distribution and produces −13 dB sidelobe levels in both the *E*- and *H*-planes. It is possible to design tapered distributions using non-uniform power dividers and non-resonant slots of different lengths for low sidelobe arrays. The slot arrays in ridge waveguide technology are expected to find many applications in the future, such as integration of planar antennas, filters, active devices, etc., up to terahertz frequencies.

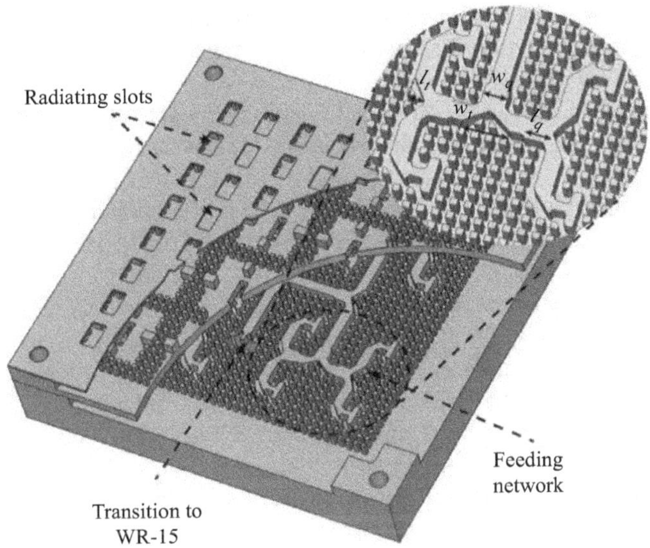

Figure 10.26 An 8 × 8 slot array in ridge gap waveguides with a corporate power divider. © IEEE 2014, reprinted from [29], with permission

References

[1] Kelly K. C. 'Recent annular slot array experiments'. *IRE National Convention Record*, pt. 1, 1957, pp. 144–151.

[2] Goebels F. J., Kelly K. C. 'Arbitrary polarisation from annular slot planar antennas'. *IRE Transactions on Antennas and Propagation.* 1961;**9**(4): 342–349.

[3] Kelly K. C., Goebels F. J. 'Annular slot monopulse antenna arrays'. *IEEE Transactions on Antennas and Propagation.* 1964;**12**(4):391–403.

[4] Ando M., Sakura K., Goto N., Arimura K., Ito Y. 'A radial line slot antenna for 12 GHz satellite TV reception'. *IEEE Transactions on Antennas and Propagation.* 1985;**33**(12):1347–1353.

[5] Ando M., Ito S., Sakurai K., Goto N. 'Suppression of reflection in a radial line slot antenna for 12 GHZ band satellite reception'. *IEEE Antennas and Propagation Symposium Digest,* 1987, pp. 898–901.

[6] Hirokawa J., Ando M., Goto N. 'Analysis of slot coupling in radial line slot antenna for DBS reception'. *IEE Proceeding H.* 1990;**137**(5): 249–254.

[7] Ando M., Numata T., Takada J.-I., Goto N. 'A linearly polarized radial line slot antenna'. *IEEE Transactions on Antennas and Propagation.* 1988;**36**(12): 1675–1680.

[8] Stutzman W. A., Thiele G. A. *Antenna Theory and Design*, 3rd edn. John Wiley and Sons, Hoboken, NJ, 2013.

[9] Davis P. W., Bialkowski M. E. 'Linearly polarized radial-line slot-antennas with improved return-loss performance'. *IEEE Antennas and Propagation Magazine.* 1999;**41**(1):52–61.

[10] Takahashi M., Takada J., Ando M., Goto N. 'Characteristics of small-aperture, single-layered, radial line slot antennas'. *IEE Proceedings-H.* 1992;**139**(1): 79–83.

[11] Miyashita H., Katagi T. 'Radial line planar monopulse antenna'. *IEEE Transactions on Antennas and Propagation.* 1996;**44**(8):1158–1165.

[12] Albani M., Cono G. L., Gardelli R. 'An efficient full-wave method of moments analysis for RLSA antennas'. *IEEE Transactions on Antennas and Propagation.* 2006;**54**(8):2326–2336.

[13] Albani M., Mazzinghi A., Freni A. 'Asymptotic approximation of mutual admittance involved in MoM analysis of RLSA antennas'. *IEEE Transactions on Antennas and Propagation.* 2009;**57**(4):1057–1063.

[14] Albani M., Mazzinghi A., Freni A. 'Automatic design of CP-RLSA antennas'. *IEEE Transactions on Antennas and Propagation.* 2012;**60**(12): 5538–5547.

[15] Hirokawa J., Ando M., Goto N. 'Waveguide-fed parallel plate slot array antenna'. *IEEE Transactions on Antennas and Propagation.* 1992;**40**(2): 218–223.

[16] Hirokawa J., Ando M. 'Sidelobe suppression in 76 GHz post-wall waveguide-fed parallel plate slot arrays'. *IEEE Transactions on Antennas and Propagation.* 2000;**48**(11):1727–1732.

[17] Deslandes D., Wu K. 'Integrated microstrip and rectangular waveguide in planar form'. *IEEE Microwave and Wireless Components Letters.* 2001; **11**(2):68–70.

[18] Cassivi Y., Perregrini L., Arcioni P., Bressan M., Wu K., Conciauro G. 'Dispersion characteristics of substrate integrated rectangular waveguide'. *IEEE Microwave and Wireless Components Letters.* 2002;**12**(9):333–335 .

[19] Bakhtafrooz A., Borji A., Busuioc D., Safavi-Naeini S. 'Novel two-layer millimeter-wave slot array antennas based on substrate integrated waveguides'. *Progress in Electromagnetics Research.* 2010;**109**:475–491.

[20] Chen X.-P., Wu K., Han L., He F. 'Low-cost high gain planar antenna array for 60-GHz band applications'. *IEEE Transactions on Antennas and propagation.* 2010;**58**(6):2126–2129.

[21] Hirokawa J., Ando M. 'Efficiency of 76 GHz postwall waveguide fed parallel plate slot arrays'. *IEEE Transactions on Antennas and Propagation.* 2000;**48**(11):1742–1745.

[22] She Y., Fujino R., Hirokawa J., Ando M., Hanatani D., Fujimoto M. 'LTCC oversized rectangular waveguide slot array antenna with air-layer in the radiating part in the millimeter-wave band'. *IEEE Transactions on Antennas and Propagation.* 2013;**61**(4):1777–1783.

[23] Xu J., Chen Z. N., Qing X. '270-GHz LTCC-integrated strip-loaded linearly polarized radial line slot array antenna'. *IEEE Transactions on Antennas and Propagation.* 2013;**61**(4):1794–1801.

[24] Kildal P. S., Alfonso E., Valero-Nogueira A., Rajo-Iglesias E. 'Local metamaterial-based waveguides in gaps between parallel metal plates'. *IEEE Antennas and Wireless Propagation Letters.* 2009;**8**:84–87.

[25] Polemi A., Maci S., Kildal P.-S. 'Dispersion characteristics of metamaterial-based parallel plate ridge gap waveguide realized by bed-of-nails'. *IEEE Transactions on Antennas Propagation.* 2011;**59**(3):904–913.

[26] Pucci E., Zaman A. U., Rajo-Iglesias E., Kildal P. S., Kishk A. 'Losses in ridge gap waveguide compared with rectangular waveguide and microstrip lines'. *Fourth European Conference on Antenna and Propagation*, 2010.

[27] Silveirinha M. G., Fernandes C. A., Costa J. R. 'Electromagnetic characterization of textured surfaces formed by metallic pins'. *IEEE Transactions on Antennas and Propagation.* 2008;**56**(2):405–415.

[28] Zaman A. U., Kildal P. S. 'Wideband slot antenna arrays with single-layer corporate feed network in ridge gap waveguide technology'. *IEEE Transactions on Antennas Propagation.* 2014;**62**(6):2992–3001.

[29] Vosoogh A., Kildal P. S. 'Corporate-fed planar 60-GHz slot array made of three unconnected metal layers using AMC pin surface for the gap waveguide'. *IEEE Antennas and Wireless Propagation Letters.* 2016;**15**:1935–1938.

Chapter 11
Manufacturing aspects

In this chapter, we will take a look at manufacturing aspects that are important for high-quality production of slotted waveguide array antennas. It is a wide subject so we will concentrate on critical parameters such as mechanical tolerances that relate to electrical performance. We will also discuss joining methods for metal materials, for example, dip brazing of aluminium. Applications at high frequencies (millimetre waves) in particular call for very high precision and special methods. An important area is the use of slotted carbon fibre reinforced plastic (CFRP) waveguides that offer light weight and thermal stability, typically required for large antenna systems for space applications. The technology is also used in some military ground and airborne radar systems.

Some of the exciting developments in low-cost fabrication using plastic materials and metallisation techniques are discussed. This area is related to the microelectronic area, particularly for high frequencies and highly integrated antenna/microwave assemblies.

11.1 Tolerance theory

When an array antenna is tested experimentally, the measured electrical performance will sometimes depart to some extent from the theoretical design due to excitation errors. The errors can be random and/or systematic. Systematic errors can in principle be corrected once the cause of the errors has been found. Random errors could be due to imperfections in the materials, tools, fixtures or other unknown source. They may be difficult to reduce at a reasonable cost. Errors due to inadequate design modelling are also important.

Several authors have treated the tolerance problem for general arrays and also for slotted waveguide arrays [1–4]. When analysing random excitation errors a common approach is to assume the errors to be uncorrelated from element to element, and with a normal distribution of the real and imaginary parts of the error. We can write the resulting voltage V_n of element n:

$$V_n = V_{n0} + \sigma V_{n0} = V_{n0}\left(1 + r_n e^{j\alpha_n}\right) \tag{11.1}$$

According to this model the phase error α_n is uniformly distributed within $(0,2\pi)$. It can be shown that the relative amplitude error r_n is Rayleigh distributed

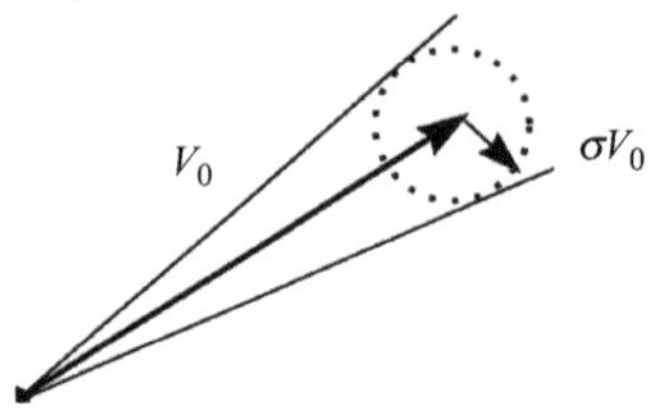

Figure 11.1 The nominal element voltage V_0 and the error voltage σV_0

while the array pattern function has a modified Rayleigh distribution; see, for example, Rondinelli [3]. The error model is represented graphically in Figure 11.1.

Assuming small errors the errors expressed in phase and amplitude become

$$\left.\begin{aligned} \sigma_{\text{phase}} &= \sigma \qquad \text{radians} \\ \sigma_{\text{dB}} &= 20\log(1+\sigma)\ \text{dB} \end{aligned}\right\} \tag{11.2}$$

As shown here the amplitude error and the phase error are both given by the same parameter σ. This can of course be questioned since in many cases the two types of error have different sources. Still, the model illustrates the level of influence from typical errors. For given or assumed phase and amplitude errors we can in practical cases write

$$\sigma^2 = \sigma_{\text{tot}}^2 = \sigma_{\text{phase}}^2 + \sigma_{\text{ampl}}^2 \tag{11.3}$$

The effect on the array gain can be estimated as

$$G = G_0/\left(1+\sigma^2\right) \tag{11.4}$$

Normally this effect is quite small. The effect on the sidelobe level is usually much more serious. The main beam power level can be written as

$$G_{\max} \propto \left(\sum_1^N V_n\right)^2 \tag{11.5}$$

while the sidelobe power level is

$$G_{\text{SL}} \propto \sum_1^N (\sigma V_n)^2 \tag{11.6}$$

Thus, the sidelobe level relative to the main beam becomes

$$\text{SL} = \text{SL}_0 + \frac{\sigma^2}{\eta N} \tag{11.7}$$

where η is the aperture efficiency:

$$\eta = \frac{\left(\sum V_n\right)^2}{N\sum V_n^2} \tag{11.8}$$

Example:

Number of radiating elements:	$N = 64$
Element error: 0.5 dB and 5°, uncorrelated =>	$\sigma_{tot} = 0.11$
Pattern function: $SL_0 = -50$ dB sidelobes,	$\eta = 0.72$ (error-free)

This gives the average sidelobe level due to the errors alone: $SL_{err}(\sim \frac{\sigma^2}{\eta N}) = -36$ dB. Sidelobe peaks will be perhaps 6–8 dB higher. Clearly, this is an unbalanced design; the tapering is too strong considering the error level (too much overdesign).

In this example the 64 elements could be the 64 slots in a single linear waveguide array. If we now take 32 *identical* slotted waveguide arrays of this kind and combine them into a 32 × 64 planar array, we would still get an error sidelobe level of about −36 dB. However, if the 32-element slotted arrays had completely independent errors (but at the same average level as before), the sidelobe average level due to errors becomes 15 dB lower, that is, at about the same level as SL_0, a much more reasonable design.

Figure 11.2 shows the calculated sidelobe level SL versus the error-free sidelobe level SL_0 for different numbers of uncorrelated elements N.

We have here discussed array element errors in dBs and degrees, while the real error character is slot positions, orientations and lengths, etc., in inches and millimetres. Furthermore, the average sidelobe level as discussed earlier is not always the most suitable parameter. A more relevant measure can be the maximum sidelobe level that is expected with, say, 95% confidence. This approach was studied by

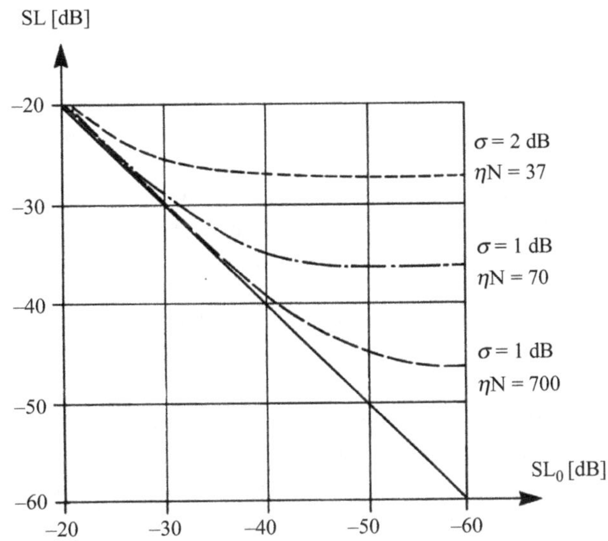

Figure 11.2 Average sidelobe level SL (achieved) versus sidelobe level SL_0 (ideal) with error σ as parameter. N is the number of uncorrelated radiating elements, and η is the aperture efficiency

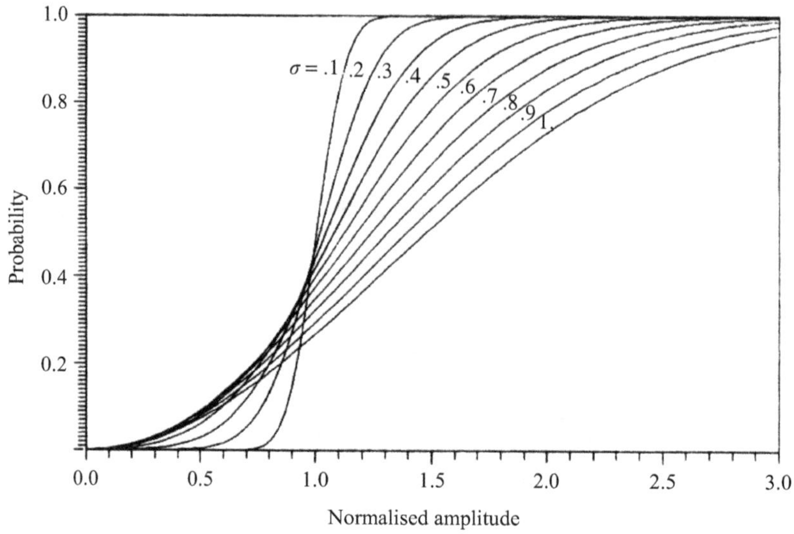

Figure 11.3 Normalised probability function for sidelobe level. Both the amplitude and the error parameter σ are normalised to the design sidelobe level. © 1982 IEEE. Adapted from [4], with permission

Hsiao [4], who derived the probability density function of the sidelobe level (see Figure 11.3).

The error tolerance problem can also be studied by direct simulation of the antenna array in question. Lee *et al.* [5] used the Monte Carlo method for determining acceptable error levels. The tolerance for each element was set inversely proportional to the nominal element weight. A Monte Carlo simulation was also used by Rengarajan *et al.* [6]. For their simulation random element error levels of 0.5 dB and 5° were assumed. The authors also made a Method of Moments analysis with assumed machining tolerances of ± 1 mil for slot positions, lengths and widths. Finally, a finite element analysis using HFSS code was made before building and testing an experimental antenna. The results from all these efforts showed good agreement.

In two early papers [7,8], the authors treated the tolerance problem for linear waveguide arrays with longitudinal slots. The mechanical errors considered were slot position errors: along waveguide, across waveguide (i.e. offset errors), and slot length errors. The errors were assumed to be independent, random and normally distributed. The error in slot position along the waveguide was found to be of minor importance compared to the errors in slot offset and slot length. Bailin and Ehrlich [7] quoted a result by Stegen [9] where the change of phase of the radiated field was measured as a function of a change in slot length off resonance. Stegen's curve is reproduced in Figure 11.4.

We find from Stegen's graph that 5 degrees of phase error results from a slot length error of about 0.5%, or less than one-tenth of one mm (X-band).

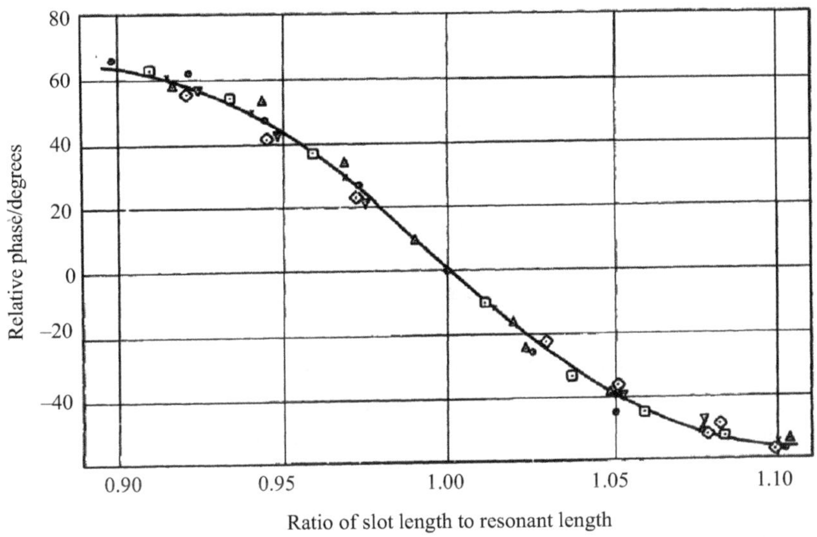

Figure 11.4 *Radiated phase as a function of change in slot length off resonance. The points on the curve indicate slot offsets, from about 0.7 to 3.9 mm. Standard X-band waveguide, slot width 1.6 mm, frequency 9.375 GHz. Adapted from Kaminov and Stegen [10]*

(We know from Chapter 5 that an error in slot offset also changes the slot resonant length as well as the slot admittance.)

The offset of a longitudinal slot is referenced to the waveguide centreline, that is, the line halfway between the inner side walls of the waveguide. This line is difficult to establish with high precision from the outside of the waveguide, especially for long waveguides. If in error the result is a systematic error, that is, a periodic offset error with a period of about one guide wavelength, giving rise to grating lobes. The artillery locating radar described in Chapter 3 (Figure 3.5) has a radiating aperture of frequency scanned non-resonant waveguides, each with 64 longitudinal slots. It has been estimated that a slight offset error in the waveguide centreline by as little as 0.04 mm would result in a 20 dB grating lobe level [11].

Rectangular waveguide sizes are standardised (EIA, IEC)[1] including tolerances and flange types [12]. Also ridge waveguide dimensions are standardised to some extent. However, for slotted waveguide arrays using ridge waveguides (e.g. to reduce the width of the waveguide) non-standard cross sections may be required. The tolerances for the critical dimensions can be analysed theoretically using HFSS or special software [13,14]. Figure 11.5 shows a simplified cross section of a ridge waveguide which is characterised by four main dimensions: width (A), height (B),

[1] EIA = Electronic Industries Association, IEC = International Electrotechnical Commission. See also Appendix.

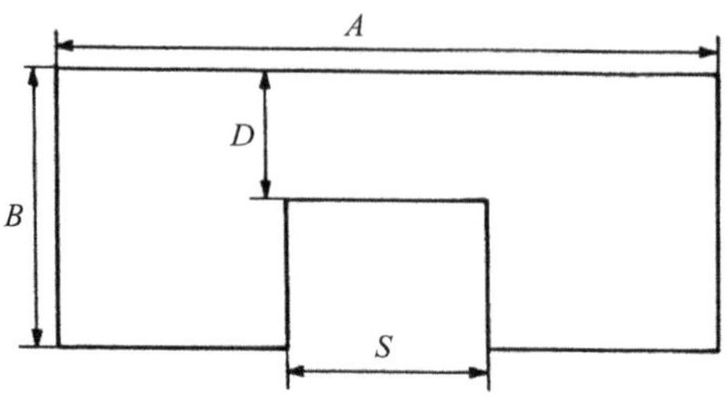

Figure 11.5 Cross section of a ridge waveguide

ridge width (S) and the separation between the upper wall and the ridge (D). From a theoretical analysis it was found that the separation (D in the figure) is the most critical and should be given the tightest tolerance. This is also the region where the electric field is concentrated.

11.2 Manufacturing methods

We will first discuss manufacturing methods using metallic materials and then continue with plastic materials. Examples of technical solutions and real arrays as presented in the open literature are also given.

11.2.1 Metallic materials

Important characteristics of materials for slotted waveguide array antennas are

- Good conductivity
- Low weight
- Corrosion resistance
- Processability

More parameters can be added to the list depending on the application. Low cost is sometimes very important. For space applications, on the other hand, thermal stability can be the critical parameter. Overviews of manufacturing methods have been presented [15,16].

Aluminium, copper and silver are typical metals used. Aluminium or aluminium alloys are widely used for frequencies up to about 30 GHz. Copper and silver are sometimes used for higher frequencies. Standard waveguides are available in many materials and can be the starting point for fabrication. Typical dimensions for the X-band WR90/R100 waveguide are inner dimensions 22.86 × 10.16 mm, wall thickness 1.27 mm and a tolerance for the inner dimensions of ±0.08 mm (±0.003 in.). For antennas a reduced height version is often preferred. Instead of starting from separate waveguides several waveguide channels can be milled out of a block of the chosen material. Die-casting and electroforming are also used for special requirements.

The traditional assembly technique for joining aluminium parts is dip brazing. The parts are kept together by fixtures, if necessary, and submerged in a bath of special salt heated to about 600°C. This is close to the melting temperature of the aluminium alloy. A filler material is added in the joints where it will melt and join the assembly. After the brazing the unit is hardened and surface treated. Anderson *et al.* [17] describe the design and fabrication of X- and Ku-band array antennas using numerically controlled machining and dip brazing. Joining the various parts by epoxy bonding has also been reported [18].

A 94 GHz linear array of 21 longitudinal slots was made in copper by photo-etching the slots on a 0.1 mm copper sheet [19]. The sheet was then brazed onto a copper block with the waveguide channel. The theoretical analysis indicated a required positional accuracy for the slot positions of 0.0004 in. (0.01 mm) and even tighter for the slot length. An array of series inclined slots was also fabricated.

In another study [20] it was shown that a linear slotted waveguide array with ten slots can be built for 94 GHz from standard WR-10 waveguide. However, the wall thickness was reduced to facilitate scaling from X-band data. The slots were milled with a machine tolerance of typically $\pm$ 0.02 mm ($\pm$ 0.001 in.).

For the machining of rectangular slots the slot ends are often rounded and a correction for this can be applied as discussed in Chapter 5, Section 5.3.3; cf. also Section 8.10.1.

Low-cost mass production of 20 and 60 GHz planar arrays was aimed for in an experimental work by Ando and Hirokawa [21]. Their findings include die-casting of waveguide channels showed promise. Laser-welding of a slotted plate to the waveguide channels caused too much distortion while bonding using a conducting adhesive worked well mechanically, but losses were too high. However, feeding adjacent waveguides in antiphase (compensated with slot offset directions) was promising since the currents in the waveguide vertical walls almost vanished.

A crucial step in the fabrication is the joining of a slot plate to the waveguide channel(s). Screws can only be considered for low frequencies. For high frequencies (millimetre waves) diffusion bonding is an interesting technique where the parts are pressed together at a high temperature causing the parts to join (no adhesive). Stainless steel is often used but it has high conduction loss. Copper is preferred for antennas and was tried by [22,23]. Also, 60% efficiency was obtained for a 94 GHz 18 $\times$ 18 slotted waveguide array. Alignment of the layers during bonding is critical. See also reference [24] and Chapter 10.

Solbach [25] reported work on linear and planar arrays at 40 and 60 GHz. The slots were photoetched on a thin (0.1 mm) copper plate. The slot lengths and widths showed a spread of about 50–100 μm, mainly due to underetching. A thinner plate or thin metallisation on a dielectric carrier was suggested to improve this. Such a dielectric carrier was used by Zhao *et al.* [26] for a ten-element slotted waveguide array at 93 GHz. They used a teflon substrate (Rogers 5880) with a thickness of 0.254 mm. The slot plate was photoetched and bonded to the milled waveguide channel using a conductive resin. A related technology is also used in stripline fed slotted arrays; cf. Section 9.7.

11.2.2 Carbon fibre-reinforced plastic materials

Plastic materials with plated conducting surfaces can be used, similar to printed circuit techniques; cf. stripline arrays, Section 9.7. The waveguide sidewalls can be formed by rows of plated-through holes. For applications requiring low weight and thermal stability CFRP is suitable. CFRP reflects microwaves and is good for reflector antennas at not too high frequencies (up to about 15 GHz), but the conductivity is not sufficient for use in waveguides. A possible solution for slotted waveguide designs is to use *metallised* carbon fibre composites.

Several metallisation techniques are described in the literature [27]. In the galvanic method a thin copper layer is chemically deposited on the composite with the aid of a catalyst. This thin layer is then reinforced by galvanic metallisation. For corrosion protection a thin layer of gold can be added. Figure 11.6 shows samples of CFRP waveguides metallised with this process. Another method is to apply a metal foil on the composite before curing. Still another method is to metallise a mandrel and then apply a layer of pre-impregnated carbon fibres (the 'prepreg'). After curing the mandrel is removed. In some cases the three methods are combined.

An X-band mobile short range radar with a metallised carbon fibre slotted waveguide array antenna is shown in Figure 11.7(a) and (b). The antenna is phase

Figure 11.6 Waveguides made of metallised carbon fibre composites (galvanic method) [28]. Courtesy of Saab AB

(a)  (b) 

Figure 11.7 Mobile radar system: (a) radar unit; (b) array antenna with metallised CFRP slotted waveguides [28]. Courtesy of Saab AB

steered in elevation. The horizontal slotted waveguide runs are fed in the centre with different slot spacings in the two halves in order to maximise the frequency bandwidth [29]. The centre feeding technique is discussed in more detail in Chapters 7 and 9.

With composites it is possible to form several waveguides in one operation, for example, to produce integrated subpanels; an example is shown in Figure 11.8. When metallised this design performs electrically as its aluminium counterpart, but with improved characteristics in terms of weight and thermal stability. The same technique has also been used for airborne applications [30].

Low weight and thermal stability are critical parameters for space applications. A metallised CFRP composite material was therefore chosen for the SAR antennas for the European Earth Resource Satellites ERS-1 and ERS-2. The C-band SAR antenna is 10 m × 1 m when deployed (Figure 3.4(b)). The antennas were manufactured by Dornier Satellitensysteme in Germany [32]. Metallised carbon fibre arrays have also been used for X-band active SAR systems, for example, in the TerraSAR-X Radar satellite [33]. The active radiating element with the dual CFRP waveguides (for dual polarisations) is shown in Figure 11.9.

Large parts of modern aircraft structures are made of CFRP panels. It has been suggested that such panels could be designed to function as, for example, radar antennas and loadbearing structures at the same time. Several studies along these lines have been published [34,35], addressing RF performance and structural performance of test panels: SWASS = Slotted Waveguide Antenna Stiffened

Figure 11.8 *An integrated ridge waveguide slotted subpanel made in metallised CFRP [31]. Courtesy of Saab AB*

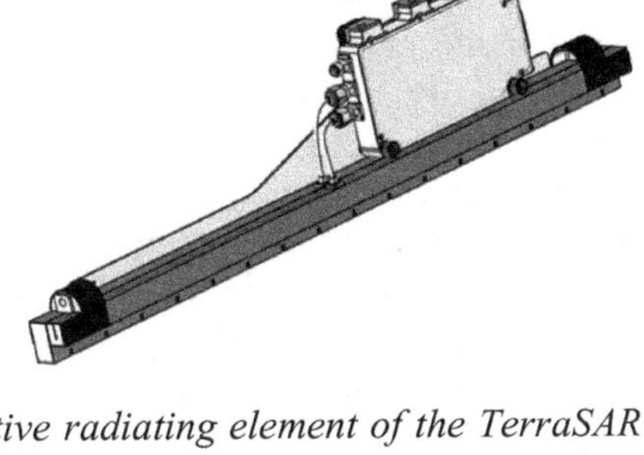

Figure 11.9 *The active radiating element of the TerraSAR array with a combination of edge slots in the rectangular waveguide and longitudinal slots in the ridge waveguide, both in CFRP technology. © 2003 IEEE. Reprinted from [33], with permission*

Structure. The tested panels were not metallised. The reported transmission losses were somewhat high and probably not acceptable for most applications. However, losses depend on the orientation of the carbon fibres in the waveguides. If the technology is successful the result could be large high-performance conformal slotted waveguide array antennas with minimum impact on drag and weight.

11.2.3 Microfabrication

Microelectronics packaging is a key element in the production of devices and systems that are becoming indispensable in our modern life. Miniaturisation is taken for granted in portable electronic devices such as mobile phones, Wi-Fi networks, automotive radar, etc. The trend towards higher frequencies makes the design and fabrication a challenging task. Circuits on ceramic substrates have been used extensively and efforts to combine this technology with antennas have been reported. The low-temperature co-fired ceramics (LTCC) technology has been studied for slotted waveguide array antennas at 79 GHz [36] and even at 140 and 270 GHz [37]. One problem is high transmission losses in the substrate material. More about the LTCC process and the materials used in [38,39].

An eight-element slotted waveguide array antenna was developed for 300 GHz using four silicon layers and a UV-sensitive epoxy-based photoresist (SU-8). The contact between layers proved to be a critical factor [40]. The slot length at this frequency is indeed small, in fact only 0.54 mm.

Photoetching of slots in a thin copper plate and brazing the plate onto a milled waveguide channel at 94 GHz has been successful [19]. It is a less complicated process, but lacks the possibility to integrate active and passive circuits in the same structure.

The PolyStrata® (literally 'many layers') process (http://nuvotronics.com) has been used for fabrication of micro-coaxial lines and components [41] and recently also for slotted waveguide arrays at millimetre-wave frequencies, 130–180 GHz [42]. According to this process a photo resist pattern is applied onto a silicon wafer. Next, a copper layer is electroplated on the wafer. Additional photoresist is applied (defining the waveguide walls) and a copper layer is again deposited, etc. About 15 layers can be built up this way. The final layer includes the slot elements. At the last step the photoresist is rinsed away through release holes. For corrosion protection the copper surface can be gold plated.

The PolyStrata® process is illustrated schematically in Figure 11.10 for fabrication of a micro-coaxial line using five layers [43]. (For a slotted waveguide array no centre conductor is needed.) Two fabricated slotted waveguide arrays for 130–180 GHz with 10 and 20 elements, respectively, are shown in Figure 11.11. The technology has been applied to slotted waveguide arrays operating up to 300 GHz [44].

The PolyStrata® process is very promising for low-cost fabrication of arrays and components at very high frequencies. Another possible fabrication method in the future might be the 3D printing technique, so-called 'Additive Manufacturing' [45,46].

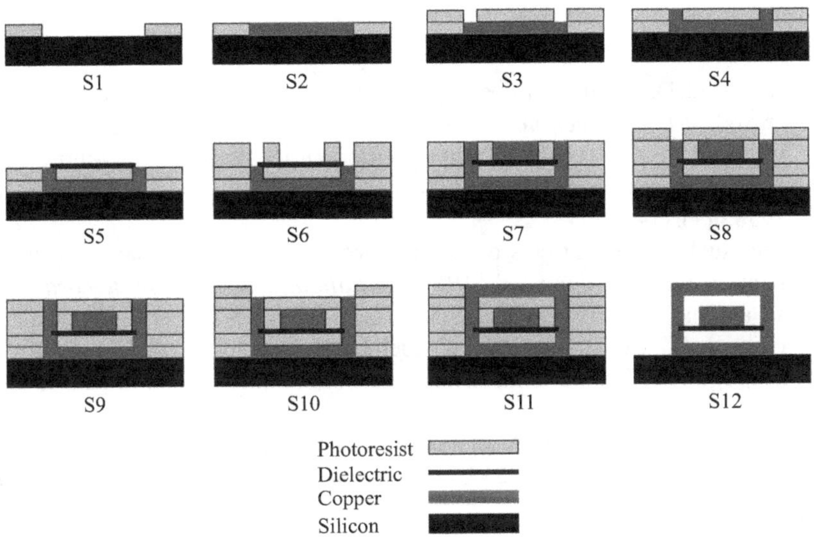

Figure 11.10 *The PolyStrata® process (five-layer) for fabrication of a micro-coaxial line. Reprinted from N. Ehsan, PhD thesis, Univ. Colorado 2010 [43]*

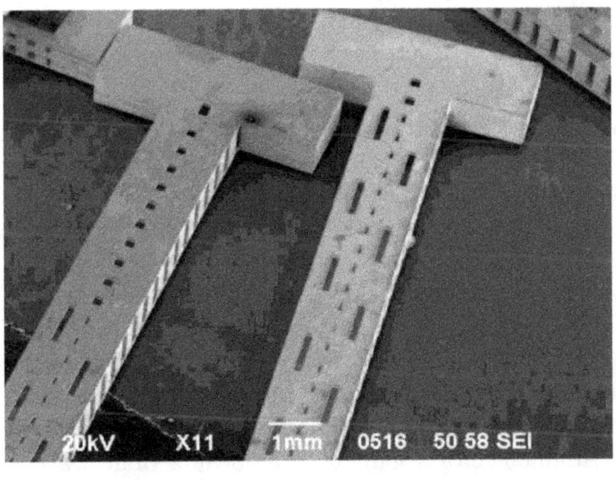

Figure 11.11 *The ends of two PolyStrata® waveguide arrays (130–180 GHz). Note the small holes that, together with the slots, are used for releasing the lossy photo resist. © IEEE 2012. Reprinted from [42], with permission*

References

[1] Ashmead D. 'Optimum design of linear arrays in the presence of random errors'. *IRE Transactions on Antennas and Propagation.* 1952;**4**(1):81–92.

[2] Elliott R. S. 'Mechanical and electrical tolerances for two-dimensional scanning antenna arrays'. *IEEE Transactions on Antennas and Propagation.* 1958;**6**(1):114–120.

[3] Rondinelli L. A. 'Effects of random errors on the performance of antenna arrays of many elements'. *IRE International Convention Record.* 1959; **1**(Part 1):174–189.

[4] Hsiao J. K. 'Constraints on low sidelobe phased array'. *IEEEAntennas and Propagation International Symposium,* Albuquerque, NM, USA, 1982, pp. 687–690.

[5] Lee J., Lee Y., Kim H. 'Decision of error tolerance in array element by the Monte Carlo method'. *IEEE Transactions on Antennas and Propagation.* 2005;**53**(4):1325–1331.

[6] Rengarajan S., Zawadski M. S., Hodges R. E. 'Waveguide-slot array antenna designs for low-average-sidelobe specifications'. *IEEE Antennas and Propagation Magazine.* 2010;**52**(6):89–98.

[7] Bailin L. L., Ehrlich M. J. 'Factors affecting the performance of linear arrays'. *IRE Transactions on Antennas and Propagation.* 1952;**1**(1):85–106.

[8] O'Neill H. F., Bailin L. L. 'Further effects of manufacturing tolerances on the performance of linear shunt slot arrays'. *IRE Transactions on Antennas and Propagation.* 1952;**4**(1):93–102.

[9] Stegen R. J. ' Slot radiators and arrays at X-band'. *IRE Transactions on Antennas and Propagation.* 1952;**1**(1):62–84.

[10] Kaminov I. P., Stegen R. J. *Waveguide slot array design*, Hughes Aircraft Company, Techn. Memo No. 348, July 1954 (available as NTIS Report AD 63600).

[11] Karlsson I. 'Applications of waveguide arrays in commercial and military radars'. *Antenna Applications Symposium,* Allerton Park, IL, USA, 1993, pp. 1–22.

[12] Brady M. M. 'A plea for clarity in waveguide designation'. *IEEE Transactions on Microwaves, Theory and Techniques.* 1965;**13**(2):247–249.

[13] Conciauro G., Bressan M., Zuffada C. 'Waveguide modes via an integral equation leading to a linear matrix eigenvalue problem'. *IEEE Transactions on Microwaves, Theory and Techniques.* 1984;**32**(11):1495–1504.

[14] Falk K. 'Admittance of a longitudinal slot in a ridge waveguide'. *IEE Proceedings H - Microwaves, Antennas and Propagation.* 1988;**135**(4):263–268.

[15] Sikora L., Womack J. 'The art and science of manufacturing waveguide slot-array antennas'. *Microwave Journal.* 1988;**31**(6):157–162.

[16] Muhs H. P. 'Mm-wave antenna'. *Microwave Journal.* 1985;**28**(7):91–194.

[17] Anderson T., Yun-Li-Ho, Michalski J. 'Design and manufacturing techniques for planar slot array antennas for a variety of radar application'. *Proceedings of the IEEE International Radar Conference Radarcon,* 1998, pp. 337–342.

[18] Singh A. K. 'A low cost, low side lobe and high efficiency non-orthogonally coupled slotted waveguide array antenna for monopulse radar tracking'. *IEEE Antennas and Propagation International Symposium,* Washington DC, USA, 2005, pp. 732–735.

[19] Rao B. R. '94 Gigahertz slotted waveguide array fabricated by photolithographic techniques'. *IEEE Antennas and Propagation International Symposium,* Houston, TX, USA, 1983, pp. 688–691.

[20] Farrar F. G. 'Millimeter wave W-band slotted waveguide antennas'. *IEEE Antennas and Propagarion International Symposium,* Los Angeles, CA, USA, 1981, pp. 436–439.

[21] Ando M., Hirokawa J. 'Single-layer slotted waveguide arrays for DBS reception and higher frequency applications'. *Electromagnetics.* 1999;**19**(1): 23–48.

[22] Hirokawa J., Zhang M., Ando M. 'Millimeter waveguide fabrication to reduce transmission loss by diffusion bonding, light-curing resin or dielectric partially-filling'. *Asia-Pacific Microwave Conference, APMC.* 2008, pp. 1–4.

[23] Zhang M., Hirokawa J., Ando M. 'Fabrication of a slotted waveguide array at 94 GHz by diffusion bonding of laminated thin plates'. *IEEE Antennas and Propagation Society International Symposium,* Charleston, SC, USA, June 2009, pp. 1–4.

[24] Ando M., Zhang M., Lee J., Hirokawa J. 'Design and fabrication of millimeter wave slotted waveguide arrays'. *4th European Conference on Antennas and Propagation (EuCAP)*, Barcelona, Spain, 2010, pp. 1–6.

[25] Solbach K. 'Some millimeter-wave slotted array antennas'. *14th European Microwave Conference*, Liege, Belgium, 1984, pp. 181–186.

[26] Zhao G., Xu H., Chen Z., Sun H., Lv X. 'A W-band waveguide slot array antenna based on print technology'. *International Conference on Microwave and Millimeter Wave Technology (ICMMT),* 2012, Vol. **3**, pp. 1–4.

[27] Jönsson D. 'Metallisation methods & applications'. *Proceedings of the ESA Symposium on Space Applications of Advanced Structural Materials,* ESTEC, Noordwijk, March 1990, pp. 35–38.

[28] Dahlsjö O., Ljungström B., Magnusson H. 'Fibre-reinforced plastic composites in sophisticated antenna designs'. *Ericsson Review.* 1987;**64**(2):50–57.

[29] Karlsson E. R. 'Waveguide element for an electronically controlled radar antenna'. US Pat. 4788 552, 1988.

[30] Brunzell S., Magnusson H. 'Lightweight CFRP waveguide array'. *Proceedings Military Microwaves*, London, UK, October 1984, pp. 305–309.

[31] Josefsson L., Derneryd A., Lagerlöf R. 'Electronic scanning with slotted waveguide arrays'. *Proceedings of the Radarcon 90 Conference,* Adelaide, Australia, April 1990.

[32] Petersson R., Bonnedal M. 'The slotted waveguide arrays of the European remote sensing satellites ERS-1 and ERS-2'. *Electromagnetics.* 1999;**19**(1): 77–89.

[33] Stangl M., Werninghaus R., Zahn R. 'The TerraSAR-X active phased array antenna', *IEEE International Symposium on Phased Array Systems and Technology,* Boston, MA, USA, October 2003, pp. 70–75.

[34] Sabat J. W. *Structural Response of the Slotted Waveguide Antenna Stiffened Structure Components under Compression.* M.Sc. Thesis, Air Force Institute of Technology, Wright-Patterson Air Force Base, OH, USA, March 2010.

[35] Callus P. J., Nicholson K. J., Bojovschi A., Ghorbani K., Baron W., Tuss J. 'A planar antenna array manufactured from carbon fibre reinforced plastic'. *28th International Congress of the Aeronautical Sciences*, ICAS 2012, Brisbane, Australia, September 2012.

[36] Bauer F., Menzel W. '79 GHz LTCC – integrated antennas', *2013 International Workshop on Antenna Technology (IWAT)*, 2013, pp. 158–161.

[37] Xu J., Chen Z. N., Qing X. '270-GHz LTCC-integrated strip-loaded linearly polarised radial line slot array antenna'. *IEEE Transactions on Antennas and Propagation.* 2013;**61**(4):1794–1801.

[38] Kulke R., Rittweger M., Uhlig P., Günner C. 'LTCC – Multilayer ceramic for wireless and sensor applications'. http://www.ltcc.de/downloads/rd/pub/10-doc-plus-engl-2001.pdf Retrieved 2017–04–02.

[39] Ullah U., Ain M. F., Mahyuddin N. M., *et al.* 'Antenna in LTCC technologies: a review and the current state of the art'. *IEEE Antennas and Propagation Magazine.* 2015;**57**(2);241–260.

[40] Wang Y., Ke M., Lancaster M. J. 'Micromachined 300-GHz SU-8-based slotted waveguide antenna'. *IEEE Antennas and Wireless Propagation Letters.* 2011;**10**:573–576.

[41] Ehsan N., Cullens E., Vanhille K., *et al.* 'Micro-coaxial lines for active hybrid-monolithic circuits'. *IEEE MTT-S International Microwave Symposium Digest,* 2009, pp. 465–468.

[42] Cullens E. D., Ranzani L., Vanhille K. J., Grossman E. N., Ehsan N., Popović Z. 'Micro-fabricated 130–180 GHz frequency scanning waveguide arrays'. *IEEE Transactions on Antennas and Propagation.* 2012;**60**(8):3647–3653.

[43] Ehsan N. *Broadband Microwave Lithographic 3D Components*, Ph.D thesis, University of Colorado, 2010.

[44] Boryssenko A., Vanhille K. '300 GHz microfabricated waveguide slotted arrays'. *39th International Conference on Infrared, Millimeter and Terahertz Waves (IRMMW-THz).* 2014, pp. 1–2.

[45] Nayeri P., Liang M., Sabory-Garcia R. A., *et al.* '3D printed dielectric reflectarrays: low-cost high-gain antennas at sub-millimeter waves'. *IEEE Transactions on Antennas and Propagation.* 2014;**62**(4):2000–2008.

[46] Nassar I. T. and Weller T. M. 'An electrically-small, 3d-cube antenna fabricated with additive manufacturing'. *IEEE Topical Meeting on Silicon Monolithic Integrated Circuits in RF Systems (SIRF)*, 2013, pp. 162–164.

Chapter 12
Outlook for the future

In this book we have analysed the design of several types of slotted waveguide array antennas, from theories and optimisation to applications and manufacturing techniques.

In this last chapter we will discuss the current status in the field and look at new technologies and new applications recently presented and researched. Still, in this short overview it is not possible to mention all the details of the evolving field; the reader is referred to the respective chapters and the references for more information.

12.1 Defence systems

The development of slotted waveguide array antennas started during the Second World War driven by the need for radar to detect and track enemy targets. The development continued during the following years, and the technology was also applied to non-military systems: communication [1], remote sensing [2,3], navigation [4], weather radar [5], etc.

The flat plate antenna succeeded the reflector antenna and became the standard antenna type in the nose of military fighter aircraft in combination with a central transmitter and receiver. The development in the microelectronic area of today has led to the point where transmit/receive modules can be miniaturised and integrated with the individual radiating elements of phased array antennas. These AESA systems (Active Electronically Steered Arrays) enable advanced signal processing and 'graceful degradation'. Since the systems are quite complex and costly they are primarily considered for the most advanced multifunction systems. Still, they are today being implemented in new fighter versions as well as in some ground and naval systems [6–8]. We also find examples where (passive) slotted waveguide subarrays are the elements in systems with the signal processing acting on the subarray level [9,10]. This passive electronically steered array (PESA) might seem as a good cost/complexity compromise, especially when the electronic scanning is mainly required in one dimension only [11].

A sometimes important aspect is the protection of the receiver circuits from (un)intentional external microwave radiation. The AESA radiating elements have a broad beam and are not protected by low sidelobes as formed in arrays with a central receiver. Looking at the other side, creating high-power radiation from slotted waveguide arrays has been discussed as a possible high-power microwave weapon [12,13].

An interesting idea is the inclusion of slotted waveguide arrays in the skin of aircraft or other vehicles: conformal arrays. The waveguides would serve as structural members and antenna elements at the same time [14]. This is to be compared with today's systems like AWACS [15] and ERIEYE [16] where a large array antenna is carried outside the aircraft body.

12.2 Millimetre waves

In 1984 a literature search revealed only three papers on millimetre-wave slotted waveguide array antennas (at 60 and 94 GHz) [17]. Today, work on slotted waveguide arrays at frequencies as high as 300 GHz (0.3 THz) [18] and even higher has been reported. An LTCC-integrated radial line slot array for 270 GHz was designed and evaluated by Xu *et al.* [19].

The benefits of going to millimetre-wave frequencies include smaller dimensions, narrow beams, large signal bandwidths, high data rates, sometimes covert communication due to atmospheric losses, etc. However, higher frequencies mean smaller antenna apertures, and hence weaker signals. This leads to a need for higher antenna gain and arrays both for transmit and receive. Typical millimetre-wave applications include automotive radar in 76–81 GHz range [20,21], planetary landing radar in 138–180 GHz [22,23], point-to-point wireless links [24], etc.

Microwave and millimetre-wave signals penetrate optically opaque materials such as composites, ceramics, concrete, wood, clothing and interact with their interior structures. Imaging techniques using various forms of modulation and signal processing have demonstrated both 2D and 3D images at 24 GHz as shown by Gahsr *et al.* [25]. In their 'microwave camera' 576 modulated sidewall slot elements were used. Applications include non-destructive testing and evaluation (NDT&E) of materials, medical imaging, airport security scanners, concealed weapon detection, etc.

12.3 Spaceborne systems

SAR stands for synthetic aperture radar, a technology used in airborne and spaceborne systems. By coherent signal processing along the flight path a large (synthetic) aperture is created in this direction. Together with a large physical aperture across the flight path very high angular resolution is possible. In one application [26] scatterometer antennas provide ocean wind data and soil moisture content. The technology is used in many remote sensing satellites such as the C-band instruments ERS-1 and ERS-2; cf. Figure 3.4(b). The ERS satellites were operational almost 20 years [27]. The major antenna technology for these systems is slotted waveguide arrays. New developments include dual polarisation, dual and triple beams, beam scanning, X-band and higher frequencies, and also millimetre waves. The new spaceborne SAR systems will provide improved high-resolution images on a global scale for monitoring the environment, wind speed, ocean currents, land mapping, climate change, etc. Together with advanced signal processing techniques, digital beam forming and interferometry, the available information content will increase at least one order of magnitude [28].

12.4 Communication

Communication based on millimetre-wave radio systems is of increasing importance. Telecom operators need high flexibility when rerouting of the network becomes necessary. Narrow beam antennas, point-to-point, are used for connecting base stations to the main switching office (backhaul). Slotted waveguide array antennas are the best solution in many cases, enabling bandwidths of about 10% and an attractive form factor. Short millimetre-wave links are more flexible and often less costly than conventional cables or fibres.

Steerable beams are desired in future high-capacity wireless local area networks. We find many interesting solutions here: adaptive beams, MIMO in WLAN (WiFi, WiGig) and also in cellular networks such as the coming 5G systems [29]. For many of these applications slotted waveguide arrays have been studied [30,31], but there are naturally also other technologies that can be considered. The low loss of waveguide antennas favours this technology.

12.5 Manufacturing

The challenges are to reduce cost and complexity. We have come a long way, for example, in integrating the feeding waveguides in the same plane as the planar slotted waveguide array [32]. Traditionally, the antenna used to be a separate unit, connected to the transmit/receive front end via coaxial cables or waveguides. At higher frequencies the active front end can be integrated with the slotted waveguide antenna using the same materials and processes, for example, LTCC, the SU-8 process or the PolyStrata® process [22,33]. One example is the LTCC substrate-integrated slotted waveguide array antenna. It is essential to master the tolerance requirements and possibly modify the design to match the available fabrication capability, especially at millimetre-wave bands [34].

12.6 Design tools and software

In the early days of slotted waveguide array antennas, say before 1960, the designs were based primarily on experimental data. In the following years electromagnetic models and software for design and analysis were used more and more. This development is reflected by several examples in this book. The first comprehensive array design method that included external and internal higher-order mode mutual coupling was due to Professor Elliott in the 1980s; see in particular Chapters 6–8.

The interest in computational electromagnetics increased even more in the 1990s and several commercial software tools suitable for analysis and design of slotted waveguide arrays became available. The computer codes employ various numerical and asymptotic techniques such as the Method of Moments (MoM), the finite element method (FEM), the mode matching technique (MM), the finite difference time domain technique (FDTD), etc. [35–38]. As computer software becomes more and more powerful the size and complexity of the structure to be analysed can be allowed to increase. The continued development in computers and software leads to better tools for design and analysis, at the same time reducing the need for expensive testing and measurements.

It shall be emphasised, however, that computers do not design antennas. It still requires a great deal of engineering efforts, for example,

- analysing system requirements
- optimising antenna configuration
- choice of materials
- manufacturing methods
- testing and verification

There exists today a plethora of computer software for design of array antennas. The following listing provides a few examples; much more information is available in the cited papers as well as on the internet.

12.6.1 HFSS

HFSS stands for high-frequency structural simulator, originally developed by Professor Z. Cendes around 1980 [39]. The code was further developed by Ansoft Corporation and is today marketed by ANSYS [40]. It is a FEM and includes an integral equation solver (MoM) which is convenient for slot array problems.

12.6.2 XFdtd

The XFdtd by Remcom [41] uses the FDTD technique, first proposed in 1966 [42]. The problem geometry is simulated by a mesh which can be directly connected to lumped circuit elements. This has become attractive for simulations of subassemblies that include both MMIC circuits and antenna parts.

12.6.3 CST Microwave Studio

CST Microwave Studio [43] is another software tool, also based on FDTD, for analysis of high-frequency devices such as antennas, filters, couplers, planar and multi-layer structures. It has adaptive meshing and includes both a frequency domain solver and a time-domain solver.

12.6.4 WASP-NET

WASP-NET [44] combines several methods: MM, the FEM and the MoM for design of slotted waveguide arrays including feed networks.

12.6.5 SWAN™

The software SWAN™ [45] is a powerful CAD tool for the design and analysis of very large slotted waveguide arrays. It is based on Elliott's theory with a number of significant extensions and improvements. It has a graphical user interface so that the user can input design and analysis parameters easily.

References

[1] Grabherr W., Huder B. 'Omni-directional slotted waveguide array antenna'. 29th *European Microwave Conference*, Munich 1999, pp. 134–137.

[2] Ekengren B. 'Ericsson SLAR – airborne radar for maritime surveillance'. *Ericsson Review.* 1979;**56**(4):151–157.

[3] Alves A. M. P., Duplat D. N., de Oliveira L. P., Hernández-Figueroa H. E. 'Aerotransported radar antenna for oil spills monitoring antennas and propagating'. *4th European Conference on Antennas and Propagation (EuCAP)*, Barcelona, April 2010, pp. 1–3.

[4] Morwing B. 'AGA-ERICON – a marine radar beacon'. *Ericsson Review.* 1981;**58**(4):180–187.

[5] Anderson T., Yun-Li-Hou, Michalski J. 'Design and manufacturing techniques for planar slot array antennas for a variety of radar application'. *Proceedings of the IEEE Radar Conference, Radarcon 98,* Dallas, TX, USA, May 1998, pp. 337–342.

[6] Brookner E. 'Phased arrays and radars – past, present and future'. *Microwave Journal.* 2006;**49**(1):1–12.

[7] Kopp C. 'Evolution of AESA radar technology'. *Microwave Journal.* 2012;**55**(8):3–12.

[8] Pendergast S. L. 'Recent advances in radar technology'. *Microwave Journal.* 2015;**58**(9):1–14.

[9] Karlsson I. 'Applications of waveguide arrays in commercial and military radars'. *Proceedings of the 1993 Antenna Applications Symposium,* Allerton House, IL, USA. pp. 1–22, 1993.

[10] EMS White Paper *Passive phased arrays for radar antennas.* EMS Technologies Inc., Norcross, GA, USA, December 2005.

[11] Lewis D. J., Lee J. R., McCarty D. K. 'A single-plane electronically scanned antenna for airborne radar applications'. *Phased-Array Antenna Symposium*, Polytechnic Institute of Brooklyn, Farmingdale NY, 2–5 June 1970.

[12] Coburn W., Litz M., Miletta J., *et al. A Slotted-Waveguide Array for High-Power Microwave Transmission*. Army Research Laboratory, Rpt. MD 20783-1197, January 2001.

[13] Baum C. E. *Sidewall Waveguide Slot Antenna for High Power.* Sensor and Simulation Note No. 503, Univ. New Mexico. August 2005.

[14] Callus P. J., Nicholson K. J., Bojovschi A., Ghorbani K., Baron W., Tuss J. 'A planar antenna array manufactured from carbon fibre reinforced plastic'.

28th International Congress of the Aeronautical Sciences, ICAS 2012, Brisbane, Australia, September 2012.

[15] Skillman W. A. 'Environmental effects on airborne radar performance'. *IEEE International Microwave Symposium,* 5–10 June, 2011, Baltimore MD, USA, pp. 1–4.

[16] Ahlbom S., Andersson P., Lagerlöf R. 'A Swedish airborne early warning system based on the Ericsson ERIEYE radar'. *Ericsson Review.* 1995;**72**(2): 54–63.

[17] Solbach K. 'Some millimeter-wave slotted array antennas'. *Proceedings of the 14th European Microwave Conference,* 1984, pp. 181–186.

[18] Wang Y., Lancaster M. J. 'A micromachined centre-fed slotted waveguide antenna for mm-wave applications'. *IEEE MTT International Microwave Workshop Series on Millimetre wave Wireless Technology and Applications (IMWS)*, 2012, pp. 1–3.

[19] Xu J., Chen Z. N. Qing X. '270 GHz LTCC-integrated strip-loaded linearly polarized radial slot array antenna'. *IEEE Transactions on Antennas and Propagation.* 2013;**61**(4): 1794–180.

[20] Hirokawa J., Ando M. '76 GHz post-wall waveguide fed parallel plate slot arrays for car-radar applications'. *Proceedings of the IEEE Antennas and Propagation Symposium,* July 2000, pp. 98–101.

[21] Menzel W. 'Antennas in automobile radar'. In *Handbook of Antenna Technologies*, Springer Verlag, Berlin Heidelberg 2015.

[22] Cullens E.D., Ranzani L., Vanhille K. J., Grossman E. N., Ehsan N., Popovic Z. 'Micro-fabricated 130–180 GHz frequency scanning waveguide arrays'. *IEEE Transactions on Antennas and Propagation.* 2012;**60**(8):3647–3653.

[23] Samoska L., Kangaslahti P., Pukala D., Sadowy G., Pollard B., Hodges R. 'A G-band 160 GHz T/R module concept for planetary landing radar'. *Proceedings of the 36th European Microwave Conference*, Manchester UK, September 2006, pp. 757–760.

[24] Frenzel L. 'Millimeter waves will expand the wireless future'. *Electronic Design*, March 6, 2013.

[25] Gahsr M. T., Abou-Khousa M. A., Kharkovsky S., Zoughi R., Pommerenke D. 'Portable real-time nicrowave camera at 24 GHz'. *IEEE Transactions on Antennas and Propagation.* 2012;**60**(2):1114–1125.

[26] Magnusson P., Di Salvo M., Scarchilli C. 'A dual polarised slotted waveguide antenna for satellite based wind scatterometet instruments'. *7th European Conference on Antennas and Propagation (EuCAP),* Gothenburg, Sweden, April 2013, pp. 3449–3452.

[27] van't Klooster. 'ERS-1, European remote-sensing satellite was launched 20 years ago'. *21st International Crimean Conference on Microwave & Telecommunication Technology,* September 2011.

[28] Moreira A. 'A golden age for spaceborne SAR systems'. *20th International Conference on Microwaves, Radar and Communication (MIKON),* 2014, pp. 1–4.

[29] Dussopt L., Bouayadi O. E., Luna J. A. Z., Dehos C., Lamy Y. 'Millimeter-wave antennas for radio access and backhaul in 5G heterogenous mobile

networks'. *9th European Conference on Antennas and Propagation (EuCAP)*, Lisbon, Portugal, April 2015, pp. 1–4.

[30] Ohira M., Miura A., Ueba M. '60-GHz wideband substrate-integrated-waveguide slot array using closely spaced elements for planar multisector antenna'. *IEEE Transactions on Antennas and Propagation.* 2010;**58**(3): 993–998.

[31] Wen Y., Wang B., Ding X. 'Wide-beam SIW-slot antenna for wide-angle scanning phased array'. *IEEE Transactions on Wireless Propagation Letters.* 2016;**15**:1638–1641.

[32] Zhang M. *Full-Structure Analysis and Design of Single-Layer Slotted Waveguide Arrays with their Practical Realisation*, Doctoral Dissertation, Tokyo Institute of Technology, Dept. of Electrical and Electronic Engineering, March 2008.

[33] Wang Y., Ke M., Lancaster M. J. 'Micromachined 300-GHz SU-8-based slotted waveguide antenna'. *IEEE Transactions on Wireless Propagation Letters.* 2011;**10**:573–576.

[34] Xu J., Chen Z. N., Qing X., Hong W. '140-GHz SIW LTCC antenna array using a large via-fenced and slotted dielectric loading'. *7th European Conference on Antennas and Propagation (EuCAP),* Gothenburg, April 2013, pp. 2861–2864.

[35] Mittra, R. (ed.). *Numerical and Asymptotic Techniques in Electromagnetics*, Springer Verlag, Berlin, 1975.

[36] Itoh, T. (ed.). *Numerical Techniques for Microwave and Millimeter-Wave Passive Structures.* Wiley Interscience, New York, 1989.

[37] Peterson, A. F., Ray, S. L., Mittra, R. *Computational Methods for Electromagnetics.* Wiley-IEEE, December 1997.

[38] Weiland T., Munteanu I. 'State of the art and the future in modelling RF components'. *1st European Conference on Antennas and Propagation (EuCAP),* Nice, France. November 2006.

[39] Cendes Z. 'The development of HFSS'. *USNC-URSI Radio Science Meeting.* Fajardo, Puerto Rico, USA, June 2016. pp. 39–40.

[40] http://www.ansys.com. Retrieved March 2018.

[41] http://www.remcom.com. Retrieved March 2018.

[42] Yee K. S. 'Numerical solution of initial boundary value problems involving Maxwell's equations in isotropic media'. *IEEE Transactions on Antennas and Propagation.* 1966;**14**(3):302–307.

[43] https://www.cst.com. Retrieved March 2018.

[44] http://www.mig-germany.com. Retrieved March 2018.

[45] http://www.swan-soft.com. Retrieved March 2018.

Appendix

Standard waveguide frequency bands

Standard rectangular waveguides are listed with their common designations and useful frequency ranges. The waveguide inner dimensions are given in mm.

Band letter*	Waveguide designation**		Frequency range (GHz)	Inner dimensions (mm)
	EIA WR-	IEC 153-2		
L	650	R 14	1.15–1.72	165.10 × 82.55
	510	R 18	1.45–2.2	129.54 × 64.77
	430	R 22	1.7–2.6	109.22 × 54.61
	340	R 26	2.2–3.3	86.36 × 43.18
S	284	R 32	2.6–3.95	72.14 × 34.04
	229	R 40	3.3–4.9	58.17 × 29.21
C	187	R 48	3.95–5.85	47.55 × 22.15
C	159	R 58	4.9–7.05	40.39 × 20.19
	137	R 70	5.85–8.2	34.85 × 15.80
	112	R 84	7.05–10.0	28.50 × 12.62
X	90	R 100	8.2–12.4	22.86 × 10.16
Ku	62	R 140	12.4–18.0	15.80 × 7.899
	51	R 180	15.0–22.0	12.95 × 6.477
K	42	R 220	17.6–26.7	10.668 × 4.318
Ka	28	R 320	26.5–40.0	7.112 × 3.556
Q	22	R 400	33.0–50.0	5.690 × 2.845
U	19	R 500	40–60	4.775 × 2.388
V	15	R 620	50–75	3.759 × 1.880
E	12	R 740	60–90	3.099 × 1.549
W	10	R 900	75–110	2.54 × 1.27
F	8	R 1200	90–140	2.032 × 1.016
	6	R 1400	110–170	1.651 × 0.8255
G	5	R 1800	140–220	1.295 × 0.6477
	3	R 2600	220–330	0.8636 × 0.4318

*Several systems for band letter designations exist [1,2]. This table is adapted from [3].

**EIA = Electronic Industries Association, New York; IEC = International Electrotechnical Commission, Geneva.

References

[1] Brady M. M. 'A plea for clarity in waveguide designations'. *IEEE Transactions on Microwave Theory and Techniques*. 1965;**13**(2):247–249.
[2] Lewin L. 'Letter symbols to designate microwave bands'. *IEEE Transactions on Microwave Theory and Techniques*. 1964;**12**(5):551.
[3] https://en.wikipedia.org/wiki/waveguide_(electromagnetism)#References. Retrieved 25 Oct 2017.

Index

www.ingramcontent.com/pod-product-compliance
Lightning Source LLC
LaVergne TN
LVHW080643210926
848485LV00003B/314

* 9 7 8 1 6 1 3 5 3 1 8 9 1 *